HEALTH EFFECTS OF
FOSSIL FUEL BURNING

HEALTH EFFECTS OF FOSSIL FUEL BURNING
Assessment and Mitigation

RICHARD WILSON
Department of Physics and
Energy and Environmental Policy Center,
Harvard University

STEVEN D. COLOME
School of Public Health and
Energy and Environmental Policy Center,
Harvard University

JOHN D. SPENGLER
School of Public Health and
Energy and Environmental Policy Center,
Harvard University

DAVID GORDON WILSON
Department of Mechanical Engineering,
Massachusetts Institute of Technology

BALLINGER PUBLISHING COMPANY
Cambridge, Massachusetts
A Subsidiary of Harper & Row, Publishers, Inc.

This book is printed on recycled paper.

International Standard Book Number: 0–88410–714–0

Library of Congress Catalog Card Number: 80–20305

Printed in the United States of America

Library of Congress Cataloging in Publication Data

Main entry under title:

Health effects of fossil fuel burning.

  Bibliography: p.
  Includes index.
  1. Coal—Combustion—Hygienic aspects. 2. Air—Pollution—Toxicology. 3. Coal-fired power plants—Environmental aspects. 4. Air quality management. I. Wilson, Richard, 1926-
RA577.C63H4   363.7'392   80–20305
ISBN 0–88410–714–0

CONTENTS

LIST OF FIGURES

LIST OF TABLES

PREFACE

Coal and oil are the primary energy sources all over the world. The supply of easily available coal is 30 times the supply of easily available oil. We therefore expect that coal use will increase as oil runs out. In the United States coal use has once again returned to the levels of the 1920s and 1940s. Energy plans call for an increase in the output of our coal mines from the present six hundred million tons a year to one and a half billion tons by the turn of the century.

Coal has always been known as a dirty fuel and for hundreds of years governments have tried to control its use. However, attempts to *measure* the adverse effects on health are comparatively recent. In this book we discuss efforts to distinguish which of the many emitted pollutants cause adverse health effects and how we can describe them numerically.

The information, voluminous though it is, is imprecise and inadequate, and to a considerable extent will always remain so. Nonetheless decisions will and must be made. All over the world, from Tokyo to Pittsburgh, air pollution has been reduced in the last 20 years. This progress is undeniable, but has been accomplished primarily by reducing the concentrations of pollutants near the sources and dispersing the pollutants over a large area. It is an open question whether the lower concentrations of pollutants which are now widespread, and to which millions of people are exposed, are serious health hazards.

This leaves us a dilemma: if adverse health effects are still present at lower pollutant levels, thousands of people would be affected; yet if these effects do not exist, or if we control the wrong pollutants, we would be doing so at fantastic expense. We hope to set the stage by providing a logical framework for discussion of these questions without the emotional polarization that is all too frequent.

We have concentrated on sulfur oxides, nitrogen oxides and particulates. In our early chapters we show the trends in emissions and the resulting concentrations. We then proceed to show how these pollutants cause obvious adverse health effects at the high concentrations that were all too frequent in the past but are now rare. This leaves the question of how to use this data—often laboratory or animal data and often fragmentary—to describe or predict effects at the lower ambient levels now common.

It has been suggested that air pollution in the U.S. may contribute 1–3 percent of the total death rate. To measure such effects directly requires extremely large populations. The *prospective* epidemiologic studies unfortunately have study populations that are too small to detect this effect. Therefore, we are then led to discuss retrospective studies of mortality. There have been several such studies which suggest a correlation between air pollution and mortality. However, statistical studies cannot prove a *causal* relationship between air pollution and mortality. We discuss several of these studies and their inevitable weaknesses. It is currently impossible to correct for differences in occupational exposures, cigarette smoking, and "socioeconomic" factors. Although a causal connection cannot be demonstrated it cannot be disproved either.

There are many strategies for controlling air pollution. The later chapters cover technical control options, operational controls, and siting methods. We discuss these and emphasize the importance of generic solutions to reduce all pollutants since we are not sure which, *if any*, have the most serious effect on human health. The choice of control strategy is often now made on political and not technical or economic grounds. We discuss how an emission fee might be used as an incentive for optimum choice.

We hope that this book will provide a perspective to understand the dilemma facing us as we turn to coal, and more important how to structure an important policy problem where the uncertainties are great.

ACKNOWLEDGMENTS

We would like to thank the many people who have helped bring this book into being. We cannot recognize everyone who has helped during the various stages of this project, but in particular we thank: Ralph Sacco for his review of global sulfur cycles; Robin Charo for her review of the carbon dioxide question; Phil Sisson for classifying various health effects studies; Mike Sylvanus for reviewing several health studies; and Pie W. Chen and Ben Chang for their contributions in atmospheric modeling and taxation proposals.

We draw special attention to the contributions of Drs. Stanley V. Dawson and John S. Evans who, through a project for the U.S. Office of Technology Assessment (OTA 1979), provide much of the inspiration and many of the examples and thoughts that appear in Chapters 4 and 6.

The text and tables were produced by Cynthia Robinson and Lissette Williams under the direction of Diane Rolinsky. Text and reference editing was performed competently by Nancy Harman, Hugh Donahue and Kathryn Lottes.

1 INTRODUCTION

1.1 SOME HISTORIC NOTES

As early as the ninth century, A.D., "sea-coales" were found on the northeast coast of England and burned as fuel. It was soon noticed that noxious fumes were produced upon burning, and the coal and the fumes became intolerable in the city of London. At that time there was still plenty of wood in the country so that during his reign of 1272–1307, Edward I, an early environmentalist, banned the use of "carbone marino" (sea coal) from London with the following proclamation:

> Restrictions on the use of kilns in London. The King to (the Sheriff) of Surrey, greetings. As a result of the serious complaint from the prelates and magnates of our realm, who frequently come to London for the good of the commonwealth by our order, and from the citizens and all the people who dwell there and at Southwark, we have learned that, whereas previously the makers of kilns in the aforesaid city and village and their neighbourhood were in the habit of using brush-wood or charcoal for their kilns, they are now again, contrary to their usual practice, firing them with and constructing them of sea-coal, from which is emitted so powerful and unbearable a stench that, as it spreads throughout the neighbourhood, the air there is polluted over a wide area, to the considerable annoyance of the said prelates, magnates, citizens and others dwelling there, and to the detriment of their bodily health. Wishing to take precautions against this kind of danger, and to

provide for the safety of the prelates and magnates, citizens and others of our faithful, we instruct you to have public proclamation made in the aforesaid village of Southwark that all those who wish to operate kilns in this same village or its neighbourhood should make them in the customary fashion from brushwood or charcoal and should henceforth make absolutely no use of sea-coal in constructing these same kilns, under pain of heavy forfeiture. And see that this order of ours is inviolably observed hereafter throughout the afore-said village and its neighbourhood. Attested by the King at Carlisle, 12 June. In like manner an order is issued to the mayor and sheriffs of London. Attested as above [Plantaganet 1307, as translated by Sir W. Hawthorne 1978, p. 485].

Edward I's successor, Edward II (1307–1327) had persons tor-tured who were found fouling the air with coal smoke. Following the reign of Edward III, Richard II (1377–1399) adopted a more moder-ate position and sought to restrict the use of coal through taxation. Two reigns later, Henry V (1413–1422) established a commission to regulate the entry of coal into London [Chambers 1977].

By the sixteenth century, the population had increased and the woodlands cleared so that coal burning became a necessity. The court physician to Queen Elizabeth I took out a patent to remove the sulfur [Armytage 1961], but, as we all know, this process is neither cheap nor reliable even today, four centuries later. One hun-dred years later the problems were already very bad. The diarist John Evelyn wrote:

It is this horrid smoke, which obscures our churches and makes our palaces look old, which fouls our clothes and corrupts the waters so that the very rain and refreshing dews which fall in the several seasons precipitate this impure vapour, which with its black and tenacious quality, spots and contaminates whatever is exposed to it [Evelyn 1661].

Burning coal produces particulate matter and sulfur oxides in the smoke, along with nitrogen oxides and their derived particles and other trace substances. For instance, it is one of the principal ways in which mercury and other heavy elements enter the atmosphere. The long-term health hazards of low levels of these metals is unknown. The amount of radioactive material—radium and thorium—is enough to make the problem of long-lived (greater than 500 years) radio-active waste from coal burning comparable to the waste from a nuclear power plant.

It is instructive to speculate on the public perception of the rela-tive problems of sulfur and other pollution. More attention has been paid to sulfur than to nitrates; it is not entirely clear why. Sulfur certainly produces an odor; when Dante saw sulfur miners in Sicily melting the sulfur out of the ore he was inspired to write *The Inferno*. Since then Hell has been composed of fire and brimstone, not fire and nitric acid. We are reminded of the prayer common in some Victorian churches in Britain: "From Hell, Hull, and Halifax, Good Lord deliver us." Whether the association of Hull and Halifax with Hell was due to air pollution is uncertain.

The Industrial Revolution was accompanied by very few pollution controls. People erected factories where the water was available for mills. The smoke released from the chimney stacks, and from the coal-burning workers' homes nearby, often was trapped in a temperature inversion. The wealthier citizens escaped the cities either by living on the tops of hills or by using weekend houses in the country. From the beginning of the current century, however, steady improvement has been made in pollution control. The first method adopted, primarily by industry, was dispersion by tall chimney stacks. For example, in a classic text, Herington [1920] says:

> It is quite true that perhaps 60 percent of the fly ash goes up through the stack. This ash is of such light fluctuant nature that it is dissipated over a wide area before precipitation occurs and no trouble can be expected from this source, although the . . . tonnage put out . . . seems great.

The next improvement was to install devices that eventually suppressed 99.5 percent by weight of the particulates, leaving mostly small aerosols to escape to the atmosphere. Unfortunately sulfur and nitrogen oxides, being gases, also escape collection by those devices designed to collect particles.

After a disastrous air pollution episode in London in 1952, when 4,000 more people died in a week than usual, Great Britain adopted a clean air act that forbade the burning of soft coal in the center of cities. Before this time, most houses were heated by soft coal, and the domestic burners used low chimney stacks and had no particulate, sulfur, or nitrate suppression. Even as late as 1950 it was reported that in Glasgow three tons of soot fell per acre per year. Although the problem was not generally as bad in the United States, there are a few cities—Pittsburgh, for example—where it was as bad.

The combined effect of locating industry and power stations in unpopulated areas, introducing tall stacks, and banning combustion of soft coal in the center of cities led to a clean-up in Britain that has been impressive and obvious. Until recently, most experts considered that these measures were all the pollution control that is necessary. New health and property damage assessments, and a general rise of standards and expectations, suggest that maybe we should do more.

The disagreement among scientists about whether or not a health hazard remains at present levels of air pollution poses a major public policy dilemma. It is therefore worthwhile to try to understand the disagreement in some detail. There is less disagreement about data than about the proper conclusions to be drawn therefrom.

On the one hand, the traditional view is that if the concentration of a pollutant can be brought below a well-defined threshold level, there will be no adverse health effects. The belief in the existence of a threshold is deeply held. The body seems to show an ability to detoxify itself so long as poisons are not present in large quantities. It is probable, of course, that individuals will show a variation in sensitivities, but insofar as these thresholds of sensitivity are genetically determined and pollution is affecting the young, we might expect that natural selection would gradually weed out of the human race those with low thresholds. All that remains, then, is to determine the threshold.

On the other hand, it is now known that many problems, cancer in particular, are determined more environmentally than genetically. Some theories of cancer induction suggest a random process resulting in the linearity of cancer incidence with the concentration of carcinogen.

This difference in preconceived views is not merely a matter of age, but is partially a matter of training. A physical scientist tends to think in terms of probabilities; an individual atom, γ ray, or molecule can set in motion a chain of microscopic and macroscopic events in the body that leads to morbidity and death. But the probability that any one pollutant molecule sets such a chain into motion must be miniscule—one in 100 million or less. A physical scientist will not automatically think in terms of a threshold, although he is of course willing to concede that other mechanisms, such as repair of damaged cells or detoxification, can under some circumstances lead to a threshold, just as the random motion of many molecules with forces among them can produce the well-defined cooperative phenomena of the melting of solids and the boiling of liquids.

This view of the physical scientist has had a great deal of impact, particularly in theories of carcinogenesis, because cancer by now is believed to have a microscopic and cellular origin. It is far less clear that the argument should apply to the breathing difficulties that are more characteristic of severe air pollution. A biologist or physician, on the other hand, deals with matter in bulk; he sees basic large effects on the body as well as the power of the body to heal itself. The concept of a threshold comes naturally to him.

The problem is that we are looking in the gray area between the microscopic ideas of the physical scientist and the macroscopic ideas of the biologist. In this book we will not resolve the resulting discrepancy but will carry these ideas through as bounds on our possible interpretation of the data.

It would be nice if we could distinguish definitively between the two limits of a firm threshold model, and a proportional incidence model by direct observation or experiment. This is not possible, largely because the effects of air pollution on death rate are delayed, and causality is hard to establish. It is not possible to prove a threshold even for tobacco cigarette smoking, which kills five to ten times as many Americans (approximately 300,000 per year) as the worst estimates of air pollution. We know from the work of Doll and Peto [1976] that smoking four cigarettes a day gives about one-tenth the incidence of lung cancer among British physicians as smoking 40 cigarettes a day, yet we cannot directly tell whether or not one cigarette a week, or living with a smoker and inhaling stale nicotine, is harmful. We can compare directly a smoker with his neighbor who is exposed to similar amounts of other pollutants, so that an epidemiologic assessment is possible. We cannot easily find large matched groups of those exposed to air pollutants and those not so exposed. Much more serious is that as we mitigate the known and obvious effects of air pollution, we are led to disperse the pollutant sources and dilute the pollutants, making the task of comparison harder. We have little hope that we will ever have much more reliable data. We are therefore forced to use animal tests, tests of metabolism, theoretical ideas, and crude census surveys to decide what an effect might be.

In Chapter 6 we will produce evidence that suggests that 50,000 deaths per year might be attributed to air pollution in the United States. If these indications that 50,000 Americans die every year are correct, the effort to reduce this pollution must be commensurate. It is worth perhaps $50 billion per year if we can reduce it completely. On the other hand, the number of victims of air pollution could be zero, and all of this money would be wasted. Moreover, the rising costs of electricity from pollutant control could give secondary adverse health effects. This has been stressed by McCarroll [1979] in the following paragraph.

> All these health costs (due to increased electricity costs) are real and must be balanced against any anticipated benefit from reduced SO_x levels achieved by costly scrubbing of stack effluents, or by burning of expensive desulfurized fuel. Since no detectable excess mortality can be found ascribable to SO_2 or sulfates at presently prevailing levels, money used to reduce these levels still further might better be used for other public health purposes.

In the following chapters we hope to present this controversy logically so that decisions can be more soundly based.

We will see that the discussion has resolved around three major contributors to the pollution: particulates, sulfur oxides (SO_x) and nitrogen oxides (NO_x) or colloquially, rocks, socks, and knocks.

1.2 SCOPE OF THE PROBLEM

It is useful and instructive to look at some very crude and elementary studies of air pollution surveys before we go into detail on the better, more complete ones. This will give us perspective and an indication of the size of the effects we are discussing and the type of problems we address in the detailed studies.

We first present a graph (Figure 1–1) of the number of deaths in Oslo, Norway, plotted against the weekly mean concentration of SO_2 [Lindberg 1968]. The statistics are poor, but nonetheless a linear relation fits well and, if the relationship could be extrapolated, shows that the death rate would be doubled at a concentration of 1 part per million parts (ppm) as compared with the U.S. air standard. Even at the level of the U.S. primary air-quality standards, 0.03 ppm of SO_2 in air, the line predicts a 3 percent increase in the death rate. This then is suggestive evidence that air pollution causes mortality; but it is very far from proof of causality. A correlation has been shown, but such numbers never prove causality. A physical scientist asks if the SO_2 concentration is the correct independent variable. Obviously, particulate, NO_x and SO_x concentrations vary together with the SO_2 concentrations. But there are also other more obscure possibilities. The air-pollution concentrations may be higher as the nights become longer. The long nights may cause individual depression and illness or an increase of automobile accidents.

Figure 1—1. Relationship between total number of deaths for 156 winter weeks in Oslo, Norway and SO_2 concentration. Equations predict number of deaths at different air pollution levels. (Adapted from W. Lindberg, *General Air Pollution in Norway*, Oslo, 1968: Smoke Damage Council.)

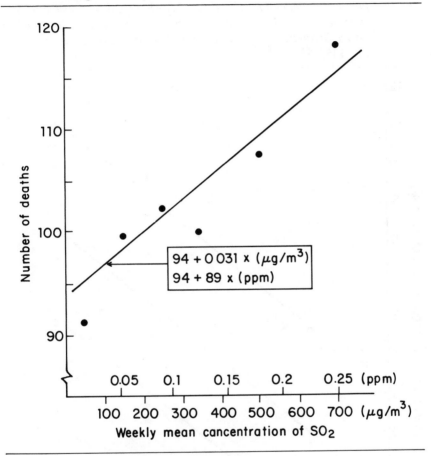

In these data we have related the death rate to air pollution at a constant position in space (Oslo) varying with time in a correlation called a time-series study.

Another easily understood study has been presented in Japanese work [see Nishiwaki et al. 1971]. Figures 1—2, 1—3, and 1—4 present bronchitis incidence as a function of sulfation index (total SO_x) by considering several Japanese cities. Again a correlation is observed that remains even after correction for smoking habits. Two features

Figure 1–2. Correlation between the prevalence of chronic bronchitis and sulfur oxide concentration by smoking history (as measured by the lead-candle method) for several Japanese cities. Curve a, 11–20 tobacco cigarettes per day; curve b, 1–10 cigarettes per day; curve c, nonsmokers. (Adapted from Nishiwaki et al., 1971, *Atmospheric Contamination of Industrial Areas Including Fossil Fuel Stations and the Method of Evaluating Possible Effects of Nuclear Power Stations*, Vienna: International Atomic Energy Agency, IAE-SN-145/16.)

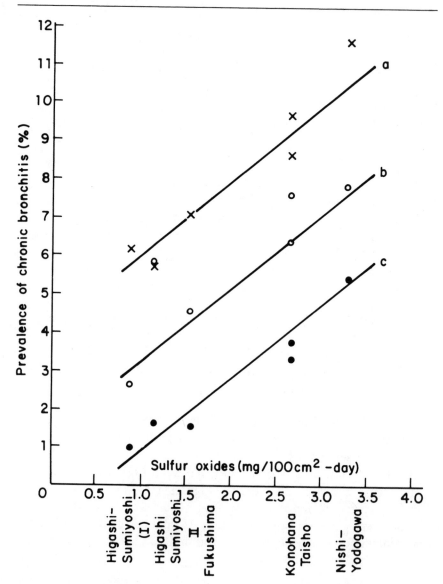

Figure 1–3. Correlation between the prevalence of chronic bronchitis and sulfur oxide concentration (lead-candle method) in Japan without separation of smokers and nonsmokers. (Adapted from Nishiwaki et al., 1971, *Atmospheric Contamination of Industrial Areas Including Fossil Fuel Stations and the Method of Evaluating Possible Effects of Nuclear Power Stations,* Vienna: International Atomic Energy Agency, IAE–SN–145/16.)

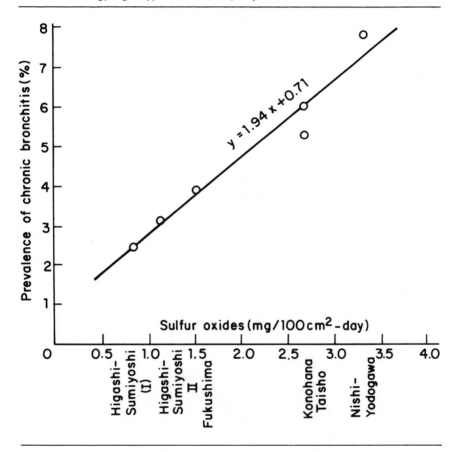

are obvious: for present levels (near the U.S. air quality standards) the effects are smaller than tobacco cigarette smoking, and the data suggest a threshold for nonsmokers. But again there are questions. Are the measurements of sulfate precipitation adequate? Is SO_x the correct index for air pollution? Are there more important differences among the Japanese cities than the air pollution index?

Figure 1−4.   Correlation between the prevalence of chronic bronchitis in men and women and sulfur oxide concentration in Japan. Men show greater prevalence of chronic bronchitis but the strength of association between sulfur oxide pollution and chronic bronchitis is similar for men and women. (Adapted from Nishiwaki et al., 1971, *Atmospheric Contamination of Industrial Areas Including Fossil Fuel Stations and the Method of Evaluating Possible Effects of Nuclear Power Stations*, Vienna: International Atomic Energy Agency, IAE−SN−145/16.)

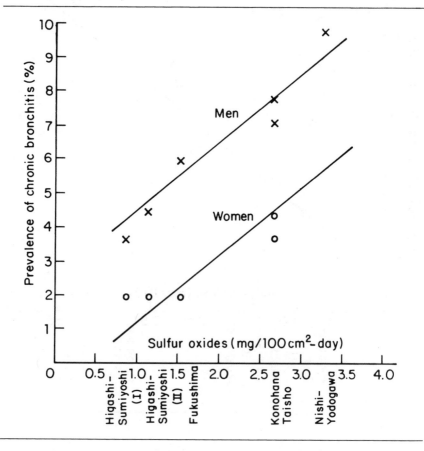

This type of study relates the death rate over a long time to an air pollutant index by varying the position in space. It is called a cross-sectional study because it uses different cross-sections of the population in space.

One of the first attempts to proceed beyond the simple correlation was made by [Winkelstein 1967] using mortality rates in Buffalo, New York. He realized that (1) wealthy people live longer than poor people, and (2) wealthy people move out of polluted city centers into the less polluted suburbs. He therefore tried to relate the mortality rate simultaneously to two variables—particulate concentration as an air pollution index, and income level. His results are shown in the two-dimensional plot of Figure 1−5. Finch and Morris [1977] fitted Winkelstein's data to an equation

$$SMR = -0.51 + (0.6 \pm 0.08)\ 10^4/I + (0.007 \pm 0.001)P$$

where

SMR = the standardized mortality ratio

I = the family income in dollars

P = the concentration of total suspended particulates (TSP) in $\mu g/m^3$.

We can see at once that if we had merely made a fit of the mortality rate to air pollution, we would be taking a section up the diagonal of this curve and finding about double the effect that we derive for air pollution when both variables are included.

The studies of Winkelstein by themselves do not answer all the questions we posed as we looked at the Japanese data. The most obvious is, if we can halve the calculated health effect by including income level as a confounding variable, are there other variables that can also reduce the calculated health effect and bring the calculated effect to zero? We will address this question in detail in Chapter 6.

1.3 POSSIBLE CHANGES IN AIR QUALITY

There are two major trends that affect air quality in opposite directions. On the one hand there is considerable emphasis on meeting air quality standards by control of emissions, and on the other there is a national energy plan to double the use of coal, the most polluting of fossil fuels, within 20 years [DOE 1976].

The present emphasis is on meeting the air quality criteria adopted by the U.S. Environmental Protection Agency (EPA) under the Clean Air Act of 1970. When these criteria were adopted, they were set at

Figure 1–5. Relationships among air pollution levels, economic status, and total mortality. This figure is produced from data of Winkelstein [1967] and shows the interaction between the independent variables of air pollution and economic status. A lack of data in four cells is indicated as ND.

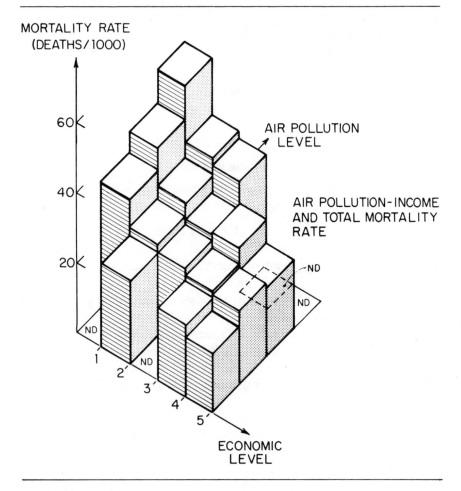

levels below which no health effect had been proved. The criteria are listed for concentrations of SO_2, NO_2, ozone (O_3), and TSP in Table 1–1. In 1970 there were many areas where the concentrations violated one or more of these criteria. For obvious reasons monitors and samplers were placed in areas where air quality had traditionally been bad—mostly in urban areas of the eastern United States.

Table 1–1.   U.S. ambient air quality standards.

Pollutant	Primary (To protect human health)		Secondary (For public welfare)
	$\mu g/m^3$	ppm	$\mu g/m^3$
Sulfur dioxide			
Annual arithmetic mean	80	0.03	—
24-hour maximum[a]	365	0.14	—
3-hour maximum[a]	—	—	1,300
Particulate matter			
Annual geometric mean	75	—	60 — guideline only
24-hour maximum[a]	260	—	150
Nitrogen Oxides (i.e., NO_2)			
Annual arithmetic mean	100	—	100
Photochemical Oxidants			
1-hour maximum	235	0.12[b]	—

a. Not to be exceeded more than once per year.

b. Revised upward from 0.08 ppm in 1979.

Source: *Federal Register* 36: 8185 (30 April 1971); ibid. 38: 25678 (14 September 1973); Title 40 CFR Part 50.

By 1975 there was already considerable improvement. Out of 210 million people, 46 million were living in areas where the criteria for TSP were exceeded—down from 74 million in 1970. The EPA reports similar improvement for SO_2 with only a few areas in violation of the standards. The monitoring for O_3, a major photochemical oxidant, is only just beginning on a large enough scale to judge improvements.

Although some of the improvement is by suppression of particularly bad sources, a large part is by dispersion of sources to the suburbs and rural areas, and the air quality, although not always measured, is almost certainly worsening in these areas. There has been a steady increase in the burning of fossil fuels, and a steady increase in the release of sulfur oxides and nitrogen oxides. This increase is likely to continue because of the increased emphasis on burning of coal.

2 PARTICULATE MATTER AND ATMOSPHERIC DISPERSION

In this chapter and the next we discuss the emissions of the major air pollutants that are believed to cause some adverse health effects. In principle the quantities of pollutant emitted are easy to measure. It is a tautology that adverse health effects are related to concentrations of air pollutants and not directly to the emissions, so we also describe how to measure concentrations and give some numbers about their distributions. However, because an adequate monitoring system does not exist to fully characterize air pollutant concentrations, we also briefly describe how air dispersion calculations may be used to calculate concentrations from known emission sources. We also describe how the anthropogenic (caused by human activity) airborne particulates are related to the natural ones.

2.1 PARTICULATE MATTER SUSPENDED IN THE ATMOSPHERE

The most obvious effect of burning coal in the preceding centuries was the production of noxious black smoke caused by the emission of particles, often of unburned carbon. These soot particles soiled clothes and reduced visibility. Even now, as shown in Figure 2−1 a and b, haze is the most obvious effect of air pollution. Haze may be caused by emissions from many different sources both natural and

15

Figure 2–1. (a) Lines of equal visibility of June 30, 1975. A large hazy area covers much of the midwestern United States on that day. High b_{scat} areas correspond to areas of low visibility and high sulfate concentration. (b) Satellite photograph of the United States on June 30, 1975 showing the area of low visibility in (a) (from W.A. Lyons and R.B. Husar [1976], "Visible images detect a synoptic-scale air pollution episode," *Mon. Weather Rev.* 104: 1623–1626.)

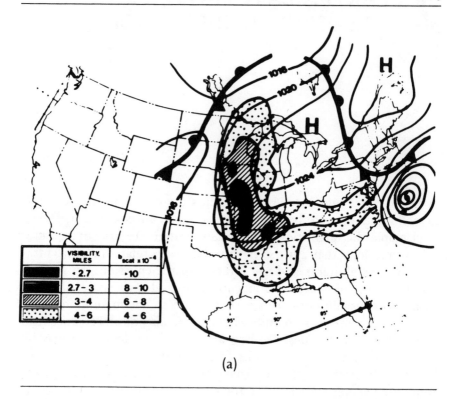

(a)

anthropogenic. On the surface weather map for June 30, 1975 are plotted isopleths of visibility. A large area of reduced visibility occurs on the back side of a high-pressure area. The area of most impaired visibility covers portions of Iowa, Missouri, and Arkansas. The particulate burden is so severe that it can be seen as a large hazy air mass scattering light back into space as photographed by the Synchronous Meteorological Satellite (SMS).

Although the absorption and scattering of light alert the observer to the presence of air pollution, the major adverse health effects arise from those particles that enter the lungs. The large, visible

Figure 2–1.   continued

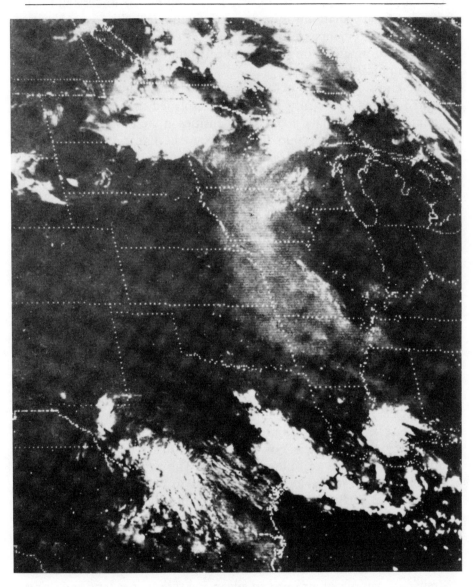

(b)

particles fall to the ground and will not travel long distances; the smaller ones can travel long distances and enter the lungs. In discussion of particulates, therefore, it is essential to understand the details of particulate measurement. Unfortunately the vast majority of our measurements are of mass concentrations and do not provide detailed information on size, chemical composition, pH, or perhaps other important properties.

The emission density of total suspended particulates (TSP) in the United States by county in 1978 is displayed in Figure 2−2. The counties with the highest emission densities in the East are the industrialized areas of the Ohio River Valley, the Great Lakes, and the Tennessee Valley, as well as the refinery and industrial counties of eastern Pennsylvania and New Jersey. In the sparsely populated western states, high emission densities are primarily associated with smelter operations. Stationary fuel combustion emissions and industrial processing emissions are approximately equal and are the major national sources of primary particles.

2.1.1 Particulate Measurement

The dust bucket was a popular particulate measurement technique in Britain during the first half of the twentieth century. A small kitchen mop-style bucket is placed outdoors for one month. The weight of the collected dust is then divided by the area of the bucket and the results expressed as grams per square centimeter (g/cm^2) per month or tons per square mile per month. A more recent text on air pollution describes this method as continuing to be a useful technique for measuring the "dirtiness" of air [West 1968]. It is not at all clear, however, what this technique measures. The particles collected will be relatively large, having sufficient mass to fall into the bucket by gravitational settling. Some implications of soiling potential may be developed with this method; but information on the all-important fine particles is completely lacking because these particles settle very slowly and comprise a minor fraction of the collected mass.

The coefficient of haze (COH)[1] is measured by automatically drawing a sample of air onto an advancing filter-paper tape and using

---

1. A COH unit is that quantity of particulate matter that produces an optical density of 0.01 when measured by light transmission at 400 nm and when compared to the transmission of dust-free filter paper that is taken as 100 percent transmission.

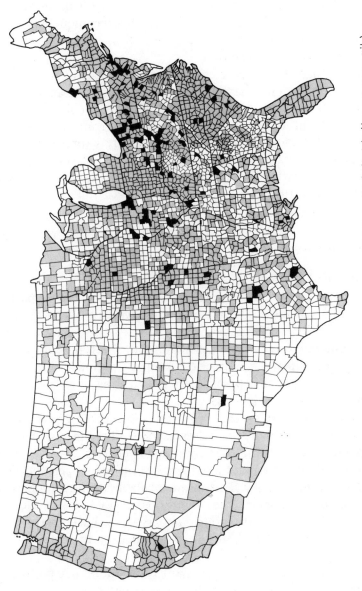

Figure 2–2.   TSP annual emissions patterns by county, 1978, in tons per square mile.

optical equipment, which essentially measures the "blackness" of the collected material. The optical density of the collected material is measured and expressed as COH units per 1,000 linear feet of air drawn through the filter [West 1968]. This technique and a similar one, the smoke index, do not correlate well with the measure of total particulate mass concentration.

The standard reference method currently used for measuring TSP is with a high-volume sampler. A known flow rate of air is drawn through a filter for a 24-hour period. From the collected mass and total volume of air sampled, particulate mass concentration is calculated. Total suspended particulates (TSP) are expressed in units of micrograms per cubic meter ($\mu g/m^3$). The TSP measure is dominated by the larger particles and the technique does not provide separate information on fine particles.

2.1.2 Ambient Concentrations of TSP

The standard method for measuring the particulate content of the outdoor air has been to weigh the total amount of all particulate matter retained on a filter paper through which ambient air has been forced by a high-volume air pump for a 24-hour period. However, this measurement technique defines only the mass per unit volume of particles present in the air over the sampling period. There are instrumental techniques that permit size classification of ambient aerosols, but these are not routinely used.

The U.S. Environmental Protection Agency (EPA) in 1979 installed 100 dichotomous samplers in the United States. These instruments measure both the coarse fraction (particles 2.5 to 15 μm) and the fine fraction (particles less than 2.5 μm) of an aerosol. The fine fraction is roughly equivalent to the inhalable (respirable) particulate fraction discussed later. The locations that EPA set as the highest priority for siting dichotomous samplers were those areas where the air-quality standard had not been attained or in populated urban areas.

In addition, there have been numerous studies using a variety of instruments and analytical techniques to describe the size and the chemical and elemental composition of urban and rural aerosols.

The TSP data averaged over 24 hours reported from about 4,008 monitoring stations operating in 1977 are stored in the National

Aerometric Data Bank (NADB) for use in assessment of nationwide progress in achieving and maintaining the national ambient air quality standards (NAAQS). These high-volume sampling locations were operated by federal agencies, state agencies, and local air pollution control agencies (Table 2–1). In addition to this monitoring network, many utilities, paper companies, smelting companies, and others operate high-volume sampling networks. Most of this data, however, is not readily available, nor is it centrally stored.

Data from the network of high-volume samplers may be of limited value in assessing true population exposures to suspended particulates for a number of reasons. First, there is no standard location for these monitors; some are at ground level near roads, whereas others are on rooftops of multistory buildings. Second, the quality control procedures have only recently been implemented by EPA [EPA 1977b]. Third, and perhaps most important, the monitor's proximity to certain pollution sources may provide samples that are unrepresentative of general human population exposure. Nevertheless, even while efforts to improve the ambient air-quality monitoring programs continue through EPA's Office of Environmental Monitoring and Surveillance this data base serves to indicate compliance status as well as trends in total suspended particulate levels.

Of the 4,008 stations reporting TSP values in 1977, only 2,699 stations had a sufficient number of samples per quarter to allow a valid annual average to be calculated. From these stations, 40 percent, or 1,970, reported a violation of the secondary TSP annual standard ($60\,\mu g/m^3$); 17 percent, or 456 of these stations, reported violations of the national primary annual standard ($75\,\mu g/m^3$). Of

Table 2–1. Total suspended particulate monitors in the United States operated by federal, state, and local agencies.

Agency	Total Suspended Particulate Monitors
Federal	90
State	2,588
Local	1,330
Total	4,008

Source: Environmental Protection Agency, *National Air Quality, Monitoring, and Emission Trends Report* EPA–450/2–78–052–December, Washington, D.C.: EPA, 1978c.

all those stations reporting at least minimum data, 36 percent, or 1,424, showed violations of the 24-hour secondary TSP standards (the second highest value not to exceed 150 $\mu g/m^3$); 8 percent, or 314 monitoring stations, reported violations of the 24-hour primary standards for TSP (the second highest value not to exceed 260 $\mu g/m^3$). Table 2–2 reports the number of stations that were violating the national primary or secondary standards for TSP. Of 50 states and the District of Columbia, 42 showed violations of the primary ambient air standard for TSP in at least one of the monitoring locations.

Using 1977 TSP data from the NADB, the number of air-quality control regions where the national annual primary air-quality standard was not achieved in at least one sampling site is shown in Figure 2–3. But this figure may be misleading. A local source may cause a violation at a particular site in an otherwise unpolluted area. EPA has recognized this problem and allows states to subdivide Air Quality Control Regions (AQCRs) into attainment and non-attainment sections.

Using the same 1977 NADB TSP data, the AQCR attainment status for the 24-hour maximum NAAQS is shown in Figure 2–4. As

Table 2–2. National summary of total stations reporting data and number reporting violations of air quality standards for total suspended particulate matter, 1977.

Data Record and Standard Exceeded	Number of Stations	% Sites Exceeding NAAQS
Valid annual data[a]	2,699	
Annual secondary	1,070	40
Annual primary	456	17
At least minimal data[b]	4,008	
24–hour secondary	1,424	36
24–hour primary	314	8

a. Must contain at least five of the schedule 24–hour samples per quarter for EPA-recommended intermittant sampling (once every 6 days), or 75% of all possible values in a year for continuous instruments.

b. At least three 24–hour samples for intermittant sampling monitors.

Source: Environmental Protection Agency, *National Air Quality, Monitoring, and Emission Trends Report* EPA–450/2–78–052–December, Washington, D.C.: EPA, 1978c.

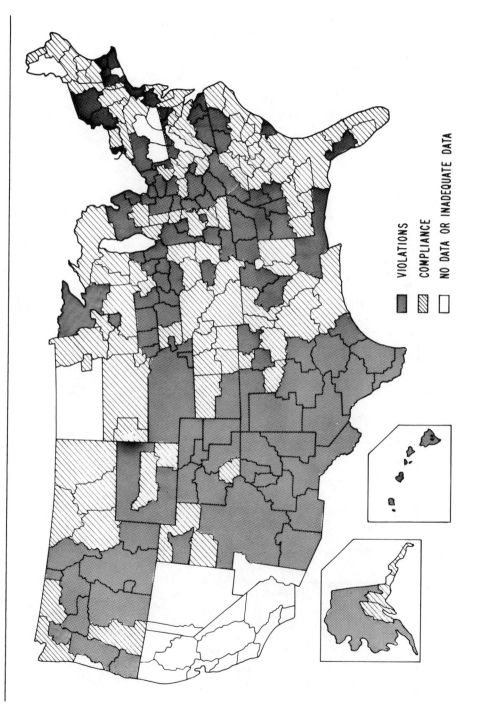

Figure 2–3. Status of compliance in 1977 with annual ambient air quality standards for particulate matter ($>75 \ \mu g/m^3$).

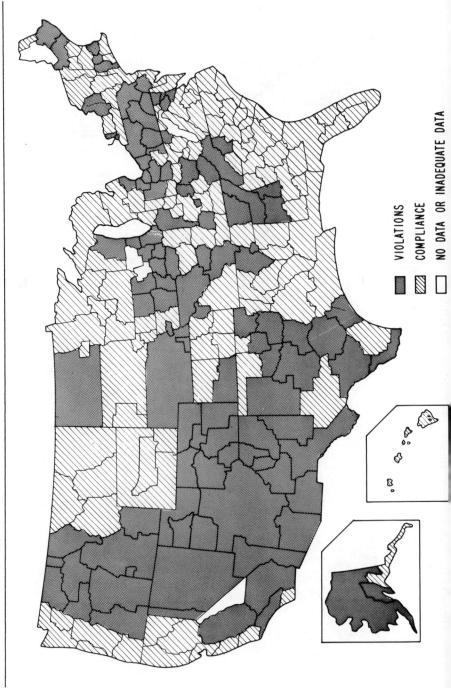

Figure 2-4. Status of compliance in 1977 with 24-hour ambient air quality standards for particulate matter ($> 260 \ \mu g/m^3$).

for the annual primary standard, a violation of the 24-hour NAAQS for TSP at one location does not necessarily imply a higher health risk for the entire population of that area. The health implication even for those living near a site in violation is not clear. Many TSP measurements are strongly influenced by resuspension of surface dust containing, for the most part, larger particles. This apparently causes numerous violations in the arid regions of the northwestern and southwestern United States. Populations living in attainment areas but exposed to TSPs high in trace metals, for example, might have a considerably high health risk.

The geographic displays of attainment status are only one way of conveying the extent of the TSP pollution problem. To indicate the severity of ambient exposures to TSP, the ninetieth percentile concentration of the 24-hour measurements was examined for all 4,008 sites in the 1977 NADB. Concentrations of TSP and other air pollutants have widely been reported to be lognormally distributed [Larsen 1971] except for the high and low ends of the distribution [Mage and Ott, 1978]. The ninetieth percentile was chosen as being a reasonably stable indicator to represent a TSP level that is exceeded on approximately 36 days of the year. This value serves as an indicator of severe TSP pollution exposure.

In Figure 2-5 the AQCRs having at least one monitoring station whose ninetieth percentile for TSP exceeds 260 $\mu g/m^3$ are displayed. The AQCRs in Montana, Arizona, and New Mexico have a large number of monitoring sites for a relatively sparse population. A number of these sites are near smelters. Hence, the high levels do not necessarily imply high population exposure to TSP. In addition, windblown soil is contributing in part to the higher levels in these states. In the Northeast and East, the elevated TSP concentrations reflect the higher density of industrial and urban emissions. In these cases, the high levels (in Pennsylvania, Ohio, New Jersey, New York, Connecticut, and Massachusetts) indicate a larger population exposed to peak TSP concentrations.

Of the nation's 254 AQCRs only 20 had ninetieth percentile TSP concentration less than 100 $\mu g/m^3$. 154 AQCRs had ninetieth percentile values in at least one site between 100 and 200 $\mu g/m^3$. From these data, we see that the great majority of the U.S. population is exposed to ambient TSP concentrations exceeding 100 $\mu g/m^3$ for at least 36 (0.1 of 365 days) days of the year.

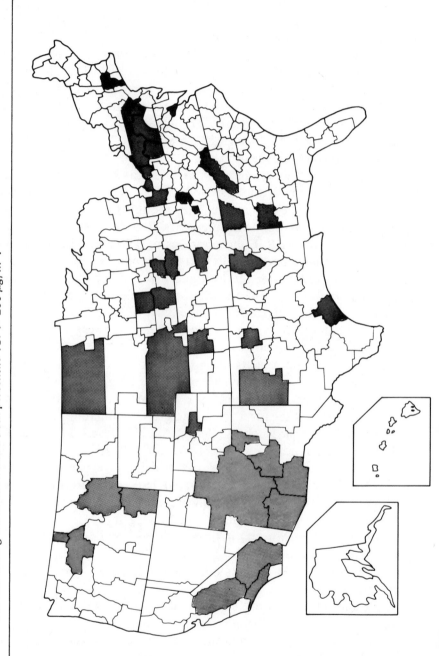

Figure 2-5. Severity of exposure to particulate matter using 90th percentile concentration of the distribution of 24-hour samples at any site within an Air Quality Control Region. Shaded areas indicate AQCRs in which at least one monitoring site in 1977 had a 90th percentile TSP > 260 $\mu g/m^3$.

2.1.3 Historic Trends in TSP Levels

The sampling methods for atmospheric particulate exposure were not well developed prior to the systematic high-volume filter measurements that commenced in the early 1960s. Dust buckets and settling jars were used as soot indicators. Ludwig, Morgan, McMullen [1971] used calculations of dustfall to plot the trends in six cities. Figure 2–6 indicates that settlable dust has been reduced a factor of two in Pittsburgh, Cincinnati, Chicago, and New York. Ludwig notes that there were many changes over this period that would decrease fugitive dust from roads and agricultural sources. It is difficult to interpret dustfall in terms of industrial controls, although many sources in the late 1950s and throughout the 1960s were being controlled, which helped to eliminate the larger particles in cities. The trends analysis by Lynn et al. [1976] of the total filtered particulate matter in several cities suggested that urban values have decreased by approximately 20 percent over the decade of the 1960s. Examination of the expanded data set for TSPs from high-volume samplers shows that for 2,707 sites the median concentration has remained about 60 $\mu g/m^3$ between 1972 and 1977. The geometric mean concentration over this period has decreased by approximately 8 percent. A slight decrease in the ninetieth percentile values is observed over this period (Figure 2–7). Reduction of the peak TSP levels reflects the

Figure 2–6. Trends in settlable dust in six U.S. cities from 1925 to 1965 (Ludwig, Morgan, McMullen 1971).

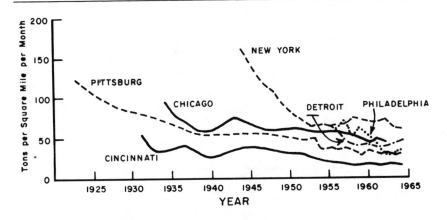

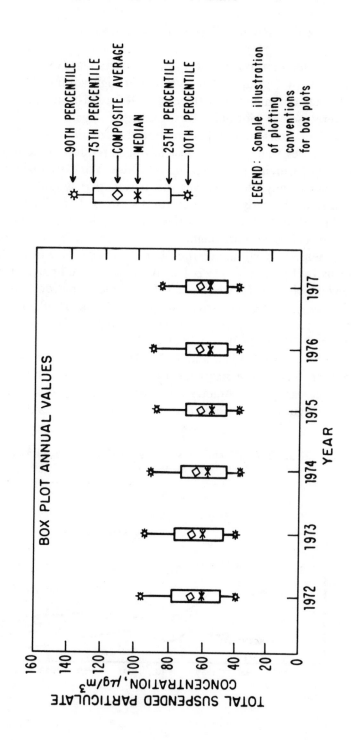

Figure 2–7. Nationwide trends in annual mean TSP concentrations from 1972 to 1977 at 2,707 sampling sites. (EPA, 1978c.)

control strategy of reducing emissions from larger stationary sources. In addition, displacing sources to rural regions and building new sources with taller stacks will decrease peak concentrations.

In view of the almost 50 percent reduction in TSP emissions between 1970 and 1977, the rather modest 8 percent decrease in the composite average annual TSP levels is disappointing. This indicates the large contribution of area sources and other nonanthropogenic sources to TSP levels that are not counted in the emission base.

The overall trend in improvement from 1972 through 1975 was followed by a reversal in some regions in 1976. Despite this one-year reversal, in 1977 60 percent of the sites showed continued improvement from the 1972 to 1977 levels. For those sites with NAAQS annual TSP violations, 77 percent nonetheless displayed long-term improvements. Approximately 25 percent of these sites that were in violation reported their lowest annual values in 1977. EPA offers the explanation that the short-term reversal in 1976 was due to unusually dry weather, resulting in wind-blown dust that may have contributed to elevated TSP levels throughout the Central Plains, Far West, Southwest, and Southeast.

2.1.4 Respirable Particles

In discussing fine particulates, the air quality criteria document for particulate matter states: "Most of the available studies on the effects of particulate air pollution . . . do not specify particle size, and this document is limited to treating particulate matter as a whole, and to considering the effects which are generally associated with the presence of particles in the air" [HEW, 1969]. The problem with omitting consideration of particle size is that this criteria document not only forms the basis for uniform federal particulate standards but has in practice been the basis for standard measurement techniques that produce the future data base. Regulation setting, by accepting the limitations of the available measurement techniques while adopting them as standard procedure, does not stimulate those studies necessary to relate particle size to effects. This can set in motion a circular dilemma that institutionalizes inadequate or inappropriate measurement and regulation.

This problem is addressed in an annual report of the President's Council on Environmental Quality:

Present ambient air-quality standards and monitoring of suspended particulates . . . are concerned only with the total weight of airborne particulate matter. But total weight is at best only a crude indicator of trends in the kinds of particulates that are most important to human health. By itself such measurement is inadequate for many scientific purposes and therefore may be inadequate as a regulatory guide [CEQ 1975].

There is currently no accepted convention either on measurement or on nomenclature. Various phrases are used: fine particulates (those below 2.5 μm in diameter in some definitions or below 1 μm in other definitions); mass respirable particles or total respirable particles, defined by various criteria to be those particles collected by the human lung.

The lack of chemical characterization of ambient particulates further limits our understanding of the biological and ecological effects of this class of pollutants. No summary exists that provides detail on the chemical composition and size distribution of emissions from primary particulate sources [Fennelly 1975]. Although considerable research has taken place on the total mass composition of particulate coal emissions, little characterization has been done on the chemical composition as a function of particle size [Bolton et al. 1973].

The lack of specification of particle size in the current standards and monitoring methods has long been recognized as a serious shortcoming. EPA has recently defined inhalable particles as being less than 15 μm in diameter. Dichotomous samplers will be deployed to 300 urban and rural locations to gather the necessary background information on ambient concentrations. In defining what an inhalable particle is, EPA had to consider the mechanisms which lead to the deposition of particles in the respiratory system.

Particle deposition in the respiratory tract is discussed in an article describing EPA's definition of inhalable particles [Miller et al. 1979]:

There are five mechanisms by which particle deposition can occur within the respiratory tract. These mechanisms involve interception, electrostatic precipitation, impaction, sedimentation, and diffusion; the latter three are the most important of these mechanisms [Stuart 1976; NAS 1977a]. Inertial impaction of inhalable particles is the principal mechanism for large particle deposition in the respiratory tract, acting on particles ranging from a few micrometers to greater than 100 μm in diameter. Sedimentation is one of the main mechanisms of deposition of inhalable particles having diameters of 0.5

to 2.0 μm, whereas diffusional deposition is important for particles less than
0.5 μm in diameter. The relative predominance of these mechanisms, with
respect to deposition in the head, the conducting airways, and the gas-
exchange areas has been studied [Heyder and Davies 1971; Lippmann and
Albert 1969; Morrow 1972; Lippmann, Albert, and Peterson 1971].

The deposition of particles within specific regions can be influenced by
changes in respiratory flow rate, respiratory frequency, and tidal volume.
Thus, the activity level of the individual and the route of breathing can significantly alter regional, as well as total, respiratory tract deposition of inhalable particles. Deposition in the conducting airways can be altered by
physiological or pathological factors. Lippmann, Albert and Peterson [1971]
have shown that deposition in the conducting airways is greatly enhanced for
asthmatic and bronchitis patients and is higher than normal in cigarette smokers who inhaled 1-5 μm particles.

Using the equation developed by the Task Group on Lung Dynamics
[1966] for the probability of deposition of particles within the head during
nose breathing, the probability is essentially one that particles $\geq$ 13.2 μm
are retained in the head during normal respiration (i.e., nasal breathing with
an inspiratory peak flow rate of 15 l/min).

Figure 2-8 shows that deposition of monodisperse aerosols in the head during inhalation via the nose is essentially 100 percent for particles $\geq$ 10 μm
with average inspiratory flow rates on the order of 30 l/min, i.e., flow rates
corresponding to moderate exercise [NAS, 1977a]. However, during mouth
breathing the nasal passages are bypassed, increasing the fraction of particles
of a given size entering the trachea. For example, Figure 2-9 indicates that
overall head deposition of 10 μm particles is only 65% when breathing by
mouth at an average inspiratory flow rate of 30 l/min [NAS, 1977a]. Thus,
this method of breathing provides increased deposition in the conducting airways and gas-exchange areas. Furthermore, the data of Figure 2-9 indicate
that, even under these inhalation conditions, only a small percentage (< 10%)
of particles > 15 μm would penetrate to the trachea. Various studies indicate
that the deposition site within the nasal airways may be a crucial factor in
determining the likelihood of possible adverse health effects because mucociliary transport may be rapid in some areas and relatively slow in others
[Lippmann 1970b; Fry and Black 1973]. All available data demonstrate that
direct health effects from inhalable particles > 15 μm are primarily restricted
to the upper respiratory tract.* Thus 15 μm would be a reasonable particle
size cut-point to include in the design of a sampler which would differentiate
particles deposited in the upper vs. lower respiratory tract.

---

*It should be noted, however, that a small number of large aeroallergen particles of the
order of 25 μm aerodynamic diameter have been found in the deep lung parenchyma, and
thus, the possibility exists that a direct contact mechanism may be operative in the genesis
of pollenic asthma; see Michel et al. 1977.

Figure 2-8. Deposition of monodisperse aerosols in the head during inhalation via the nose vs $D^2 F$, where D is the aerodynamic equivalent diameter (μm) and F is the average inspiratory flow (l/min). The inspiratory flows in the individual studies of this composit range from 5-60 l/min. The heavy solid line is the International Commission on Radiological Protection Task Group Deposition model. (NAS, 1979a, p. 120).

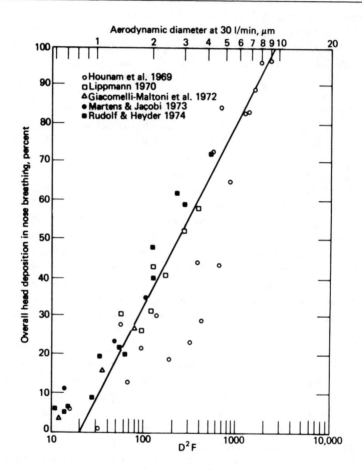

Conducting airway deposition includes deposition both by impaction in the larger airways and by sedimentation in the smaller airways. While it would be desirable to have a sampler that had a cut-point that could delineate a particle size range that would primarily be associated with deposition in the conducting airways, the tremendous variability among individuals in this region prevents such a refinement. Palmes and Lippmann [1977] identified a *characteristic airway* parameter which relates to the average size of an individual's

Figure 2-9. Deposition of monodisperse ferric oxide aerosol in the head of nonsmoking healthy males during mouthpiece inhalation as a function of $D^2 F$, where D is the aerodynamic equivalent diameter (μm) and F is the average inspiratory flow (l/min). An eye-fit line describes the median deposition between 10 and 80 percent. (NAS, 1979a, p. 119). The fitted line is extrapolated to 15 μm.

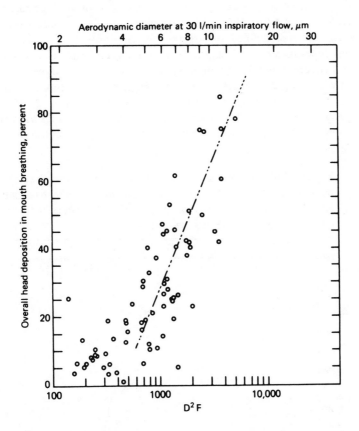

bronchial tubes. Deposition curves for isopleths of this parameter [NAS 1977a] illustrate the variation in conducting airway deposition associated with particles in the 2–10 μm range. For example, in nonsmoking healthy males, deposition of 5 μm particles varies between 33 and 77%, depending upon an individual's characteristic airway dimension value. On the other hand, conducting airway deposition of 5 μm particles ranges from 30 to 95% for cigarette smokers. The range of conducting airway deposition values is less variable for particles < 2 μm or > 10 μm, due in large part to the fact that conducting airway deposition is slight for particles < 2 μm and is nearly complete for particles > 10 μm. Therefore, there is no *standard* conducting airway deposition curve, and hence, there appears to be no clear basis for establishing a particle size range which is exclusively restricted to the conducting airways.

For mouth breathing at 30 l/min average inspiratory flow rate, a maximum gas-exchange area deposition of approximately 50% is associated with a particle size of $\sim$ 3 μm in nonsmoking healthy males (Figure 2–9). When the route of breathing is nasal, a maximum deposition of about 25% occurs with 2.5 μm particles, with a nearly constant deposition of 20% for all particles between 0.1 and 4 μm. Deposition patterns in the gas-exchange areas of the lung are not well defined for cigarette smokers and for individuals with chronic lung disease. At first glance it would appear that a cut-point anywhere between 2 and 3 μm would reflect particle deposition primarily associated with the gas-exchange areas of the lung, since deposition in the head is slight (5–10%) for particles in the 2–3 μm size range (Figure 2–10). However, conducting airway deposition is much more variable for 3 μm particles (11–40% deposition) than for 2 μm particles (5–22% deposition) [NAS 1977a, pp. 611–612].

Figure 2-10. A comparison of the deposition in the alveolar region by mouth and nose breathing as a function of diameter. Lippmann and Albert utilized the data of several investigators to form these curves by eye-fit. (NAS, 1979a, p. 124).

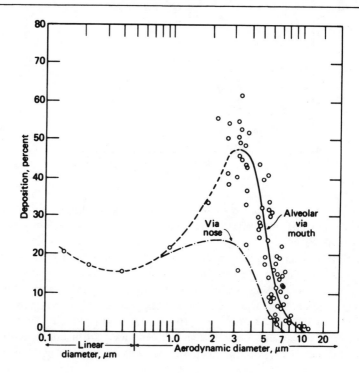

It is hoped that the EPA's network of dichotomous samplers will provide more information on the sources and composition of suspended particulate matter. The dichotomous sampler may become the reference instrumental method for an inhalable or fine particulate standard should either be promulgated.

Limited field studies have indicated that the total dichotomous mass concentration is approximately 60 percent of the TSP concentrations as measured by the high-volume sampler, and the fine fraction is typically 50 percent of the dichotomous mass [Pace and Mayer 1979]. This relationship is not consistent at all times of the year and may not be consistent at all locations (as shown in Figure

Figure 2–11. Monthly mean concentration ($\mu g/m^3$) for co-located high-volume and dichotomous samplers in Topeka, Kansas. Mean values represent 8 days or more of sampling per month. Dichotomous concentration is approximately 60 percent of TSP concentration. Fine fraction is approximately 50 percent of dichotomous concentration except when the ground is snow-covered in January and February 1979. (Reproduced with permission from J.D. Spengler, "Comments on 'Atmospheric dispersion modeling: A critical review' by D.B. Turner," J. Air Poll. Control Assn. 29(9). 1979.

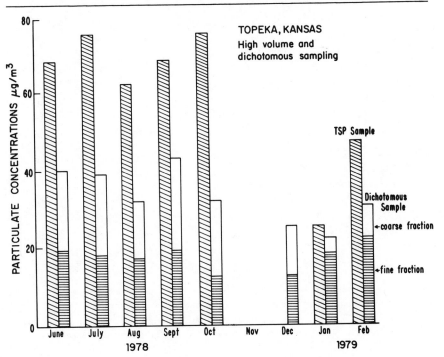

2-11). Figure 2-11 compares high-volume and dichotomous monthly mean samples from Topeka, Kansas. The pattern is altered in the winter months of January and February 1979. The ground was covered with snow, so the contribution of windblown or suspended dust to the coarse fraction was greatly reduced.

The distribution in size of small particles emitted into the atmosphere from natural and anthropogenic sources is shown in Table 2-3. In part, this distribution is reflected in the dichotomous sampling results in St. Louis, Missouri (Figure 2-12). Crustal material dominates the coarse size fraction, whereas sulfates dominate the fine fraction [Pace and Mayer 1979].

Table 2-3. Estimates of particles smaller than 20 μm radius emitted into or formed in the atmosphere.

Type of Particle	Quantity Emitted in Atmosphere (10^9 kg/yr)
Natural	
Soil and rock debris[a]	100–500
Forest fires and slash-burning debris[a]	3–150
Sea salt	(300)
Volcanic debris	25–150
Particles formed from gaseous emissions	
Sulfate from H_2S	130–200
Ammonium salts from NH_3	80–270
Nitrate from NO_x	60–430
Hydrocarbons from plant exudations	75–200
Subtotal	773–2200
Man-made	
Particles (direct emissions)	10– 90
Particles formed from gaseous emissions	
Sulfate from SO_2	130–200
Nitrate from NO_x	30– 35
Hydrocarbons	15– 90
Subtotal	185–415
Total	985–2615

a. Includes unknown amounts of indirect man-made contributions.

Source: NAS, Committee on Airborne Particles, Committee on Medical and Biological Effects of Environmental Pollutants, *Airborne Particles*, Baltimore: University Park Press, 1979a.

Figure 2-12. Source contribution at the Regional Air Pollution Study sites in St. Louis as estimated by elemental composition of particles. (T.G. Pace and I.L. Mayer, "Preliminary characterization of inhalable particulates in urban areas," 72nd Annual Meeting, Air Pollution Control Association, Cincinnati, Ohio, June 24–29, 1979, Fig. 3.)

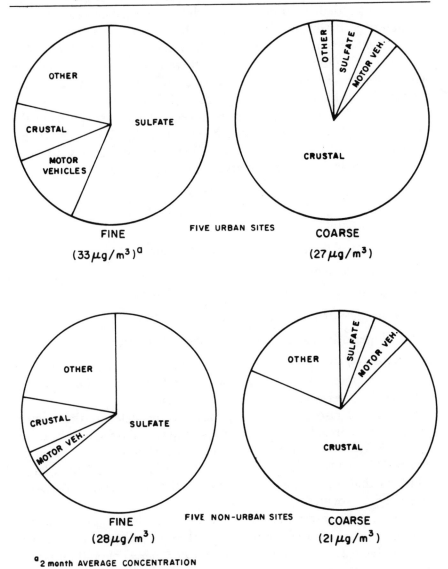

2.2 KNOWN PROPERTIES OF FINE PARTICLES

Particles with diameters of less than one μm (approximately 4/100,000 of 1 inch) originate principally from combustion and condensation of gases. Larger particles generally derive from abrasive processes, for example, natural erosion and industrial grinding. Particles smaller than 0.1 μm move in the air much like gases and are characterized by diffusional forces and brownian, or random, motion as they collide with gas molecules. These very small particles quickly agglomerate following collision and thereby grow in size. Particles with diameters larger than 10 μm have sufficient mass so that they may settle out from the supporting gas [HEW 1969].

The most important factor in removal of large particles from the lower atmosphere, or troposphere, is gravitational settling. In the particle range of approximately 1–17 μm the "speed of sedimentation," or terminal settling velocity, is given by Stokes' law

$$v \; = \; \frac{d^2\,\rho g}{18\,\eta}$$

where (in cgs units)

   v = terminal settling velocity (usually expressed in cm/sec);

   ρ = particle density (g/cm^3)

   g = acceleration caused by gravity (981 cm/sec^2)

   d = particle diameter (cm); and

   η = kinematic viscosity of air (poises = g/cm-sec)

It should be noted that the atmospheric residence time, which is roughly proportional to the inverse of the terminal settling velocity, varies as the inverse square of particle diameter.

Stokes' law also governs the removal of particles by the electrostatic or centrifugal suppressors in use today. These suppressors act by increasing the acceleration of the particles above that caused by gravity. The constant g in Stokes' equation is replaced by a larger value, so that the settling velocity onto the catching area is greater than gravity provides.

However, the settling velocity still varies as the square of the particle diameter, and whereas the removal efficiency for the total mass of particles may be impressively high (99.5 percent), and all visible smoke may be removed, the removal efficiency for the all-important respirable particles is much lower.

Gravitational fall, governed by Stokes' law, is the way in which large particles are removed. Small particles may be removed by agglomeration to larger particles, but the most important mechanism for removal of small particles with diameters of less than a few micrometers is wet removal. Particles can function as nuclei for water vapor condensation. Scavenging of small particles by small droplets within clouds is believed to be the primary removal mechanism for this size particle, a process that is called rainout. The water droplets will begin falling as rain and in a process called washout will accumulate particles that are larger than 2 μm in diameter and that lie in the path of droplet descent. The chemical properties of the individual particles will also affect removal processes, with the more chemically reactive particles growing through agglomeration. For these reasons, particles with the longest atmospheric residence time will be those that are least reactive, that are smallest, and that do not form nuclei.

The length of time required for particles to settle is an important factor in their potential health impact. Wind-blown dust is responsible for almost half of the total mass of particles in the air, but the large particle size and rapid settling velocity assure a short-term and localized impact. Other sources, such as transportation, which emits only 1 percent of particulates, have much greater atmospheric impacts because of the long residence time of their smaller particulate emissions [Fennelly 1975]. The mass, chemical composition, and particle size distribution of a polluter's emissions should be analyzed to assess the potential long-range impacts of the particulate emissions.

As was seen above, measurement techniques based upon the total mass, which is directly proportional to particle volume, lose important information on fine particles. Although particle volume is largely accounted for by the larger particles, the actual frequency distribution of particulates is skewed toward the submicrometer particles. Because of this, the submicrometer particles account for a large fraction of the total surface area of all particulates (Hidy and Burton 1974]. Because particle surface area is important for chemical and biological activity, caution is required in interpreting the significance of the mass concentration of particulate emissions and atmospheric aerosols.

The fact that most of the surface area, in contrast to the mass, is contained in the smaller particles is especially significant in the condensation of volatile metals and liquids on atmospheric particles.

Because condensation is, in part, a surface-mediated process, it would be expected that condensed volatiles would be found predominantly on the smaller particles. Confirmation of this has occurred in studies on fly ash that show that toxic volatile elements such as lead, manganese, cadmium, thallium, chromium, arsenic, nickel, and sulfur increase in concentration with decreasing particle size [Fennelly 1975; Natusch, Wallace, and Evans 1974]. Table 2−4 classifies the major chemical species associated with atmospheric aerosols by size. The normally bimodal chemical species appear in both the coarse and fine size ranges.

The small anthropogenic primary particles, with long residence times and considerable surface area, are uniquely produced in combustion processes. Considerable energy is required to split relatively small particles into smaller units, and mechanical processes can rarely produce particles smaller than several micrometers. The energy of high-temperature combustion, however, is capable of producing submicrometer particles. Secondary small particles are formed from condensation and chemical transformation of natural and anthropogenic gases. Figure 2−13 schematically displays the distribution of the three principal modes of atmospheric aerosols. The principal processes contributing mass in each mode and the removal mechanisms are shown.

There has been little study of the total emissions of the various elements. However, in Table 2−5 we make a partial and approximate list. The values on this list are very variable. For example, some coals contain mercury; others contain very little. It is, however, interesting that fossil fuel burning is responsible for half the release of mercury

Table 2−4.   Classification of major chemical species associated with atmospheric aerosols from EPA 1979.

Normally Fine	Normally Coarse	Normally Bimodal	Variable
$SO_4^=$, C (soot), Organic (condensed vapors), Pb, NH_4^+, As, Se, H^+	Fe, Ca, Ti, Mg, K, $PO_4^=$, Si, Al, organic (pollen, spores, plant parts)	NO_3^-, Cl^-	Zn, Cu, Ni, Mn, Sn, Cd, V, Sb

Figure 2-13. Idealized schematic of an atmospheric aerosol surface area distribution showing principal modes, main sources of mass for each mode, and the principal processes involved in inserting mass in each mode and the principal removal mechanism. (Reproduced with permission from K.T. Whitby and B. Cantrell, "Atmospheric aerosols—characteristics and measurements," International Conference on Environmental Sensing and Assessment, IEEE, Las Vegas, September 14–17, 1975.)

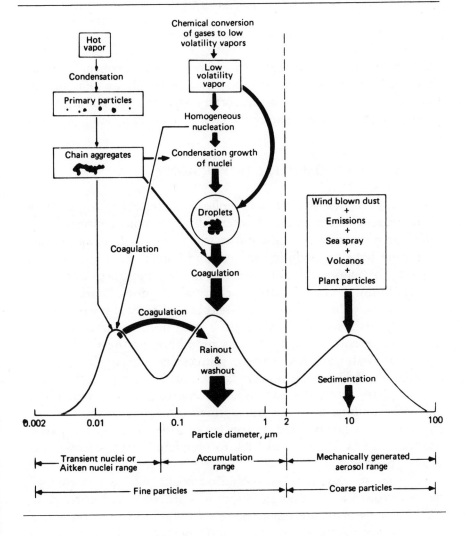

Table 2–5.   Some toxic chemicals emitted in 1973
[Wilson and Jones 1974].

	Tons/year
Mercury [Joensuu, 1971]	1,000
Vanadium (emitted by oil burners)	200,000
Nickel	100,000
Beryllium	1,000
Radium and thorium [Chou and Earl 1972]	10^{-4}
	(or 100 curies/yr)

into the biosphere. The reader should not be led to conclude, by the shortness of this list, that these and other chemicals are unimportant. Indeed they may be the most important aspects of air pollution. The brevity indicates unfortunately the lack of detailed knowledge.

2.3   ATMOSPHERIC DISPERSION

The transport and dispersion of pollutants emitted into the atmosphere prevents the buildup of potentially hazardous concentrations. The first mitigation step, therefore, was to establish an appropriate separation between polluting sources and humans breathing near the ground. The natural dilution occurring in the atmosphere was a sufficient pollution control strategy for several centuries. Unfortunately, this is not now the case. In urban areas as well as rural towns, sources and receptors are close. Often there is insufficient separation, for example, between an airport and its neighbors, or the traffic and a pedestrian walkway. Even where sources have been placed in the countryside and have tall chimney stacks, problems may still exist.

Now there are many such sources emitting into a common air mass. The pollutants can be transformed into other chemical species that remain in the atmosphere for days. Carried along in the mass of air, they can be transported back to populated areas, affecting millions of people.

In order to understand air-pollution health effects and controls, we must discuss atmospheric dispersion processes numerically. We rely on mathematical estimations of air pollution concentrations to design control strategies, set new source performance standards, prescribe chimney heights, test source compliance with ambient air quality standards, and prevent significant deterioration of air quality. There are numerous other applications of air pollution dispersion modeling. These include the modeling of photochemical reactions and transport, visibility impacts, and regional transformation and transport of sulfates.

The models used in air pollution studies are usually variations of three types. Proportional models have been used to estimate the amount of emission reductions needed to achieve a proportional reduction in ambient concentrations. This type of modeling was applied extensively in the 1960s and early 1970s as local and state air pollution control agencies devised fuel sulfur regulations and transportation control plans.

At the opposite end of the scale are the numerical simulation models using derivations of the Nevier–Stokes equation for fluid flow. This model requires specification of boundary as well as initial conditions across the spatial array of interest. Concentrations are estimated by a stepwise finite differentiation scheme. This type of model has been extensively used in photochemical modeling, where the intermediate concentrations of a number of reactive pollutants are needed.

The third and most widely used formulation of atmospheric dispersion is predicted on the turbulent structure of the lower atmosphere having a Guassian or normal statistical distribution. That is, the eddies or turbulence in the air act on the emitted pollution in such a way that the concentrations will be distributed normally around a peak center-line concentration.

2.3.1   Dispersion from a Single Chimney Stack

The basic formula for the spread of pollution from a stationary source was developed by Sutton [1932]:

$$C = \frac{Q}{2\pi u \sigma_y \sigma_z} \exp\left(\frac{-y^2}{2\sigma_y^2}\right) \left\{ \exp\left(-\frac{(z-h)^2}{2\sigma_z^2}\right) + \exp\left(-\frac{(z+h)^2}{2\sigma_z^2}\right) \right\}$$

where

C = the concentration of pollutant in quantity per unit volume ($\mu g/m^3$);

Q = the quantity emitted per unit time by a chimney stack considered to be at the origin or coordinates. The coordinate axes are chosen as shown in Figure 2−14 so that the wind travels with velocity u along the x axis; the z axis is vertical and xy plane ($z = 0$) is the ground;

σ_y and σ_z = the standard deviations of the normal distribution densities of the dispersion. Their values depend upon the weather conditions, reflecting the mixing and turbulence of the lower atmosphere, and vary with x;

h = the height of the plume, not simply the stack height because hot exhaust gases usually make the plume rise even after leaving the stack, although adverse meteorologic conditions can cause downwash.

As written, this equation represents the fact that there is no loss of matter; as the pollutant leaves the stack it is transported downwind and dispersed in the vertical and horizontal directions. The second term in the bracket is required to maintain conservation of mass. This is the "reflection" term for the portion of the plume that has mathematically intercepted the ground. The uses of this equation are discussed in Turner [1979]. In order to use it, values of the constants σ_y and σ_z and h must be determined from observations or direct measurement of atmospheric variables. Pasquill [1961, 1962] listed several weather conditions that bear his name and these can be used to categorize mixing conditions and hence the dispersion coefficients σ_y and σ_z. Models are also used to calculate plume rise. Turner [1972] lists various weather conditions for the Pasquill classification schemes as shown in Table 2−6. Pasquill A is very turbulent air; Pasquill F is very stable air found only at night. Turner lists various formulas for these; Table 2−7, from Rogers and Gamertsfelder [1971], is an average of weather conditions at several nuclear power plant sites. The atmosphere is usually (40−60 percent of the time) found in a neutral condition (D stability).

For long-term average conditions we can measure the wind direction and speed over a year, or we can assume that over a year the wind will blow from each direction in turn. This is equivalent to rotating the coordinate system around the z axis or integrating over y. Then we can average over wind direction to find

$$\bar{C} = \frac{Q}{(2\pi)^{3/2}\, ux\sigma_z} \left\{ \exp\left[\frac{-(z-h)^2}{2\sigma_z{}^2} \right] + \exp\left[\frac{-(z+h)^2}{2\sigma_z{}^2} \right] \right\}.$$

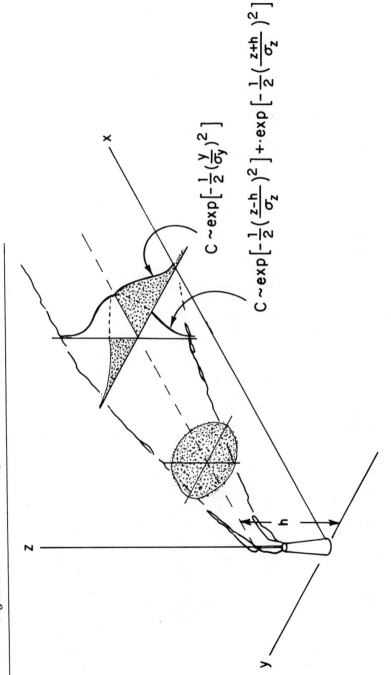

Figure 2–14.   Coordinate system for Gaussian plume dispersion.

Table 2-6. Key to stability categories.

Surface Wind Speed (at 10 m) $m\ sec^{-1}$	Day Incoming Solar Radiation			Night Thinly Overcast or $\geq 4/8$ Low Cloud	$< 3/8$ Cloud
	Strong	Moderate	Slight		
< 2	A	A–B	B		
2–3	A–B	B	C	E	F
3–5	B	B–C	C	D	E
5–6	C	C–D	D	D	D
> 6	C	D	D	D	D

The neutral class, D, should be assumed for overcast conditions during day or night [Turner 1972].

Table 2-7. Meteorologic data for atmospheric release calculations.

Weather Type	A	B	C	D	E	F	G
Probability of weather condition	0.019	0.081	0.136	0.44	0.121	0.122	0.08
Wind velocity, u_x (m/sec)	2	3	5	7	3	2	1
Distance, x (m)				σ_z (m)			
200	28.8	20.3	14	8.4	6.3	4.05	2.63
500	100	51	32	18	13	8.4	5.5
1,000	470	110	59	32	21.5	14	9.2
2,000	3,000	350	111	51	34	21.5	13.7
5,000	–	1,900	230	90	57	35	23
10,000	–	–	400	140	80	47	31
20,000	–	–	650	200	110	58	37
50,000	–	–	1,200	310	150	75	48
100,000	–	–	1,800	420	180	90	55

Source: T. Rogers and C.C. Gamertsfelder, *Proceedings for Environmental Effects of Nuclear Power Stations*, Vienna: International Atomic Energy Agency, 1971.

We are usually interested only in the concentrations at ground level $z = 0$. Then

$$\bar{C} = \frac{Q}{2^{1/2} \, \pi^{3/2} \, ux\sigma_z} \exp\left(\frac{-h^2}{2\sigma_z^2}\right) \, .$$

Finally we can insert values of σ_z and u from Table 2–7 and average over different weather conditions.

These formulas assume that the wind is blowing steadily in one direction. The formula is adequate to describe concentrations at a distance from the source as long as the wind is constant, particularly in direction but also in speed. Beyond 30 km the formula is inadequate. Variations in atmospheric mixing, terrain, wind speed, and direction combined to increase the uncertainty of the Gaussian formulation at long distances.

With these caveats, it is useful to discuss the qualitative features of the dispersion as given by these formulas. We see that the dispersion constants, σ_y and σ_z, increase with increasing x (Figures 2–15 and 2–16). Recent field studies have shown that σ_y and σ_z increase as a linear function of distance for the first few kilometers and then are assumed to increase as a function of $x^{½}$.

1. The ground level concentration is low when $h > \sigma_z$; this is an expression of the obvious fact that in steady winds a plume blows over the heads of persons nearby.

2. For long distances downwind the concentration experienced at the ground is virtually independent of initial stack height and plume rise. The concentration decreases in proportion to $(1/(x \, \sigma_z))$. This is approximately $(1/x)^{1.8}$.

3. The maximum ground-level concentration can be estimated by calculating the concentration at the distance where $h = \sigma_z\sqrt{2}$. Setting the plume height above the ground equal to 2 times the coefficient of vertical dispersion maximizes the gaussian plume equation. At this point the approximate concentration may be calculated as

$$C_{max} = \frac{0.117Q}{u\sigma_y \sigma_z}$$

In Figure 2–17 the distance to maximum concentration, expressed as the maximum of the expression Cu/Q, is displayed as a function of atmospheric stability (classes A through F) and effective stack height. For a given windspeed and direction, emission rate, and

Figure 2–15. Standard deviation, σ_y, in the crosswind direction as a function of distance downwind. (Reproduced from D.B. Turner, *Workbook of Atmospheric Dispersion Estimates*, Washington, D.C.: HEW, 1969.)

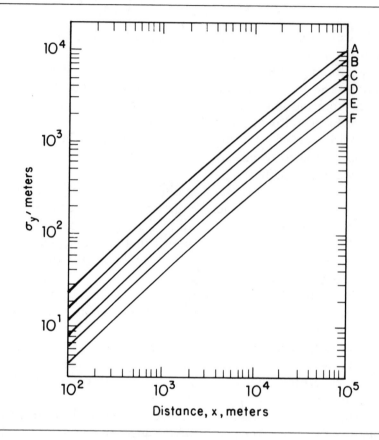

effective stack height, the location of expected maximum concentration occurs further from the source as the atmosphere becomes more stable. The expected concentration experienced at the ground under the plume decreases with increasing distance to maximum impact.

Under a constant atmospheric mixing condition the location of maximum impact increases as stack height increases. The longer transport time and the further separation from the ground allows more dilution and hence lower concentrations.

The variation of C_{max} with h has been stressed by many English authors [Clarke, Lucas, and Ross 1970; Ross and Frankenberg 1971]

Figure 2-16. Standard deviation, σ_z, in the vertical direction as a function of distance downwind. (Reproduced from D.B. Turner, *Workbook of Atmospheric Dispersion Estimates*, Washington, D.C.: HEW, 1969.)

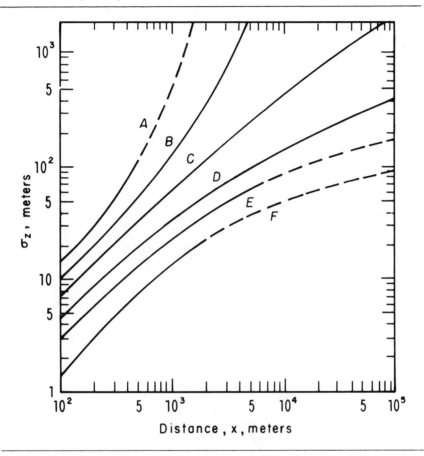

and is the basis for reliance on tall chimney stacks to reduce ground-level concentration. Indeed, if the aim is to ensure that nowhere in the vicinity of a power station are the concentrations above a certain value, as defined by air quality standards, or requirements to prevent significant deterioration of air quality, then the use of a tall stack is certainly beneficial. However, if the concern is for impact of second-ary pollutants that build up downwind, or widespread exposure to low levels of pollution, then taller stacks are not the solution.

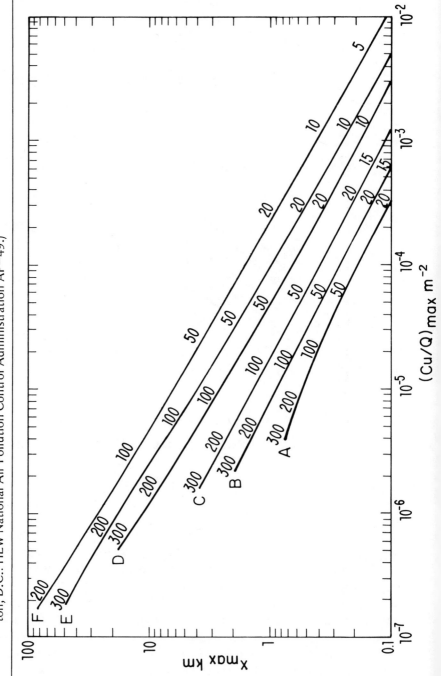

Figure 2–17. Distance of maximum concentration (x_{max}) and maximum Cu/Q as a function of stability (A through F) and effective height (meters) of emissions. (Reproduced from *Air Quality Criteria for Particulate Matter*, Washington, D.C.: HEW National Air Pollution Control Administration AP–49.)

At larger distances, where the Gaussian model is inadequate, the predicted pollutant concentrations are small. If the pollutant concentrations are below a threshold for adverse health effects, this inadequacy does not matter. If the health hazard is directly proportional to dose, without a threshold, or if ambient pollutant concentrations are above the threshold, then this formula is no longer adequate to suggest procedures for protection of public health, and we have to proceed further as suggested in the next sections.

We have estimated, and illustrate in Figure 2–18, the ground level concentrations averaged over wind directions for various weather conditions of Table 2–7. We assume a production of 300 tons of SO_2 per day—about that from a large power station with no sulfur suppression, and a wind of 7 m/sec, with varying direction. For Pasquill B conditions, the vertical dispersion is large, and concentrations are high at 1,500 m from the source, even with an effective stack height of 250 m. For Pasquill C conditions, the vertical dispersion is less; with a stack height of 250 m the concentration is not high until 5,000 m from the source (although it is now higher than for Pasquill B conditions). For the Pasquill C conditions we plot two curves, one with a 125 m effective stack height and the other with 250 m. At large distances the two curves are the same. The use of a taller stack enables the pollution to skip over the shorter distances. Fortunately, as shown in Table 2–7, Pasquill B and C conditions typically occur only one-quarter of the time, so the SO_2 concentrations averaged over all weather remain less than the air quality standard.

2.3.2 Large-Scale Transport and Dispersion

When considering dispersion of pollutants much beyond 30 km, the Gaussian model described in the preceding section is not appropriate. Additional factors must be considered for long-distance transport and dispersion of pollutants. In this section we first present an elementary didactic model to illustrate how other properties of pollutant transport and removal must be considered in calculating pollutant concentrations at some distance from the source. The impetus for developing such models is the relatively new health concern for the exposure of large populations to low concentrations of pollutants. This is in contrast to the belief that at sufficiently low concentrations no one will be harmed.

Figure 2-18. The use of the dispersion formula and coefficients σ_z from Table 2-7; the concentrations are averaged over all possible wind directions for the weather conditions shown.

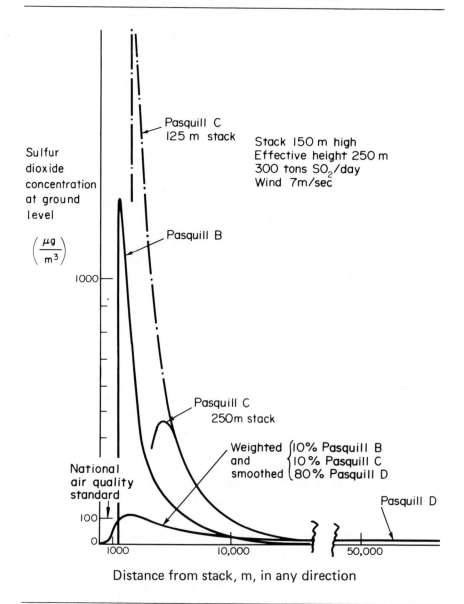

Because many of the pollutants we are concerned with are re-moved slowly from the atmosphere, they can be transported by large-scale weather features such as high-pressure anticyclones over long distances, exposing a large number of people. Fortunately, the atmosphere eventually cleanses itself. Reaction of gases such as SO_2 and NO_2 to form other gases or aerosols, absorption of gases by con-tact with particles and surfaces, and absorption by water result in a relatively short lifetime for these gases in the atmosphere. The half-life of SO_2 in the atmosphere is less than a day or two. Particles are removed by dry deposition processes as they strike the ground and by wet removal processes such as rainout and washout. The wet re-moval processes are considered more important for the particles of less than 10 μm because they are not effectively removed by gravity. For the particles between 1 and 10 μm collection by raindrops (washout) falling through a polluted layer under clouds is an effec-tive removal process. For smaller particles of less than 1 μm the in-cloud removal mechanisms (rainout) are believed more important. Particles act as condensation nuclei for cloud droplet formation. The hydroscopic salts such as sulfates, sodium chloride, and nitrates are good condensation nuclei. Cloud droplets range in size from 10 to 100 μm and in number concentrations of 100 to 1,000 drops per cubic centimeter. Continental clouds have a higher number concentration, but with droplets of smaller size, than do clouds formed in maritime air masses. This is because there are many more particles over continents to serve as condensation nuclei. In fact, some of the highest concentrations of cloud drops have been meas-ured in clouds forming in polluted air masses. Ultrafine aerosols are scavenged by cloud droplets. Molecular diffusion is effective in driv-ing these small particles toward the growing cloud drops.

Although we know that these wet processes occur, we have only limited knowledge of their overall and relative removal rates for par-ticles. It will be seen that attempts to parameterize these processes in long-distance transport models are still rather simplistic. One of the problems is that on the average only one out of ten clouds actu-ally produces precipitation. Hence, particles can go through the cloud formation process many times before being removed by falling raindrops or growth by agglomeration to a size falls by the force of gravity.

We saw in Section 2.3.1 that at short distances of up to 30 miles (50 km) the concentration averaged over a long time decreases almost as the square of the distance from the stack ($1/x^{1.8}$). However, pollutants are often trapped between the ground and an inversion layer. Under these conditions, the pollutant no longer disperses vertically, but only horizontally. The concentration then falls more slowly with distance ($1/x$). The height of the inversion layer is variable but is typically 1 km (3,000 ft.).

In order to build the didactic model, we assume that two-thirds of the 33 million tons of total U.S. SO_2 emissions [Altshuller 1973], or about 22 million tons, are emitted yearly in an area 1,500 km north to south and 2,000 km east to west, roughly covering an area from Chicago to St. Louis and from Boston to Washington. We assume that 20 percent of the SO_2 converts into sulfates, and that these sulfates stay in the air until the next rainfall or change in air mass. We will also assume that the sulfates quickly mix in the air up to 1.5 km. Mixing in the eastern United States can extend to 3 km and more on occasion. Measurements of pollutant transport caused by vertical mixing suggests that pollutants are more or less distributed between the ground and 1.5 km.

We further assume that the concentration of sulfates rises between air mass changes and falls to zero every seven days. Twenty-two million tons of SO_2 emitted per year equals 420,000 tons each seven days. If 100 percent of the SO_2 converted to sulfates, the weight would increase to 630,000 tons in seven days from the additional weight of oxygen. With only 20 percent of the SO_2 converted into sulfates, this is reduced to 126,000 tons or $\simeq 1.310 \times 10^{11}$ micrograms (μg) in the seven-day period. This is produced in a volume $(1.5 \times 10^6 \text{ m}) \times (2 \times 10^6 \text{ m}) \times 1,500 \text{ m} = 4.5 \times 10^{15} \text{ m}^3$. The sulfate concentration then rises from zero to approximately 30 μg/m^3 just before the next air mass moves in. The overall average is approximately 15 μg/m^3, which is about the measured average value over parts of the northeastern United States.

There is indeed a high concentration of sulfate in the heavily industrialized region of the northeastern United States. As we shall discover in Section 3.1.8, a high washout of acid rain also occurs in this region. Unfortunately, no one measured carefully before coal burning commenced a century ago, and there has been no experimental or accidental stoppage of sulfur-bearing fuels for a long enough time to tell whether this sulfate concentration comes from industrial and utility sources. We believe, therefore, that there is a prima facie case that these high concentrations do come from industrial and utility sulfur emissions, and that the onus is on any polluter to prove otherwise.

These calculations indicate that estimates of long-range transport depend on accurate knowledge of mixing heights, air mass movement, SO_2 source strength, SO_2 conversion rates, and SO_2 and sulfate removal rates.

This simple model provides an interesting regulatory strategy to be tested. To lower effectively the population exposure to long-distance transported pollutants may require national control or siting strategies. If we wish to disperse pollutants, we should locate major sources at the edges and downwind of a populated region. Yet a glance ahead at Figure 3−2 shows extensive emissions along the Ohio Valley, in the middle of the populated northeastern United States. If we wish to control sources, then cleaner fuels or more stringent emission limits should be required of sources upwind of populated regions.

2.3.3   Computer Calculations for Large-Scale Mixing

The two simple approaches to diffusion calculations in Section 2.3.1 and 2.3.2 have given us perspective and important public policy indications; namely, that tall stacks can reduce local high concentration levels, and that long-range effects also can be important. The simple models can also lead us to those factors that are important in more precise calculations.

As an example, SO_2-to-sulfate conversion rates have been reported to range from 0.2 to 3 percent per hour. Even if the conversion rate is as low as 0.5 percent per hour, it is high enough to produce a great deal of sulfate from an elevated source of SO_2.

We see this as follows: 50 percent of the time the atmosphere may be considered to be neutrally stable (Pasquill D) with a typical wind velocity of 7 m/sec (Table 2−7); then the plume from a 200 m high stack will travel 20,000 m before the spreading plume strikes the ground. Ten hours later, just half of the plume has come in contact with the ground. Then, based on this simple gaussian assumption, considerable conversion will have taken place. The point illustrated is that there are two competing processes removing SO_2 from the plume, dry deposition at the ground (expressed in units of velocity) and conversion to SO_4. We conclude that under typical conditions there is sufficient time for substantial conversion to sulfates.

To be more precise, we can take detailed computer models of the transport of these pollutants over large distances. There are several models, and these models, using known emissions data, can be used to predict pollutant concentrations to be compared with the measured concentrations [Nochumson 1978; OECD 1977; Meyers et al. 1978; Hidy et al. 1977; Johnson, Wolf, and Mancuso 1978; Mills and Hirata 1978; Lavery 1978]. These calculational procedures are still under development, but general features can be outlined here.

Two of these computer programs that have been widely tested are those of OECD [1977] and Meyers et al. [1978]. Typically, a trajectory for a puff of pollutant is calculated in time steps (3 hours in OECD and 6 hours in Meyers et al.). The various parameters included in the calculations are in Table 2−8.

The OECD program predicted annual average concentrations of sulfates at 31 measuring stations in 8 countries. There are some parameters that are adjusted to fit the data. Nonetheless, the agreement of the observed and the calculated SO_2 and SO_4 concentrations shown in Figures 2−19 and 2−20 is good. Because of this agreement between prediction and measurement, the OECD program can be used as a first approximation for prediction and interpolation between measurement stations for long-term mean concentration.

Because the Meyers transport and conversion model is similar to the OECD program and uses similar constraints, it can also be used. Figure 2−21 is a computer graphic output for the Brookhaven model. The calculated annual SO_4 concentrations reproduce the elevated concentrations observed over the northeastern section of the United States.

Table 2−8.   Parameters used in long-range transport.

	OECD 1977	Meyers et al. 1978
Deposition rate of SO_2 (cm sec^{-1})	0.8	3.4
Deposition rate of SO_4 (cm sec^{-1})	0.2	0.23
Transformation rate (% hr)	0.75	0.55
Mixing height	1,000 m	1,000 m
Primary sulfate (% of emitted SO_2)	0.05	0.02

Figure 2–19. Observed ($\bar{q}_{obs}$) versus estimated ($\bar{q}_{est}$) annual mean concentration of SO_2 for 31 European stations in 1974. Proportional, ideal, relationship is represented by the solid line. (From *The OECD Programme on Long Range Transport of Air Pollution: Measurements and Findings*, Paris: OECD, 1977.)

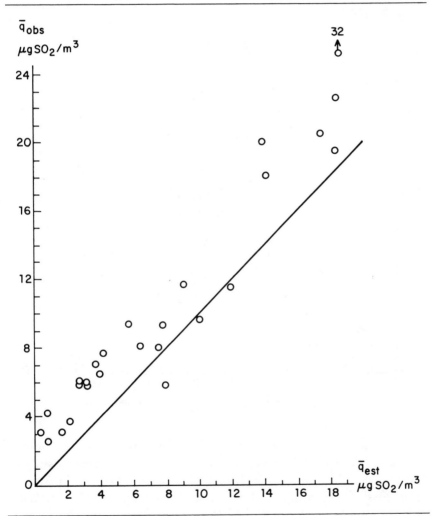

Figure 2–20. Observed $(\bar{s}_{obs})$ versus estimated $(\bar{s}_{est})$ annual mean concentration of SO_4 for 31 European stations in 1974. Proportional, ideal, relationship is represented by the solid line. (From *The OECD Programme on Long Range Transport of Air Pollution: Measurements and Findings*, Paris: OECD, 1977.)

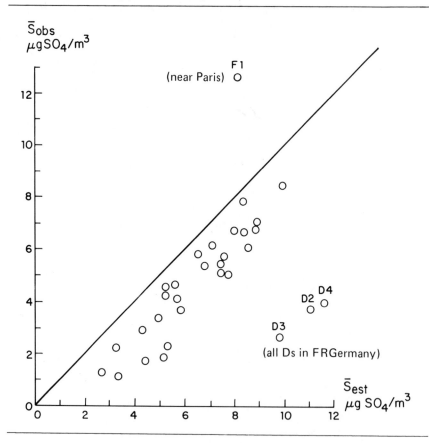

Nitrates are also transported over long distances, and contribute to acid rainfall over the northeastern United States from fossil fuel burning in Ohio and Pennsylvania, and to acid rainfall in Sweden from fossil fuel burning in Germany and England. However, even less is known about potential health effects from particulate nitrates than is known about sulfates. Polycyclic aromatic hydrocarbons do not transport over very long distances because they tend to break down in ultraviolet light to more innocuous compounds.

Figure 2–21. The predicted annual sulfate concentrations in 1990 from the Brookhaven long-distance transport model. Emissions are from major coal-burning facilities and include area sources with emission controls assumed to remove 85 percent of the SO_2. (Adapted from R.E. Meyers et al., *Constraints on Coal Utilization with Respect to Air Pollution Production and Transport over Long Distances: Summary*, Upton, N.Y.: Brookhaven National Laboratory Report, 1978.)

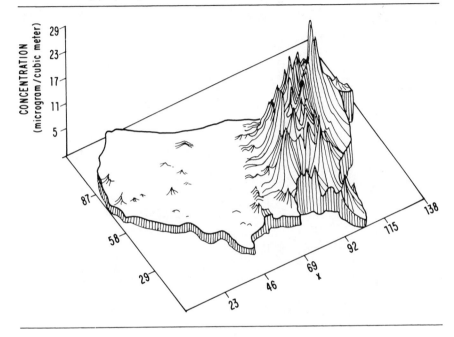

2.3.4 Predictions of Pollutant Concentrations

The traditional purpose of air pollution dispersion modeling is to predict concentrations of the pollutant from known sources and to test compliance with standards or Prevention of Significant Deterioration (PSD) regulations. Ultimately, dispersion models can be used as part of an estimate of health damage expected from single or multiple sources of pollution by calculating population exposures. The critical component that is needed is an appropriate model for the health damage to be used with the dispersion model.

In the next four chapters we will discuss the damage function that expresses the probability of death or sickness (M) as a function $f(c)$ of the concentration (c) of pollutant. The total mortality or morbid-

Table 2-9. Projected SO$_4$ air quality impact for 1985.

This table is created with emission data from 1,088 pollutant sources with a total emission of 21.855 million tons of SO$_2$/year.

EPA Region	Exposure to Conc. at or Above 9.0	Exposure Total	Concentration				Population	Area
			Average	Pop. Wt.	Maximum	Minimum		
1	0.000	70.6	3.85	5.20	6.97	1.45	13.57	161.1
2	73.867	228.7	7.61	7.92	14.96	4.61	28.89	141.7
3	266.760	316.1	10.97	11.68	27.88	6.05	27.06	309.0
4	166.668	279.0	6.96	7.00	24.69	0.06	39.87	947.7
5	444.302	540.0	7.20	10.71	27.88	0.70	50.41	827.9
6	0.000	26.1	0.74	1.13	8.31	0.00	23.11	1405.9
7	19.737	50.2	2.43	4.14	13.41	0.43	12.12	725.0
8	0.000	4.1	0.51	0.65	1.58	0.11	6.25	1469.2
9	0.000	8.2	0.26	0.30	2.20	0.00	27.11	972.0
10	0.000	1.4	0.19	0.21	0.66	0.02	6.87	628.3
Summary	Σ = 971.335	Σ = 1524.4					Σ = 235.26	

List of States by EPA Region

Region 1	Connecticut	Maine	Massachusetts	New Hampshire	Rhode Island	Vermont
Region 2	New Jersey	New York				
Region 3	Delaware	D.C.	Maryland	Pennsylvania	Virginia	West Virginia
Region 4	Alabama	Florida	Georgia	Kentucky	Mississippi	North Carolina
	South Carolina	Tennessee				
Region 5	Illinois	Indiana	Michigan	Minnesota	Ohio	Wisconsin
Region 6	Arkansas	Louisiana	New Mexico	Oklahoma	Texas	
Region 7	Iowa	Kansas	Missouri	Nebraska		
Region 8	Colorado	Montana	North Dakota	South Dakota	Utah	Wyoming
Region 9	Arizona	California	Nevada			
Region 10	Idaho	Oregon	Washington			

Units
Exposure: (Million Persons) $\times$ (Microgram/Cubic Meter)
Concentration: Microgram/Cubic Meter
Population in millions, area in thousand square kilometers

Source: R. E. Meyers et al., [1978].

ity rate becomes $M = \int f(c) \times P(c) \, (dc)$ where $P(c) \, dc$ is the number of people who are exposed to a predicted concentration between c and $c + dc$; $\int_0^\infty P(c) \, dc = P$ (the total population).

In one model, where the probability of death is proportional to the concentration $f(c) = ac$, and where a is a constant, then $M = a \int c P(c) \, dc$. We define $\int c P(c) \, dc$ to be the total integrated exposure. We see that the expression for the total mortality has two factors: the coefficient of air pollution a, and the integrated exposure.

We can also define the population-weighted average concentration c. For past and present health measurements we can use actual measurements, but for attribution of health effects to specific sources prediction is needed.

A useful result of any computer calculation, therefore, is to derive the quantities $\int c \, p(c) \, dc$ and $\bar{c}$. The simple calculations based upon Sutton's equation (Section 2.3.1), give a logarithmically infinite result for the first quantity; as we have already noted, these calculations are appropriate only for distances up to about 30 km. Beyond this distance the long-range transport is important, and the mechanism for washing pollutants out of the air provides a way of preventing the integral from diverging.

For the United States these integrals have been computed using the Brookhaven model by Meyers et al. [1978] using various assumptions on emissions and the conversion rates in Table 2−8. In Table 2−9 we show sulfate calculations for major stationary sources expected to exist in the United States in 1985, with reasonable assumptions about stack gas suppression and sulfur content of the fuels. The total SO_2 emitted from all sources was assumed to be 22 million tons. This can be compared with the 1963 and 1968 emissions of 23 and 33 million tons, respectively. The population exposure to sulfates is calculated by multiplying the concentration averaged over a 32×32 km grid by the population in that grid. This has been done for each of the 10 EPA regions and then aggregated for the continental United States. The total aggregated exposure to sulfates is estimated as $1,520 \times 10^6$ persons-micrograms per cubic meter. We return to these estimates in Section 6.8.

The development and improvement of models to estimate the population exposures to pollutants both locally and regionally should be promoted. When combined with actual measurements they complement our understanding of basic transport, conversion, and removal processes in the atmosphere. When coupled to health damage functions they provide a quantitative basis for judgment.

3 GASES AND ORGANIC MATTER AS AIR POLLUTANTS

Coal combustion discharges gases and particulates to the atmosphere. The primary gases are oxides of sulfur, nitrogen, and carbon. Particulate or fly-ash emissions contain a variety of trace metals, organic matter, and radionuclides. This organic and inorganic material can undergo a variety of chemical reactions as it is transported and dispersed in the atmosphere. Among the secondary coal emission products are sulfate and nitrate compounds. Table 3–1 presents the gaseous components and concentrations of the constituents of ordinary, polluted air. Table 3–2 summarizes coal-related direct emissions and their conversion products.

Of the host of emission products from coal combustion, attention is focused on the health effects of sulfur dioxide (SO_2), sulfates, nitrogen dioxide (NO_2), ozone (O_3), and fine particulates. Population exposure to these substances is likely to increase with the expansion of coal use. However, exact increases are difficult to estimate. We do not yet know how much emissions will be controlled, with respect to both new source performance standards for future facilities and emission requirements for older facilities. The modeling in Section 2.3 gives us a reasonable prescription for calculation once these emission increases are known. Carbon monoxide has not been included in this review because the overwhelming population exposure is from mobile sources (automobiles).

63

Table 3–1.   Gaseous composition of the ambient atmosphere[a].

Gas	ppm[b]	µg/m³ at STP
nitrogen	780,000	
oxygen	209,000	
argon	9,300	16,000,000
carbon dioxide	315	630,000
neon	18	16,000
carbon monoxide	0.03–0.14	35–160
krypton	1.1	4,100
methane	1.4	900
nitrous oxide	0.26	480
helium	5.1	850
hydrogen	0.5	40
xenon	0.086	470
butane		150
ethane		120
propane		90
ethylene		45–800
nitrogen dioxide	0.02–0.07	40–130
acetylene		40
nitric oxide		40
propene		30
isobutane		30
sulfur dioxide	0.02	53
ozone	0.02	40
ammonia		10
formaldehyde		10
trans-2-butene		3
cis-2-butene		3
methylacetylene		2
dichlorofluoromethane		3
dichlorodifluoromethane		3.5
methyl chloride		1.1
fluorotrichloromethane		0.3
hydrogen sulfide		0.07
vinyl chloride		2.3–70

a. Mean values; for most of these pollutants, mean values may be somewhat higher or lower in different parts of the world.

b. Composition of dry, pollution-free air. Air may contain 1,000–50,000 ppm water vapor, with a normal range of 10,000–30,000 ppm.

Source: F. Sawicki, "Chemical composition and potential genotoxic aspects of polluted atmosphere," in *Air Pollution and Cancer in Man*, ed. by U. Mohr, D. Schmahl, and L. Tomatis, Lyon: WHO, 1977, IARC Scientific Publications No. 16, p. 134.

Table 3–2. Some gaseous and particulate substances from coal combustion[a].

Substance[b]	Toxicity		Sources of Pollution	Environmental Standards	Comments
	Acute	Chronic			
Particulates					
Total suspended particulates	With SO_2 in episode conditions contributes to mortality and morbidity	Pulmonary irritation, chronic obstructive and restrictive lung disease	Soil erosion, natural volcanos and fires, industrial activity, fossil fuel combustion: coal and oil, secondary atmospheric conversion of gaseous compounds	260 $\mu g/m^3$ — 24 hr. max. 75 $\mu g/m^3$ — annual ave.	A very broad class, undifferentiated by particle size or chemical composition
Sulfates	Increased respiratory disease; breathing difficulty in asthmatics	Respiratory disease and increased mortality suspected	Conversion of SO_2 to sulfates in the atmosphere, therefore primary sources are SO_2 emissions from coal and oil combustion. Smelters, kraft paper mills, sulfuric acid plants also produce sulfates. Natural sources: H_2S emissions, volcanos, sea salt		
Nitrates, nitrites	Increases infant susceptibility to lower respiratory infection because of conversion of nitrates to nitrites	May combine with amines to form carcinogenic nitrosamines, also mutagenic and teratogenic. Nitrites a direct animal carcinogen	Conversion of NO_2 to nitrates and nitrites in the atmosphere; therefore primary sources are NO emissions from fossil fuel combustion, fertilizer production, munition production, chemical plants, and auto and industrial emissions		

(Table 3–2. continued overleaf)

Table 3-2. continued

| Substance[b] | Toxicity | | Sources of Pollution | Environmental Standards | Comments |
	Acute	Chronic			
Organic matter	Unknown for many compounds. Specific toxicity for others	Long-term is potentially carcinogenic and mutagenic	Fossil fuel direct and indirect use—combustion, refining, plastics, tars, coking, chemical production		Higher concentrations likely to be associated with nondirect combustion of fuels
Arsenic (oxide forms)	Effects large to small depending on form and route of exposure; rarely seen	Carcinogen, and teratogenic cumulative poison	Weathering, mining and smelting, coal combustion, pesticides, detergents	0.05 mg/liter drinking water	
Beryllium	Short-term poison at high concentrations, especially toxic by inhalation	Long-term systemic poison at low concentrations; carcinogenic in experimental animals	Industrial; combustion of coal, rocket fuels	$0.01\ \mu g/m^3$ hazardous air pollutant	
Cadmium	Very toxic at high concentrations to animals and aquatic life. Toxic by all routes of exposures	Possible carcinogen, cumulative poison; associated with hypertension, cardiovascular disease, kidney damage	Weathering; mining and smelting, especially of zinc; iron and steel industry; coal combustion; urban runoff; phosphate fertilizers	0.010 mg/liter drinking water 40 $\mu g/1/$day proposed effluent standard (withdrawn)	Chronic cadmium poisoning resulting in illness and death has occurred in Japan, where cadmium mobilized by mining contaminated daily diet.

Cadmium (*continued*)			0.05 mg/liter drinking water	Margin of safety — measured levels of cadmium in renal cortex compared to threshold for renal dysfunction — is low: 4 to 12.5
Chromium	Hexavalent form most harmful; skin and respiratory tract irritant	Carcinogenic; workers engaged in manufacture of chromium chemicals have incidence of lung cancer, no evidence of risk in nonoccupational exposure	No chromium now mined in U.S. Emissions from industrial processes, including electroplating, tanning, dyes; coal combustion	
Mercury	Methyl mercury and mercury fumes very toxic; other forms of variable toxicity	Methyl mercury very toxic, cumulative poison; affects central nervous system	Weathering, volcanoes, mining and smelting, industrial, pharmaceuticals, coal combustion, sewage sludge, urban runoff, fungicides	0.002 mg/liter drinking water; maximum of 2,300 grams mercury in emissions from stationary sources; 20 µg/1/day proposed effluent standard (withdrawn)	Environmental pollution leading to contamination of fish and shellfish caused illness and death in Japan; contamination of fish in U.S. has caused closure of waters to commercial fishing
Selenium	Soluble compounds are highly toxic	Probable carcinogen; also essential for life	Natural, mining and smelting, industrial process, coal combustion	0.01 mg/liter drinking water	Interacts with other metals, increasing or decreasing toxicity

Table 3–2. continued

Substance[b]	Toxicity		Sources of Pollution	Environmental Standards	Comments
	Acute	Chronic			
Gases					
Sulfur dioxide	Increased respiratory impairment—morbidity and mortality—in combination with particulates	Increased respiratory disease and decreased respiratory function with particulates	Sulfur contained in fossil fuels, smelters, volcanoes	365 $\mu g/m^3$ — 24 hr. max. 80 $\mu g/m^3$ — annual mean	Coal combustion presently represents between 60% and 70% of U.S. SO_2 emissions
Nitrogen dioxide	Increase respiratory infections	Changes suspected in lung function; emphysema	Nitrogen fixation in high temperature combustion, and from nitrogen contained in fossil fuels: coal, oil, gasoline combustion	100 $\mu g/m^3$ — annual mean	Organically bound fuel nitrogen is a more important component for coal NO emissions than for the other fossil fuels
Carbon monoxide	Behavior changes, nausea, drowsiness, headaches, coma, death	Increase risk of coronary heart disease—arterial sclerosis suspected	Incomplete combustion of fossil fuels: coal, oil, gas, gasoline	40 mg/m^3 — 1 hr. max. 10 mg/m^3 — 8 hr. max.	
Ozone	Increased respiratory infection, eye irritation, headaches, chest pain, impaired pulmonary function	Unknown	Photochemical reactions involving hydrocarbons, nitrogen oxide, and other compounds in lower atmosphere; reaction of atomic oxygen and oxygen in upper atmosphere	0.12 ppm — 1 hr. max.	

| Aromatic hydrocarbons | Fatigue, weakness, skin paresthesias (> 100 ppm) | Irritation, leukopenia and anemia. Certain compounds are mutagens and carcinogens | A broad class of compounds naturally evolved from organic material, and from the evaporation and combustion of fossil fuels and other organic industrial chemicals | non-methane HC 160 $\mu g/m^3$ – 3 hr. | Higher concentrations likely from less efficient and smaller boiler operation. Higher concentrations possible proximate to coal conversion facilities. Standard designed for photochemical oxidant control |

a. This table is provided merely to indicate some of the substances which are potential environmental hazards along with some information on each regarding toxicity, sources, standards, etc. It is not to be interpreted as definitive.

b. The substance listed is not necessarily the form in which it becomes a potential environmental threat. In some cases the oxide or some metabolite, rather than the substance itself, is the culprit.

Although total emissions of most gaseous and particulate matter have increased with time, the geographic distribution of these emissions has changed as industrial sources, processes, and fuels have changed. The control of stationary and mobile sources, those existing as well as new ones, resulted in a reduction of emissions of some pollutants, and reduction in ground-level concentration of many more. However, since 1970, the general increase in fossil fuel burning has slowed the trend of the late 1960s, and as of the mid-1970s the emissions of all pollutants—for which the U.S. Environmental Protection Agency (EPA) has set air quality criteria—began increasing. Figure 3–1 plots the annual emissions of the five pollutants HC, SO_x, NO_x, CO, and TSP for the years 1970 through 1976 [EPA 1977a].

Although they are very difficult to project, emissions of SO_x, NO_x, and total suspended particulates (TSP) are expected to increase further as existing oil-fired electric generating plants and industries convert to coal and as new coal-fired sources come on line. The magnitude of emission increases depends on the assumptions for control of existing sources, in particular on whether existing power plants are required to retrofit with SO_2 scrubbers.

The nationwide emission estimates for 1977 by source category are shown in Table 3–3. This table will be referred to in subsequent discussions on specific pollutants.

3.1 SULFUR CONCENTRATIONS
AND EMISSIONS

Anthropogenic sulfur pollution arises mostly from the burning of sulfur in fossil fuels to form sulfur dioxide (SO_2), which is released as a gas. It is known that the SO_2 in the atmosphere comes almost exlusively from pollution sources. In 1977, as seen in Table 3–3, over 90 percent of all anthropogenic SO_2 was from combustion sources. For those sources with short chimney stacks, the concentrations of SO_2 can reach high levels near the source.

During this century, and in the United States particularly since the passage of the Clean Air Act Amendments of 1970, there have been major attempts to reduce high concentrations of SO_2. In urban areas, burning of gas and low-sulfur oil has replaced coal; and in more rural areas tall chimney stacks that disperse emissions more efficiently

Figure 3−1.   Trends in U.S. emission of five major pollutants during the early 1970s. [Data from EPA 1977a].

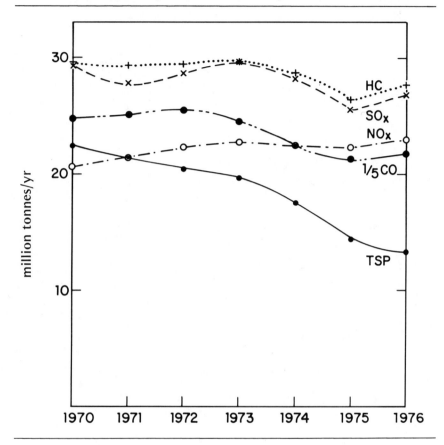

have been built by electric utilities. Between 1964 and 1971 the annual arithmetic mean of SO_2 concentrations was reduced by 50 percent at 32 sites studied by the Environmental Protection Agency [EPA 1973a]. Between passage of the amendments to the Clean Air Act of 1970 and an assessment at the end of 1974, national SO_2 concentrations were reported to have fallen an additional 25 percent [CEQ 1975]. Figure 3−2 displays regional emission densities of SO_2 in the United States. Substantial shifts in emission patterns are expected as existing gas-fired sources in the South convert to coal and as new coal sources are built in the Midwest.

Table 3–3. Nationwide emission estimates, 1977 (EPA 1978c) (10^6 metric tons/year).

Source	TSP	SO_x	NO_x	VOC	CO
Transportation	1.1	0.8	9.2	11.5	85.7
Highway vehicles	0.8	0.4	6.7	9.9	77.2
Nonhighway vehicles	0.3	0.4	2.5	1.6	8.5
Stationary fuel combustion	4.8	22.4	13.0	1.5	1.2
Electric utilities	3.4	17.6	7.1	0.1	0.3
Industrial	1.2	3.2	5.0	1.3	0.6
Residential, commercial, and institutional	0.2	1.6	0.9	0.1	0.3
Industrial processes	5.4	4.2	0.7	10.1	8.3
Chemicals	0.2	0.2	0.2	2.7	2.8
Petroleum refining	0.1	0.8	0.4	1.1	2.4
Metals	1.3	2.4	0	0.1	2.0
Mineral products	2.7	0.6	0.1	0.1	0
Oil and gas production and marketing	0	0.1	0	3.1	0
Industrial organic solvent use	0	0	0	2.7	0
Other processes	1.1	0.1	0	0.3	1.1
Solid waste	0.4	0	0.1	0.7	2.6
Miscellaneous	0.7	0	0.1	4.5	4.9
Forest wildfires	0.5	0	0.1	0.7	4.3
Agricultural burning	0.1	0	0	0.1	0.5
Coal refuse burning	0	0	0	0	0
Structural fires	0.1	0	0	0	0.1
Miscellaneous organic solvent use	0	0	0	3.7	0
Total	12.4	27.4	23.1	28.3	102.7

Note: A zero indicates emissions of less than 50,000 metric tons per year.

Source: *National Air Quality, Monitoring, and Emission Trends Report* EPA-450/2-78-052, Washington, D.C.; EPA, 1978c.

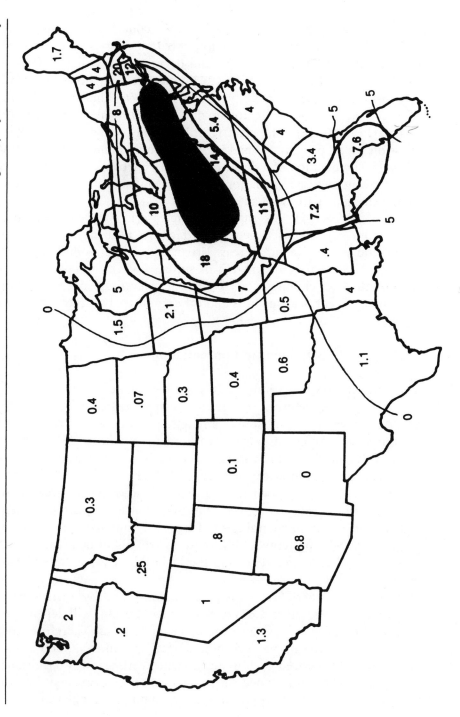

Figure 3–2.   SO$_2$ emission density for the United States in the mid-1970s. Values are g/m^2/yr. [From OTA 1979].

Despite the large decrease in urban SO_2 concentrations since the early 1960s, the nationwide emission of SO_2 increased by 45 percent from 1960 to 1970 [NAS 1975] and decreased only slightly (8 percent) from 1970 to 1976 [EPA 1977a]. Figure 3−1 shows the trend of emissions for SO_2 for the period 1970−1976. Almost 90 percent of the increased SO_2 emission is due to additional electric power plants [EPA 1973a]. The electric utility industry increased its SO_2 emission by 12 percent, from 15.7 to 17.6 million metric tons [EPA 1977a]. The paradox of a decline in urban SO_2 concentration with a simultaneous increase in nationwide SO_2 emission is accounted for by the changing pattern of sources.

Since the early 1970s the major increase in SO_2 emissions has come from new electrical generating facilities utilizing tall chimney stacks and located in rural areas. Urban SO_2 levels are high only in the areas immediately downwind of sulfur emitters, and particularly those with short chimney stacks. High SO_2 levels are not noticed in rural areas except near smelters and other industrial sources.

3.1.1 SO_2 Ambient Concentrations

The ambient concentrations of SO_2 are determined by (1) the density of emission sources; (2) source characteristics such as stack height, exit velocity, and source strength; (3) the local meteorology; (4) the local topography and surrounding buildings; (5) reaction rates of SO_2 in the plume; and (6) the removal rates because of precipitation, deposition at the surface, and other reactions. These factors interact in such a way that in urban and industrialized areas with high densities of SO_2 emissions, the SO_2 concentrations are much higher than in the surrounding rural area. It is quite common to find gradients in SO_2 patterns within these metropolitan or urban areas, with a central core area reporting the highest SO_2 values, as stylized in Figure 3−3.

In locations where SO_2 emissions are dominated by a single or a few point sources, the pattern of SO_2 concentrations will depend on topography, meteorology, and source characteristics. Therefore, the concentration patterns may be asymmetrical, and the temporal distribution may be skewed to low mean values with a few intermittant high peaks. These differences in concentration patterns may be important to the effect experienced in exposed human populations.

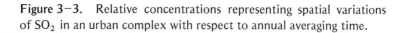

Figure 3-3. Relative concentrations representing spatial variations of SO_2 in an urban complex with respect to annual averaging time.

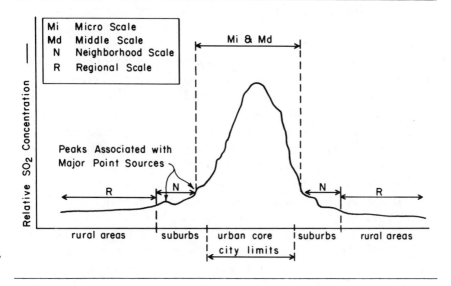

In 1970, the concentrations of SO_2 in the United States violated the primary national ambient air quality standards (NAAQS) in many areas. By 1977, there were only a few air quality control regions (AQCRs) where the desired air quality had not yet been attained. This has been achieved in urban areas by (1) restrictions on burning fuel with high sulfur content; (2) replacement of old sources in densely populated (urban) areas by new sources in less populated (rural) areas; and (3) building taller chimney stacks to disperse the gases.

There is concern that some retrenchment may occur. Regulations for use of low-sulfur fuel have already been relaxed in some states for economic reasons. Increased combution of coal in new facilities and the demand by the federal government that industrial plants and electric utilities change to burning coal from oil and gas will also increase population exposures to SO_2. Locating sources with taller stacks in a line, in the direction of the prevailing winds, has been demonstrated to lead to violations of SO_2 concentrations in areas previously well below national standards.

3.1.2 SO$_2$ Monitoring

The EPA has initiated several changes in the federal, state, and local air monitoring networks. By 1981 states will operate a selected number of sites in the National Air Monitoring Station Network (NAMS). These sites would be located in areas with the highest concentrations of pollution and of population. The monitors would serve to assess the trends and progress in reducing pollutant concentrations. By 1983, state and local agencies would be operating the State and Local Air Monitoring Station (SLAMS) network, which would be a part of the air quality implementation plan for each state. It is expected that this would mean fewer monitoring sites than currently operate; however, federal coordination of air monitoring is designed to provide better data through quality control. The trend toward reducing the number of operating stations is already apparent in the 1977 SO$_2$ data. There were almost 120 fewer monitoring sites reporting data in 1977 than in 1976 (2,365 vs. 2,482). This trend for SO$_2$ will be more pronounced, for many states terminated all or most of their 24-hour West–Gaeke bubbler sampling in 1978. SO$_2$ sampling by the West–Gaeke (pararosaniline sulfamic acid) method was subject to temperature degradation. Rather than maintaining the denser network of samplers, state and local agencies have chosen to use fewer, but more accurate, continuous monitoring instruments. Table 3–4 gives the number and type of SO$_2$ monitors operated by federal, state, and local agencies during 1976 and 1977. It should be noted that there are many more SO$_2$ monitors operated by electric utilities, paper companies, smelting companies, and others. These data, however, are not part of the national aerometric data banks. Table 3–4 is noteworthy because of the variety of sampling methods currently used to measure SO$_2$ in the atmosphere. There are some serious, unresolved questions about the direct comparability of these methods. By 1978 it was recognized that carbon dioxide (CO$_2$) could bias the calibrations of the flame photometric instruments. It was also recognized that hydrocarbons were an important interference for the pulse fluorescent instruments.

At first sight, the reduction of the number of monitoring stations seems to be a backward step. Yet air pollution concentrations now vary less dramatically across a city because peak concentrations have been reduced; and therefore fewer monitoring stations are needed.

Table 3—4. SO$_2$ instruments reporting valid annual data to National Aerometric data banks in 1976 and 1977. Table indicates the variety of instrumental methods to measure SO$_2$ concentrations in the atmosphere.

Continuous — Reporting hourly averages	*Years*	
Method	*1976*	*1977*
Colorimetric Reactions	82	95
Conductometric	20	25
Coulometric	339	365
Flame Photometric	121	105
Hydrogen Peroxide NaOH Titration	11	76
UV Pulsed Flourescent	17	—
3-Other Methods	13	10
Total continuous	603	676
Integrating — Reporting 24-hour averages		
Method		
West-Gaeke/Pararosaniline- Sulfamic Acid Reaction	1,879	1,689
Total Number of Annual Averages	2,482	2,365

Overall, SO$_2$ monitoring should improve as EPA requires quality control procedures of state and local agencies. However, the historic data are of unknown accuracy. Presumably the West–Gaeke bubbler data were subject to comparable exposures, collection, and analysis procedures by the responsible agencies. To that extent, they may still be both useful for trends analysis and relevant to health studies basing population exposures on similar methods.

3.1.3 National Status of SO$_2$ Concentrations

Of the 2,365 monitoring sites reporting data in the year 1977, only 1,355 had sufficient data within each quarter and for the entire year to produce valid data for an annual standard. Of these sites, 19 indicated a violation of the primary standard for SO$_2$ of 80 μg/m^3 or 0.03 ppm. This represents approximately 1 percent of all the stations

reporting. All 2,365 monitoring sites were eligible for use in determining compliance or violation of the daily or 3-hour standard because the criteria for using a monitoring site to determine compliance
with these secondary standards are less severe than for the annual
standard. Of these stations, 58 indicated violation of the 24-hour
primary (daily) standard. This standard is violated by the second
occurrence, within the same year, of a concentration in excess of
$365 \, \mu g/m^3$ for a 24-hour average. These 58 stations reporting violations represent only 2 percent of all the reporting stations. For the
3-hour secondary standard of $1,300 \, \mu g/m^3$, or 0.5 ppm, 30 sites
reported a violation. This represents approximately 1 percent of all
the monitoring sites (see Table 3–5).

To give a national perspective to the SO_2 emission and concentration patterns, in the Northeast the largest categories are residential space heating and power plant emissions. As a result of the
relatively large area source contributions to SO_2 emissions in the
northeastern urban areas, the expected annual averaged SO_2 concentrations might appear as they do in Figure 3–3. The central urban
core might be expected to have SO_2 concentrations many times
higher than the surrounding rural regions. In the Midwest, large electric utility sources are the major SO_2 sources. In the South, emissions from transportation, power plants, and industrial processes
dominate, whereas in the West the industrial processes emission cat-

Table 3–5. National summary of total stations reporting data and
number reporting violations of air quality standards, 1977.

Pollutant	Data Record and Standard Exceeded	No. Stations	% Sites Exceeding NAAQS
SO_2	Valid annual data[a]	1,355	
	Annual primary	19	1
	At least minimal data[b]	2,365	
	24-hour primary	58	2
	3-hour secondary	30	1

a. Record must contain at least five of the scheduled 24-hour samples per quarter for
EPA-recommended intermittent sampling (once every 6 days) or 75% of all possible values
in a year for continuous instruments.

b. At least three 24-hour samples for intermittent sampling monitors or 400 hourly
values for continuous instruments.

egory (largely smelting operations) is the largest. Over the entire nation, SO_2 problems can range in complexity from single source impacts over flat terrain to impacts from multisource urban industrial complexes located in complex terrain settings. The maps displaying the areas violating the national air quality standards for SO_2 demonstrate that the complexity of sources and concentration patterns can result in violations.

In about half the AQCRs, mostly in rural areas, there are inadequate data to judge compliance, although compliance is probable. The AQCRs violating the primary 24–hour (daily) standard are displayed in Figure 3–4. The figure shows that 29 AQCRs showed violations in 1977; 18 AQCRs had insufficient or no data; and the remaining 200 were in compliance.

3.1.4 Severity of SO_2 Ambient Exposures

The ninetieth percentile of the 1977 SO_2 data was examined for every AQCR. In Figure 3–5 we show the three ranges for the ninetieth percentile observations in each AQCR. The ninetieth percentile serves as an indicator of the severity of the SO_2 ambient concentrations. It represents the high concentrations to which the population might be repeatedly exposed. When the ninetieth percentile is based on continuous monitoring, it represents an ambient SO_2 concentration that would be experienced about 870 hours per year. When the ninetieth percentile concentration is determined by 24–hour SO_2 bubblers, then it represents an exposure level for 36 days of the year.

On the basis of the 1977 aerometric data, the urban areas around Washington, D.C., western Pennsylvania, eastern Ohio, southern Indiana, northern Kentucky, St. Louis, and Indianapolis have levels repeatedly exceeding 200 $\mu g/m^3$ with some days greater than 365 $\mu g/m^3$. Close examination of the aerometric data indicates that peak exposures are often two to five times the ninetieth percentile concentration. The large point sources in the West are responsible for the high ninetieth percentile levels in Montana, Nevada, Idaho, Arizona, and New Mexico. Throughout the Midwest and Upper Plains states, the ninetieth percentile levels are less than 100 $\mu g/m^3$, indicating well dispensed concentrations in these areas.

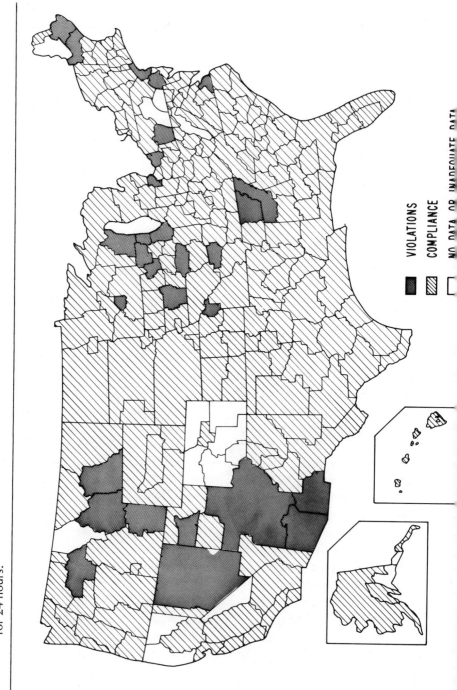

Figure 3–4. AQCRs violating primary 24-hour SO$_2$ standard, 1977. The second highest value exceeded 365 μg/m^3 for 24 hours.

VIOLATIONS

COMPLIANCE

NO DATA OR INADEQUATE DATA

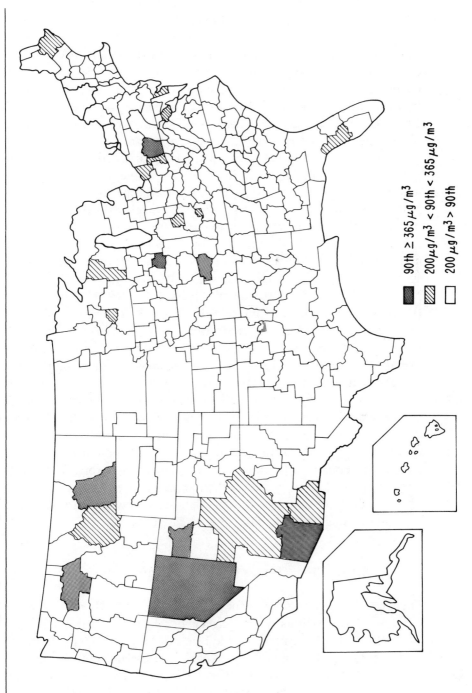

samples reported at any site within an AQCR.

90th ≥ 365 $\mu g/m^3$

200$\mu g/m^3$ < 90th < 365 $\mu g/m^3$

200 $\mu g/m^3$ > 90th

3.1.5   Historic Trends in SO_2 Concentrations

The SO_2 levels in the United States in most urban areas have decreased steadily since the mid-1960s. The trends of decreasing SO_2 concentrations have three distinct periods. From 1964 to 1969 the improvement was gradual. In the middle period, between 1969 and 1972, the improvement in most urban areas was more pronounced. Since 1973 the improvement has again become slower. The 1977 EPA trends report states: "In most urban areas, this is consistent with the switching in emphasis from attainment of standards to maintenance of air quality; that is, the initial period was to reduce pollution to acceptable levels followed by efforts to maintain air quality at these lower levels." From 1972 through 1977, annual average levels dropped by 17 percent, for an annual improvement rate of about 4 percent per year. Figure 3-6 summarizes the annual averaged SO_2 concentrations for 32 urban EPA National Air Surveillance Network (NASN) stations for the years 1964-1971. In this figure the first two periods are apparent. In Figure 3-7 the national trends in annual averaged SO_2 concentrations from 1972 to 1977 at 1233 sampling sites are displayed. The diamond symbolizes the composite overall annual average value. The X is the median value, and the various percentile points are shown in the legend.

Over the period 1970-1977, SO_2 emissions are reported to have decreased only slightly (Figure 3-1 and EPA 1978c). In 1970 the estimated annual anthropogenic SO_2 emissions were 29.8 million metric tons. By 1977 this was reduced to 27.4 million metric tons. The improvement in the ambient air quality levels for SO_2 reflects displacement of sources from urban areas to rural areas, the restriction of sulfur content used in fuels that affect low-level area sources, and the building of newer sources with taller stacks.

3.1.6   Urban Trends in SO_2 Concentrations

The air quality criteria for sulfur oxides [HEW 1970a], presented the frequency distributions for SO_2 levels in selected American cities for 1962-1967. Improvements in the levels of SO_2 in Chicago, Philadelphia, St. Louis, Cincinnati, Los Angeles, and San Francisco can be demonstrated by comparing the more recent aerometric data with historic data for the decade 1960-1970 from the Continuous Air

Figure 3–6. Average SO_2 concentrations at 32 urban NASN stations, 1964–1972. (Reproduced from National Academy of Sciences and National Academy of Engineering, Commission on Natural Resources, *Air Quality and Stationary Source Emission Control*, report prepared for U.S. Senate Committee on Public Welfare, Washington, D.C.: Ser. No. 94–4, 1975.)

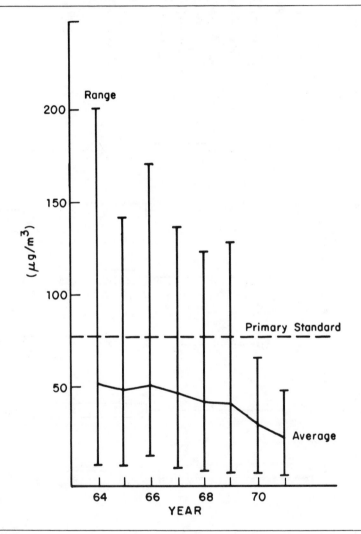

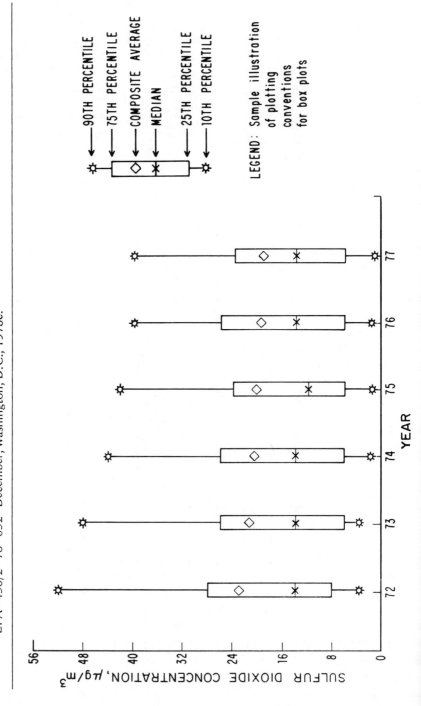

Figure 3–7. Nationwide trends in annual average SO$_2$ concentrations from 1972 to 1977 at 1,233 sampling sites. (Reproduced from Environmental Protection Agency, *National Air Quality, Monitoring and Emissions Trends* Report EPA–450/2–78–052–December, Washington, D.C., 1978c.

Monitoring Project (CAMP). In all these cities more than one contin-uously recording monitor operates. The station reporting the highest levels during the more recent period was used in order not to present a conservative estimate of the improvement. At any rate, the compar-isons are only approximate because the location of the monitors and the instrumental methods used were not similar.

In all the cities except St. Louis, the peak concentrations have been reduced. In most cities the peaks are less than one-half the ear-lier values. In Chicago the fiftieth percentile concentration dropped from 0.08 to 0.022 ppm; the primary annual standard is met, but there are still occasional violations of the 24-hour SO_2 NAAQS. In Philadelphia the fiftieth, seventieth, and ninetieth percentile concen-trations are less than one-half the earlier values. In St. Louis and Cin-cinnati, the fiftieth percentile concentrations are lower. The highest concentrations occur as frequently in St. Louis as they did before in one small section of the city. Los Angeles shows improvement in reducing the high concentrations, but the fiftieth percentile concen-tration is actually slightly higher than in the previous decade. Simi-larly, San Francisco has trimmed the peaks while having started with a very low median value.

The frequency of peak levels has been reduced in most urban areas. The steady improvement of SO_2 ambient air quality has been slowed somewhat in recent years. In 1977 only 1 percent of the SO_2 monitoring sites show violations of the annual NAAQS. In 1974 the annual mean SO_2 standard was exceeded in 3 percent of the moni-toring stations (31 of 1,030), compared with 16 percent in 1970. In 1977 2 percent of the sites reported violations of the 24-hour stand-ard. In 1974 this standard was exceeded in 4.4 percent of the report-ing stations (99 of 2,241), compared with 11 percent in 1970. Many of these sites reporting violations of the 24-hour standard are in remote areas near large point sources. In 1977 a large percentage of the U.S. population was exposed to ambient concentrations of SO_2 exceeding 100 $\mu g/m^3$ about 10 percent of the time.

3.1.7 SO_2 to Sulfate (SO_4) Conversion

The entire sulfur cycle will be discussed in Section 3.1.9, which includes details of SO_2 gas production and conversion to particulate sulfates. For the present we note that this conversion takes place in the atmosphere, and we can use computer programs to calculate sulfate concentrations from sulfur oxide emissions, provided that the conversion rate is assumed as an adjustable parameter.

Chapters 4, 5, and 6 present evidence that ambient sulfate particulates may pose a health hazard at ambient levels. To the extent that this association is real, health effects could be reduced if the conversion was inhibited. Our current understanding of the conversion is insufficient to suggest a control strategy along this line. Studies of the chemical process are incomplete [Husar, Lodge, and Moore 1978; Hidy and Burton 1974]. The dominant chemical mechanism responsible for conversion of SO_2 into sulfate in clean portions of the stratosphere is still speculative, but even less is known about the conversion processes taking place in already polluted atmospheres where the various air constituents can interreact with one another [Haury, Jordan, and Hofmann 1978; Calvert 1973]. Particularly important is information on the sulfate species formed and the rate of conversion in different atmospheres.

Sulfates exist as several chemical species that may have varying health and ecological effects. Neither the chemical species nor particle-size distribution (two of the most important parameters influencing health effects) of the sulfates are determined by using the standard techniques for sulfate analysis. Although it is quite likely that sulfates create a larger health hazard than SO_2, the lack of information on chemical species and particle size makes development of an optimal control program difficult.

3.1.8 Distribution and Trends
of Sulfate Concentrations

Whereas SO_2 concentrations are high within a few miles of the emission source, sulfate concentrations usually are not. This is because the conversion of SO_2 to sulfur proceeds sufficiently slowly that the sulfur has traveled miles downwind before conversion is complete.

Figure 3–8 shows the distribution of sulfate concentrations in the United States. Figure 3–9 shows the distributions of the acidity in rainfall in the eastern United States. The lines are lines of constant pH for (a) 1955–1956 and (b) 1972–1973; (c) shows the increase in hydrogen ion concentration. This acid rainfall can come from

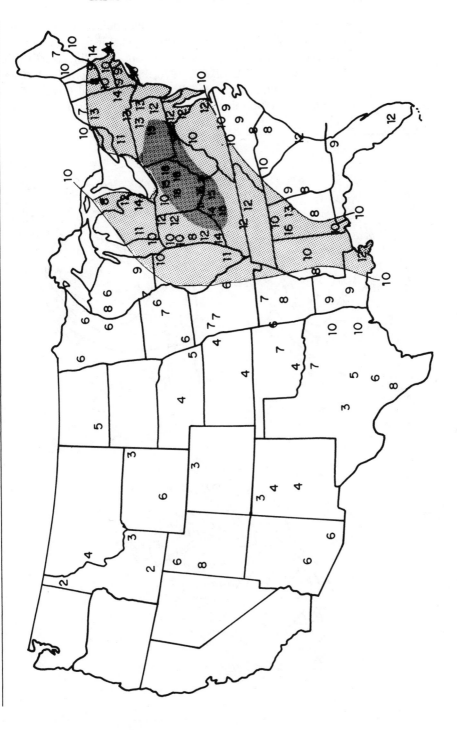

Figure 3–8.   Yearly average sulfate concentration during the mid-1970s, units in $\mu g/m^3$. From OTA (1979).

Figure 3–9.   Annual average pH of precipitation for (a) 1955–1956 and (b) for 1972–1973; (c) the average annual increase in hydrogen ion concentration in precipitation. Units are (meq/liter/year) × 10⁵. (a, b redrawn from G.E. Likens, 1976, "Acid Precipitation," *Chem. Eng. News* 54 (29); c from A.P. Pszenny, "Chemical Characteristics of Rainfall in a Boston Suburb," M.S. thesis, Boston University, 1978.)

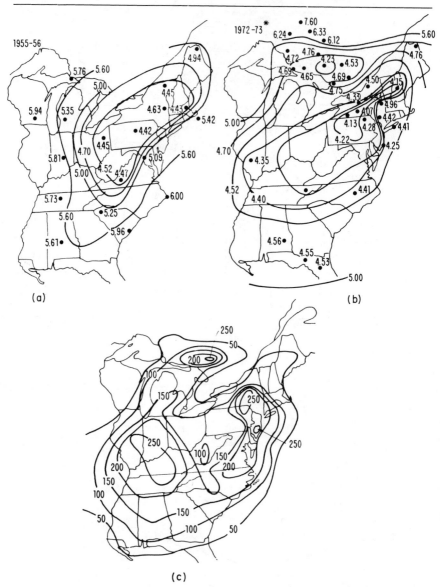

(a)

(b)

(c)

nitrates as well as from sulfates. The determinations are not precise, but they do show that the area experiencing annual average rainfall pHs below 5 in 1977 cover the entire East Coast of the United States. Sulfates have not been specified by EPA as "criteria pollutants" (those for which air pollution criteria exist) and this has reduced the incentive for collecting good data. Much of the earlier sulfate data came from the EPA's NASN, in which sulfate data are gathered primarily from the western states. However, west of the Mississippi the sulfate level is near the global average of $2 \, \mu g/m^3$ [EPA 1973a]. Annual average concentrations from 10 to 18 $\mu g/m^3$ are reported from many urban areas of the East. There are numerous days where the 24-hour average concentrations are over 20 $\mu g/m^3$ for large areas of the Northeast and Southeast. Occasionally, daily values reach $40-80 \, \mu g/m^3$ in the West Virginia panhandle and areas of eastern Ohio and western Pennsylvania. This is the region of maximum emission density for SO_2 in the United States.

Sulfate concentrations tend to be high a few hundred kilometers downwind of emission sources. Sulfates have been estimated to travel more than 1,000 km with a residence time of several days [OECD 1977; Likens and Bormann 1974]. Substantial reductions in ground-level SO_2 values within an air quality control region have not been accompanied by a reduction in sulfates in the region [Altshuller 1976]. Sulfate concentrations are high in the eastern United States, where sulfur emission densities are also high, and low in the western United States, where sulfur emission densities are low. This gross agreement is compared more precisely in Sections 2.3.2 and 2.3.3, and is also found in Europe [OECD 1977].

Either SO_2 or sulfates, independently, or synergistically, could be responsible for adverse health effects. If they varied together in space and time, then we would not need to distinguish their health effects even if we could do so. However, the differing geographical distributions of the concentrations may enable us to separate their health effects; and it is important to do so, because as the location of emission sources is changed so will the relative concentrations of SO_2 and sulfates. We will discuss in Chapter 6 evidence that sulfates (and not SO_2) may be the cause of a health effect at low levels. However, this has not been distinguished from an effect of other respirable particulates.

3.1.9 Sulfur Cycles

Even before human consumption unlocked the sulfur from coal and oil, natural processes released significant amounts of sulfur to the atmosphere. Sulfates are found in sea spray, and hydrogen sulfide (H_2S) comes from the decomposition of organic matter in swampland, tidal flats, and bogs [Hitchcock 1976]. Volcanic activity is a natural source of H_2S as well as the only direct natural source of small amounts of SO_2. Coastal H_2S production is thought to be the most important natural source of sulfur—with an estimated emission (as the element S) of 2 million tons per year. This is in contrast to an anthropogenic emission of 15.7 million tons/yr [Granat, Rodhe, and Hallberg 1976].

People have mined the element sulfur for its chemical and pharmaceutical properties for thousands of years. Today, human consumption contributes 60 percent of global emissions, mostly through the burning of fossil fuels [Granat, Rodhe, and Hallberg 1976]. Approximately 95 percent of this anthropogenic sulfur emission is in the form of SO_2 [HEW 1970a]. Of the natural sulfur, 65 percent is in the form of H_2S, and most of the balance is sea spray sulfate [Kellogg 1972].

As seen in Table 3–3, for the United States, stationary sources account for about 90 percent of the anthropogenic emissions of SO_2. Coal alone accounts for 60 percent of the SO_2 emissions [Hakkarinen 1975; MIT 1970). If the burning of coal increases at 4 percent per year until 1980 and 3.5 percent per year thereafter, with current control techniques and efficiencies, people would be contributing, worldwide as much sulfur to the atmosphere by the year 2000 as nature does [Robinson and Robbins 1968; MIT 1970]. But before the year 2000, anthropogenic sources will more than match natural sulfur emissions in the northern hemisphere.

Currently, in the industrialized regions of the northeastern United States and northwestern Europe, human sources already overwhelm emissions from natural processes; moreover, these anthropogenic emissions are likely to be in the form of small particulates that are especially dangerous. Global sulfur cycles are studied in order to compare the magnitude of sulfur from human activity with that produced by nature and to determine the sources of sulfur, sinks (places where sulfur is absorbed), and atmospheric residence of the many sulfur species [Junge 1973; Eriksson 1963; Robinson and Robbins 1968, 1970; Kellogg et al. 1972; Friend 1973; Granat, Rodhe, and

Hallberg 1976]. Most of these authors have given estimates of the magnitudes of various sources and sinks for sulfur. However, because of the paucity of measurement, particularly of natural emissions, these estimates can err by an order of magnitude or greater [Nochumson 1978]. Only the total emission of anthropogenic sulfur is known with any degree of certainty; but even this estimate could err by 25 percent [Rodhe 1978].

The largest uncertainty in the net emission of sulfur is in the emission of reduced volatile sulfur compounds such as H_2S, and dimethyl sulfide ($(CH_3)_2S$). Until better data are gathered on the emission of reduced sulfur compounds and on the concentration of different sulfur species in the air, global sulfur cycles will remain uncertain. Figure 3—10 presents a summary of the estimated fluxes for a global sulfur cycle.

However, Rodhe [1972] first proposed the use of regional sulfur budgets. Regional budgets are important because anthropogenic sulfur emissions are concentrated in only a few places on the globe, and because atmospheric sulfur residence time is relatively short. Regional budgets, where good local data are available, may provide insight into the transformation and deposition of sulfur species [Rodhe 1978].

An understanding of sulfur chemistry is important for determining not only the sources of sulfur pollution but ultimately the control of sulfur pollution. Some of the SO_2 from a stack will be deposited on the ground by dry impaction with surfaces or by entrainment in a rain droplet with subsequent rainout. What does not fall to the ground as SO_2 will be transformed into sulfate particulates, as illustrated in Figure 3—11.

Sulfates are produced either directly, in the form of natural sea sprays, or indirectly through atmospheric chemical transformation of another sulfur precursor. Of the 130 million tons of sulfates that are estimated to be produced from sea spray, only 10 percent reach land [Kellogg et al. 1972]. Whereas sulfate levels directly over the oceans may be from 2 to 5 $\mu g/m^3$ [Bulfanini 1971], values in the urban Northeast, averaged over a 24-hour period, have been recorded in excess of 50 $\mu g/m^3$. In the Boston area, the annual 24-hour averages have all been in excess of 15 $\mu g/m^3$ over the past decade [MASN 1970]. In order to account for the sulfate concentration in excess of the global background of 2 $\mu g/m^3$ we must turn to the conversion of sulfur precursors into sulfate.

Figure 3-10. Summary of global sulfur cycle with estimates for major fluxes. Values are in 10^5 tonnes/year for (F) Friend [1973]; (K) Kellogg et. al. [1972]; (J) Junge [1963]; (R) Robinson and Robbins [1970]; and (E) Eriksson [1963].

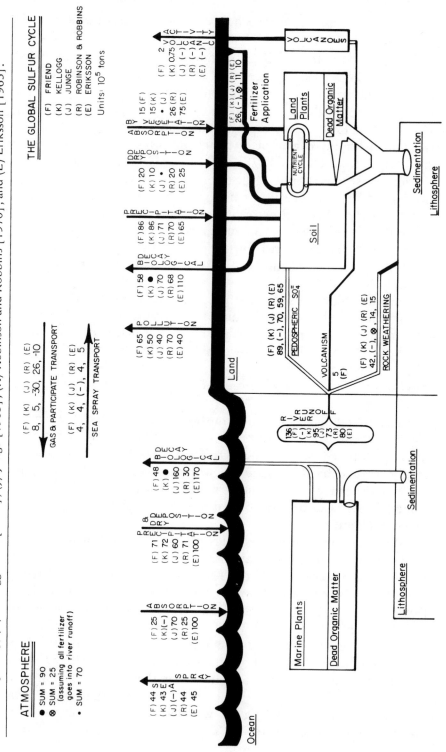

Figure 3–11.   Estimated sulfur budget for a power plant plume over a 4-day period. The sulfur gas (SO_2) deposits or converts to an aerosol. (Adapted from R.B. Husar et al., "Sulfur budget of a power plant plume," *Atmos. Envir.* 12 (1978): 549–568.)

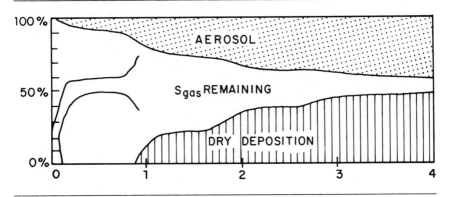

Both H_2S and SO_2 can be oxidized to sulfates. The oxidation of H_2S proceeds through a reaction in which H_2S is first converted to SO_2 [Friend 1973]:

$$2H_2S + 3O_2 \rightarrow 2H_2O + 2SO_2.$$

Because of the limited data available on the ambient H_2S levels and the possibility of its oxidation in photochemical smog, the hypothesis that H_2S contributes to urban sulfate concentrations should not be completely dismissed [Hitchcock 1976; Kellogg et al. 1972].

However, the oxidation of SO_2 is the most likely conversion mechanism and source of sulfates, at least in industrialized areas. Table 3–6 presents a brief summary of several conversion reactions and their rates. SO_2 may be oxidized using the energy of the sun to drive the reaction. In clean air this reaction is relatively slow, but in the presence of hydrocarbons and nitrogen oxides in polluted air, the reaction proceeds more rapidly [HEW 1970a; Bulfanini 1971]. The most rapid and probably most important mechanism for conversion of SO_2 to sulfates involves absorption of the SO_2 gas into water droplets. The presence of trace amounts of metal catalysts and atmospheric ammonia further speed this reaction [HEW 1970a]. A final conversion mechanism is thought to occur by absorption of SO_2 onto carbonaceous soot particles. For a more detailed presentation of the chemistry and conversion reactions the reader is referred

Table 3–6. Mechanisms by which sulfur dioxide is converted to sulfates.

Mechanism	Overall Reaction	Factors on Which Sulfate Formation Primarily Depends
1. Direct photooxidation through electronic excitation.	$SO_2 \xrightarrow{\text{sunlight, oxygen, water vapor}} H_2SO_4$	Sulfur dioxide concentration, sunlight intensity.
2. Indirect photooxidation or photochemically induced oxidation.	$SO_2 \xrightarrow[\text{hydroxyl radical (OH)}]{\text{smog, sunlight, water vapor}} H_2SO_4$ HO_2, CH_3O_2 radicals	Sulfur dioxide concentration, sunlight intensity, OH, HO_2, CH_3O_2 concentration.
2. Air oxidation in liquid droplets.	$SO_2 \xrightarrow[\text{oxygen}]{\text{liquid water}} H_2SO_4$ $NH_3 + H_2SO_4 \longrightarrow NH_4^+ + SO_4^=$	Ammonia concentration.
4. Catalyzed oxidation in liquid droplets.	$SO_2 \xrightarrow[\text{heavy metal ions}]{\text{oxygen, liquid water}} SO_4^=$	Concentration of heavy metal (Fe, V, Mn) ions or carbon particles.
4. Oxidation in liquid droplets by strong oxidants.	$SO_2 \xrightarrow[H_2O_2]{\text{ozone}} SO_4^=$	Ozone or hydrogen peroxide concentration.
6. Catalyzed oxidation on dry surfaces.	$SO_2 \xrightarrow[\text{fly ash, carbon or other particles}]{\text{oxygen, water}} H_2SO_4$	Particle concentration and surface area.

Source: Altshuller, 1980.

to several excellent articles [Husar, Lodge, and Moore 1978; Kellogg 1972; Calvert 1973].

Concentrations of sulfates are directly dependent upon SO_2 concentration, air particulates that catalyze or provide a surface for reaction, atmospheric humidity, sunlight intensity, and the SO_2 atmospheric residence time. Sulfate concentrations are known to be inversely related to wind velocity and precipitation. The most important constant for determination of long-term average concentrations is the conversion rate, which in itself varies between normal and polluted air. This constant was important in the computer models of Section 2.3.3. So far the constant has been adjusted to fit the data.

Although the chemical and physical processes leading to sulfate production are complex, with no single factor dominating [Hidy and Burton 1974], it is generally accepted that SO_2 is the primary sulfate precursor.

With coal burning producing 60 percent of anthropogenic SO_2 and with more than 90 percent of the atmospheric SO_2 derived from human activity, coal combustion may be responsible for more than 50 percent of ambient sulfate levels. This value of 50 percent could deviate up or down on the basis of at least two factors. (1) The proportion of sulfate caused by coal burning could be greater than 50 percent because conversion is strongly dependent upon the atmospheric residence time of SO_2, which is increased with the tall stacks used in coal combustion; other SO_2 sources, such as space heating, generally have shorter stacks, resulting in reduced atmospheric residence; (2) The proportion of sulfates caused by coal may be less than 50 percent if another conversion process involving H_2S is found that proceeds more rapidly than those reactions presently known.

3.2 NITROGEN DIOXIDE (NO_2) AND OZONE (O_3)

Nitrogen dioxide (NO_2) is formed from oxidation of nitric oxide (NO). NO may not be very harmful in itself, but NO_2, a reddish-brown gas with a pungent odor and toxic effects.

The major source of nitric oxide is the natural decay of biological sources, and probably 500 million tons of NO are produced per year in the United States by this means.

The anthropogenic source of NO is from any combustion process where air is used. Nitrogen and oxygen in the air combine to form NO according to the reaction

$$N_2 + O_2 \underset{\leftarrow}{\overset{\rightarrow}{\rightleftharpoons}} 2\ NO.$$

At room temperature the equilibrium for this reaction is such that there is a predominance of nitrogen and oxygen; at high temperatures the equilibrium tends to a predominance of NO. This is expressed quantitatively by the equilibrium constant

$$k = \frac{[NO]}{[NO_2]\ [O_2]} = 21.9\ exp(-43,900/RT) = 10^{-31}\ \text{at room temperature}$$

$$\simeq \frac{1}{3000}\ \text{at } 1700°C\ (= 1973°\,K)$$

where

[NO_2], [O_2], and [NO] = the concentrations of the respective molecules,

R = the universal gas constant (2 calories/gram), and

T = the temperature in degrees Kelvin.

In a flame, therefore, NO tends to be formed, and the East Coast from Boston to Washington, D.C. has the highest emission density because of mobile sources. In only four air quality regions are there any violations of the primary annual air quality standard for NO_2 (100 $\mu g/m^3$ annual average): Los Angeles, Chicago, Cleveland, and the New York–New Jersey area. In half the AQCRs there is inadequate data but compliance is probable. Measurements of nitrate ions from NASN are subject to uncertainty because of filter artifacts and volatilization problems. As a general pattern (Figure 3–12), however, nitrates appear the highest through the eastern half of Texas up through the eastern half of the Midwest and across the industrialized North from Chicago to Philadelphia.

The higher the temperature of the flame the more NO is formed; following burning the gas usually cools sufficiently rapidly that equilibrium between N, O, and NO_x is lost and high NO concentrations persist.

If there is just enough air available for combustion of air and fuel, there will be no oxygen left to combine with the nitrogen to produce NO. Of course this is a simplified discussion. We should really discuss

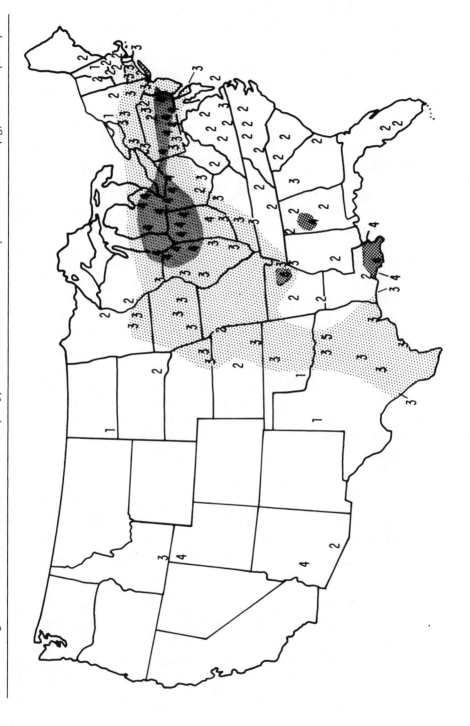

Figure 3–12.   Concentration of nitrate (NO_3) in the United States in early 1970s. Values in $\mu g/m^3$. OTA (1979).

a three-body equilibrium of N_2, O_2, and fuel but the general result that NO production is reduced is still correct.

The above discussion is about the combination of N_2 in the air and O_2 in the air. In addition to N_2 in the air, there is organically bound nitrogen in coal that adds to the ambient concentrations when burned.

There are two concerns with NO emissions. First, the NO_2 formed from NO can directly produce health problems, and second, photochemical reactions in the atmosphere can produce oxidants such as ozone (O_3).

One rapid mechanism leading to oxidant formation is a three-stage reaction where hydroxyl radical (OH) and carbon monoxide (CO) from the plume liberate atomic hydrogen, which then combines with oxygen molecules to form the hydro-peroxy radical. This product subsequently combines with NO to form NO_2 and to regenerate the OH. In a second mechanism, free-radical chains rapidly convert NO to NO_2. This mechanism is known to involve the organic vapors of polluted atmospheres and will enhance NO_2 production in such polluted atmospheres. This second reaction is especially important as the plume constituents necessary for the first reaction disperse.

The presence of sunlight on NO_2 leads within an hour or two to production of a like amount of NO and O_3. The O_3 that is produced is not a problem immediately surrounding any single power plant because O_3 is not directly emitted. However, just as sulfates pose a long-range problem, O_3 is a problem for hundreds of miles, and the source of O_3 is the combined emissions from many sources, stationary and mobile.

Figure 3–13 shows how photochemical reactions produce a complex variation of NO levels in Los Angeles. In this case, the main source of nitrogen oxides is automobile exhausts, which occur primarily during the day.

Table 3–3 gives a breakdown of NO_x emissions in the United States by source category for 1977. Electric power plants contribute slightly less to NO_x emissions than do highway vehicles. Overall, however, the combination of stationary fuel combustion contributes more than 50 percent of the annual emissions of 23 tons, whereas all transportation contributes approximately 45 percent. Increased use of coal by stationary sources in the 1980s and 1990s will further increase the relative contribution of stationary sources. The spatial distribution of NO_2 emissions is displayed in Figure 3–14. The pat-

Figure 3–13.   The variation of NO_2 and oxidant concentrations over a typical winter day in Los Angeles. (From EPA, 1971, page 6–2).

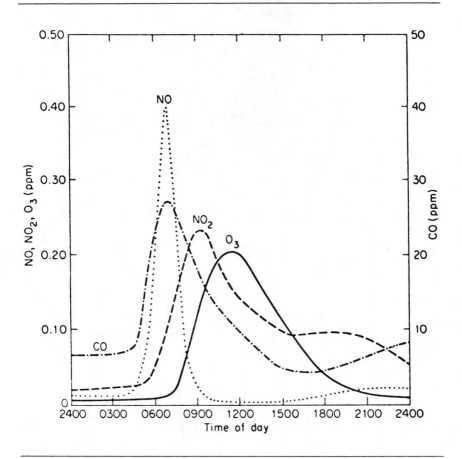

tern is similar on a regional basis to the SO_2 emission densities; that is higher in the north central and northeast than in the west and south. Differences are apparent in intraregional comparison. The populated corridor along the east coast of the United States has the high NO_x emission density. Automobiles, space heating, and power plants contribute to the NO_x emissions in this populated area.

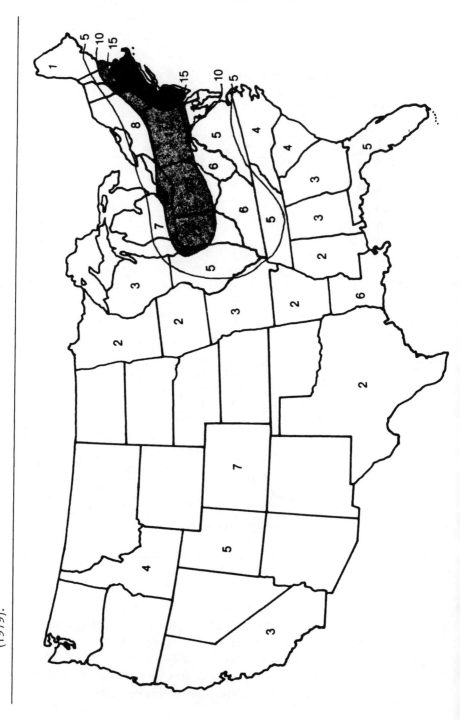

Figure 3–14. NO₂ emission density for the United States in the early 1970s. Values are g/m²/year. From OTA (1979).

3.3   CARBON DIOXIDE (CO_2)

Whenever fossil fuels (e.g., coal and oil) or nonfossil carbon fuels (e.g., wood) are burned, carbon dioxide (CO_2) is produced. Although the CO_2 does not have adverse health effects at ambient concentrations, the potential exists for temperature and climate shifts that could have serious consequences.

The CO_2 problem can be dealt with on four levels. The uncertainty in our knowledge increases with each succeeding level. These are:

1. What are the levels of atmospheric carbon dioxide?
2. What effect will increased fossil fuel use have on atmospheric levels of carbon dioxide?
3. What effect will increased atmospheric carbon dioxide have on the mean global temperature?
4. How will increased global temperature affect global climate and ecology?

In the late preindustrial period around 1850 the concentration of CO_2 in the atmosphere was about 270 ppm by volume [Stuiver 1978]. In 1973 the earth's atmosphere contained approximately 700×10^9 tonnes (1 metric tonne = 10^6 grams) of the element carbon in the form of CO_2 [Baes et al. 1976a, b]. This represents a concentration of approximately 330 ppm by volume.

The concentrations of CO_2 as measured in Mauna Loa, Hawaii, are shown in Figure 3−15. In addition to the annual variation caused by absorption and desorption of living plants of 5 ppm there is an annual increase of 0.8 ppm per year. Similar records at stations in Australia, the South Pole, Alaska, Sweden, and Long Island, New York, indicate increases of CO_2 concentration of 0.5−1.5 ppm/yr. On a global basis, this measured increase in concentration represents an increase in total carbon content of the atmosphere, as carbon dioxide, of roughly 2.5×10^9 tonnes/yr.

Two methods are used to measure the specific contribution of fossil fuel burning to concentrations of atmospheric CO_2. The first method compares ratios of carbon 13 (^{13}C) to carbon 12 (^{12}C) in tree rings and glaciers. CO_2 that is derived from fossil fuels and decaying plants is deficient in ^{13}C relative to atmospheric CO_2. The second method compares the ratio of ^{14}C to ^{12}C [Suess 1955]. Because the ^{14}C has decayed in the fossil fuels since the time they interacted with the atmosphere, this method also provides information on com-

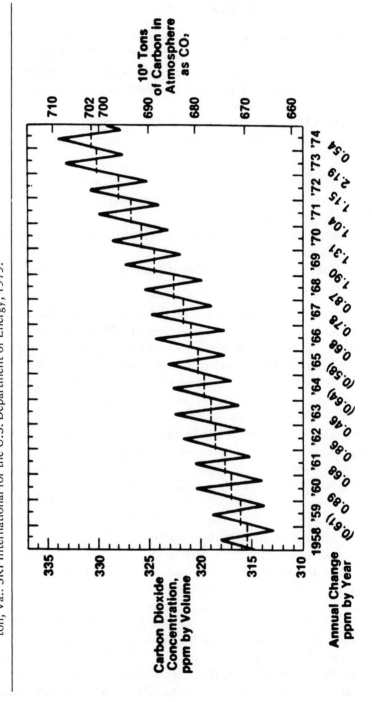

Figure 3–15. Measured CO_2 concentration at Mauna Loa Observatory. From Keeling [1976a, b] as appeared in "The long-term impact on atmospheric carbon dioxide on climate," JASON Technical Report JRS–78–07, Arlington, Va.: SRI International for the U.S. Department of Energy, 1979.

bustion-formed CO_2, but the use of atomic weapons in World War II and other bomb tests since has increased the ^{14}C content of the atmosphere and has made the measurement more difficult.

The ^{13}C/^{12}C ratio in Douglas fir trees indicates a flux from the biosphere of 120×10^9 tonnes of carbon between 1850 and 1950, or an average of 1.2×10^9 tonnes (C as CO_2)/yr [Stuiver 1978]. This is to be compared with the rapid increase in fossil fuel use between 1950 and the 1970s, which has generated a production rate of 5 to 6×10^9 tonnes (C as CO_2)/yr [Rotty 1976]. The difference, 5 to 6×10^9 minus 1.2×10^9 tonnes, would be absorbed by the terrestrial biosphere or by the surface waters of the oceans.

Current research, however, indicates that the terrestrial biosphere may be a source, rather than a sink, for CO_2 [Wong 1978]. Removal of forests (deforestation) can reduce photosynthesis (reducing CO_2 uptake) and enhance decay of the forest material (increasing CO_2 release) [Woodwell and Pecan 1973]. Estimates of CO_2 release from deforestation vary widely (Table 3–7).

Assuming that the lowest figure of 1.5×10^9 tonnes carbon/yr [Wong 1978] is correct, the annual carbon as CO_2 flux from both fossil fuel burning and deforestation would be 6.5×10^9 tonnes/yr. The difference between this number and the measured CO_2 increase of 2.5×10^9 tonnes/yr indicates removal mechanisms that are two-thirds efficient, not 50 percent as was previously assumed.

The oceans are probably the sink for removal of atmospheric CO_2. Removal of CO_2 depends on a two-step process in which atmospheric CO_2 is first absorbed in the surface layers of the ocean to be secondly transported and stored in the deeper ocean layers, where the amount of CO_2 that can be absorbed is vast. But the time constant—900 years—for mixing of the deep and shallow oceans is long compared with time spans at issue. Therefore if the first step can be saturated with higher CO_2 concentrations, the oceans would retain

Table 3–7.   Estimates of net CO_2 release from deforestation.

Tonnes $\times 10^9$ Carbon as CO_2 Released	Investigator
1.5	Wong [1978]
4.5	Bolin [1970]
4–8	Woodwell et al. [1978]

fraction of the CO_2 produced by fossil fuel combustion and urther exacerbate atmospheric CO_2 concentrations.
...ods of estimating the rate of circulation between deep and shallow oceans differ from the box models [Broecker and Li 1970] to the box diffusion models [Oeschger et al. 1975], but all give slow rates.

2.4.1 Global Temperature Changes

If atmospheric CO_2 concentrations do increase, what effect will that have upon global temperature? The geologic record offers a long history of natural climate change as a reference against which to judge the significance of CO_2-induced changes. Temperature fluctuations of the last 100,000 years have a maximum fluctuation of $8°K$ [ICAS 1974]. Over the last 200 years, this oscillation has been only $0.5°K$, with a maximum rate of change of $\pm 0.0075°K/yr$. According to Broecker [1975], we are entering a warming period. One must remember, however, the small magnitude of natural short-term trends; the current warming trend is expected to result in changes of probably no more than $0.5-1.0°K$.

Changes in CO_2 might radically alter this state of affairs. CO_2 is transparent to incoming ultraviolet radiation, but opaque to the resulting reradiation in the infrared. Because CO_2 also absorbs incoming infrared radiation, it is an integral part of the mechanism of global heat balance. Thus, without compensating charges, increased levels of CO_2 will result in higher temperatures. The decrease in the effective emissibility of infrared radiation from the earth, with resulting atmospheric heating, is responsible for the so-called greenhouse effect. Callendar [1938] was probably the first to predict significant temperature changes from anthropogenic CO_2. Mitchell [1961] suggested that just such a process is responsible for the $0.7°K$ rise in average temperature since 1850.

Direct measurement of the temperature effects of CO_2 are difficult. The underlying natural fluctuations in temperature do not provide a stable baseline against which to reference anthropogenic disturbances. Other such changes, for example as cooling from increased reflectivity of the lower atmosphere caused by particulate matter, or interference with the O_3 layer from anthropogenic pollutants, further confound the relationship between CO_2 level and temperature. We resort to models of varying complexity to predict temperature change in response to shifts in CO_2 concentration. Models by Schneider [1975], Manabe and Wetherald [1975], and Baes et al. [1976a, b] suggest an increase in temperature of 1.0 to 5.0°K per doubling of CO_2 concentration. The temperature change is likely to be more extreme in the higher latitutes. This large uncertainty is primarily due to five factors: (1) the average fraction and temperature of cloud cover have important effects on heat balance mechanisms; (2) no model can adequately treat CO_2 levels in relation to heat exchange between oceans and atmosphere; (3) if warming is sufficient to cause a significant decrease in snow and ice cover, a decrease in the albedo (reflectivity) of the earth's surface will result, amplifying any warming trend started by an increase in CO_2; and (4) increased CO_2 uptake by land biota by photosynthesis might be stimulated by high CO_2 levels.

The final uncertainty is by far the greatest: Both the oceans and fresh water could be affected by a lowering of pH from dissolved CO_2. The stability of ecosystems could be affected by changes in temperature and precipitation, or by direct sensitivity to CO_2 levels. Agricultural regions may be altered positively or negatively by changes in temperature and precipitation. The present global temperature is only 6°K different from the last ice age [IWCI 1978].

Figure 3–16, from Marland and Rotty [1977], shows cumulative CO_2 production since 1960, and projected to the year 2030 using a range of fossil fuel use and the expected increase in CO_2 concentration. 50 percent of the CO_2 is assumed to be retained by the atmosphere.

In Figure 3–17 Rotty [1977] presents a composite picture of temperature variation resulting from extremes in fossil fuel use and showing the variation produced by deforestation (±1%/yr in carbon stored in forests) and ocean absorption of CO_2 (40–60 percent of atmospheric production). The range of temperature is large and can be reduced only through further research and determination of fossil fuel production.

3.4 PARTICULATE ORGANIC MATTER (POM)

When a material containing carbon and hydrogen is burned, organic hydrocarbons are often produced. These hydrocarbons are present in

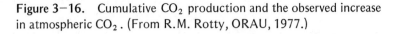

Figure 3–16.   Cumulative CO_2 production and the observed increase in atmospheric CO_2. (From R.M. Rotty, ORAU, 1977.)

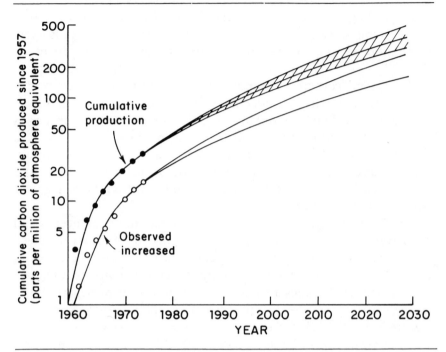

the unburned coal; they also arise from incomplete combustion, although we do not know the details of the variability of the production of hydrocarbons among various processes.

Table 3–8 shows the concentrations of benzo(α)pyrene in and near several U.S. cities [Sawicki et al. 1960]. The concentrations are high in Altoona, Pennsylvania, no doubt caused by heavy coal burning. Benzo(α)pyrene is a known potent carcinogen in animals.

Concentrations of a number of polycyclic organic compounds deposited on particulate samples observed near coal-fired power plants are shown in Table 3–9. These polycyclic aromatic hydrocarbons (PAH) are the most important of the organic materials, because many of them are known to be carcinogenic in animals or people (Section 5–3). Table 3–9 shows that some of these chemicals seem to be produced in the plume; the distribution of these compounds varies at distances downwind, although this could be a consequence

Figure 3-17. Projected atmospheric CO_2 concentrations and possible changes in the average surface temperature. (From R.M. Rotty, "Uncertainties associated with future atmospheric carbon dioxide levels," Institute for Energy Analysis, ORAU.)

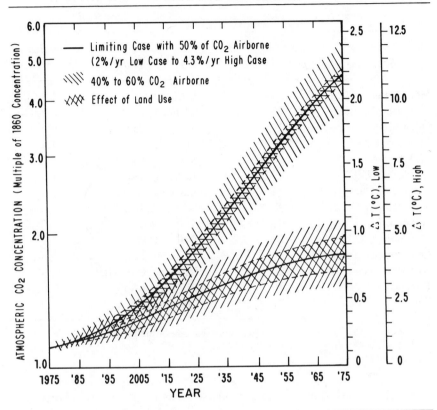

of the condensation of vapors on particulates. It is therefore desirable to measure the ambient concentrations of all of them. However, this has not been done and is probably too expensive to become a routine procedure for all PAH. For the most part, measurements of the ambient concentrations of only benzo(α)pyrene (B α P) exist [NAS 1972]. B α P is therefore used as an indicator of polycyclic aromatic hydrocarbons, and is used as the independent variable for the analysis of carcinogenesis. Because of this, only rough conclusions can be drawn. As we discover which hydrocarbons are the most important, we can then measure them.

Table 3−8.   Benzo (α) pyrene (BαP) concentrations in urban sampling sites for January through March 1959.

High BαP		Low BαP	
Urban Sampling Site	ng BαP per m³ Air	Urban Sampling Site	ng BαP per m³ Air
Montgomery, Ala.	24	Little Rock, Ark.	1.5
Indianapolis, Ind.	26	Glendale, Calif.	0.8
Des Moines, Iowa	23	San Jose, Calif.	0.6
Portland, Maine	21	Miami, Fla.	1.9
St. Louis, Mo.	54	Shreveport, La.	0.7
Charlotte, N.C.	39	Jackson, Miss.	1.2
Cleveland, Ohio	24	Las Vegas, Nev.	1.4
Youngstown, Ohio	28	Bismarck, N.Dak.	0.4
Altoona, Pa.	61	Tulsa, Okla.	1.0
Columbia, S.C.	24	Dallas, Tex.	1.4
Chattanooga, Tenn.	31	Houston, Tex.	1.6
Knoxville, Tenn.	24	Salt Lake City, Utah	0.5
Richmond, Va.	45	Burlington, Vt.	1.0
Wheeling, W.Va.	21	Cheyenne, Wyo.	1.2

Source: F. Sawicki et al., "Benzo (α) pyrene content of the air of American communities," *J. Amer. Ind. Hyg. Assoc.* 21 (1960): 443.

Emissions of B α P from various sources in the United States are shown in Table 3−10; they vary considerably with the method of burning. Hand-stoked and open fires, whether for coal, refuse, or wood, are particularly bad. There is therefore much room for improvement in reducing these emissions. It is noteworthy, however, that the larger units produced the least B α P emissions.

These concentrations of PAH have been falling since 1959. In London, Lawther and Waller [1978] mention a decline of a factor of ten between 1935 and 1965. In the United States, a drop of a factor of about five is observed from an average of 6 μg/m³ over all measuring sites to about 1 μg/m³. No figures are provided for a population-weighted annual average.

Table 3-9. Polycyclic organic compounds identified in stack and plume particulates from coal-fired power plants. [Data from Natusch, 1978; Stahley, 1976; Korfmacher, Natusch, and Wehry 1977]

| | Polycyclic Organic Compounds, ng/m³ | | |
| | In Stack | In Plume | |
Compound		0—5 miles	5—10 miles
Fluoranthene	2.7	5.0	4.4
Pyrene	16.4	60	9.0
Benzo (a) anthracene	69	232	14.4
Chrysene	48	68	10.8
Perylene	< 2	7.0	8.6
Benzo (e) pyrene	9.8	15.8	13.2
Benzo (α) pyrene	12.9	16.2	8.2
Benzoperylene	12	13	—
1, 2, 4, 5 - Dibenzopyrene			
3, 4, 7, 10 - Dibenzopyrene			
Phenanthrene			
Dimethylbenzanthracene			
Anthracene		trace	
Benzo (k) fluoranthene		concentrations	
9, 10 dimethyl anthracene			
Benzo (b) phenanthrene			
Fluorene			
Triphenylene			

It is encouraging to observe the steady downward trend in B α P concentrations in the United States. Long-term trends from the NASN network show that in 1966 the annual mean concentrations across 34 NASN urban sites was 3 μg/m³. In 1975 the mean level was less than 1 μg/m³. Figure 3—18 plots the fiftieth and ninetieth percentiles from 34 NASN sites from 1966 to 1975.

Table 3-10.   Sources of benzo (α) pyrene in the United States [modified from NAS 1972].

	Emission Factor, Lbs. BαP/10⁴ Tonnes Fuel	Total Yearly Emission, Tonnes of BαP/Year (1967)
Coal[a]		
hand-fired (residential)	1,100–2,200	420
intermediate	0.01–0.03	10
steam power	0.01–0.3	1
Oil[b]	0.1–1	2
Gas	—	2
Wood[c]	20–1,000	40
Incineration		33
Open burning		
forest and agriculture		140
coal refuse		340
miscellaneous		74

Energy content of fuels modified from NAS 1972.

a. 3.7×10^7 BTU/tonne coal.

b. 4.6×10^7 BTU/tonne oil.

c. 2×10^7 BTU/tonne dry wood; also variability in burning from von Foerster [1978].

Figure 3–18.   BαP annual trends (1966–1975) in the 50th and 90th percentiles for 34 NASN urban sites. (Adamson & Bruce, 1979)

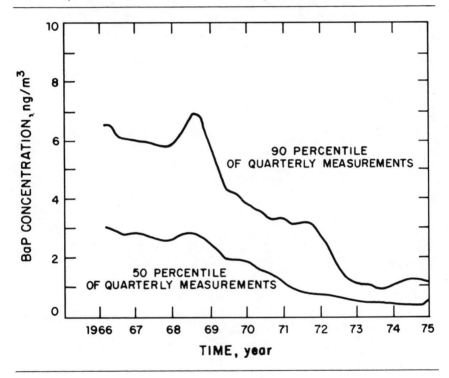

4 HEALTH EFFECTS OF AIR POLLUTANTS FROM COAL COMBUSTION

Coal combustion produces a variety of air pollutants. People will always be exposed to a mixture of several substances but experimental exposures are often limited to one or two pollutants. Yet potential synergetic effects of many pollutants may be of critical importance. Table 4-1 lists several reviews; various studies are summarized in Table 4-2.

4.1 ACUTE AND CHRONIC EFFECTS

Health effects may be divided into two classes, in one of which the effects are acute or prompt. The second class of health effects includes the chronic effects that build up slowly from continued exposure to low levels of pollutants. There is no general acceptance of the magnitude—or even the reality—of these effects. However, if real, they are large because the number of persons exposed is large, and although proof of their reality is not available, and may never be, no one can disprove the existence of a large health effect of air pollution either. It is this second class of effects that concerns us most in this book, and we will outline our arguments in Chapter 6. The low-level long-term effects, if any, may be different from the prompt effects; for example, carcinogenesis is clearly different from bronchi-

Table 4-1.   Index to recent air pollution health reviews.

Subject	Review
General	Rall [1978] Ferris [1978] American Thoracic Society [1978] Stern [1977] Lave and Seskin [1977] OTA [1979]
Sulfur oxides	NAS [1978] Colucci [1976] NAS [1975] HEW [1970a] Rall [1974] NAS [1979b]
Particulates	NAS [1979a] Perera and Ahmed [1978] NAS [1975] NAS [1972] HEW [1969]
Nitrogen dioxide	EPA [1978a] NAS [1977a] HEW [1971]
Ozone	EPA [1978b] NAS [1977c] HEW [1970b]

tis. Nonetheless, in assessing the low-level effects, we would like to keep the prompt high-level effects in mind, and look in particular for aggravation of conditions that are severe at higher levels.

Unfortunately, this goal is not completely possible to achieve. The low-level studies use large-scale statistical data, and only limited information is available for these: mortality, and cause of death. Even the major confounding variable, cigarette smoking, is not listed in the death statistics.

Any reproducibly measurable biological effect in mammals that is related to a pollutant exposure can be considered as a possible adverse health effect. The underlying assumption in this definition is that homeostasis (corrective action of body processes to maintain constancy of function) is such an important mammalian phenomenon that any measurable persistent departure of the internal environment of the body from constancy is immediately suspect. For example, running increases the pulse rate, but this change would be considered homeostatic (a prompt corrective action of the body to restore equilibrium) and, therefore, beneficial except for certain sensitive individuals such as those with heart disease. On the other hand, a pollutant that alters the resting pulse rate would immediately be suspected of producing a potentially harmful effect.

Thus implicit in the definition of an air pollution health effect is the implication that the pollutant exposure would result in any damage leading to (1) even temporary impairment of body defense mechanisms, or (2) irreversible alterations of structure, regardless of degree, for small increments can be added to existing alterations or to natural processes such as aging.

Much of the pioneering data on prompt health effects were obtained at the workplace. However, there are not many data for determining effects at lower ambient levels. In part this is because ambient levels are often below the background at the workplace, and the pollutant combinations in occupational settings may not be applicable to ambient situations. Furthermore, specific worker populations do not necessarily have the same susceptibility to pollutants that the general population has, making it difficult to generalize results.

The other sort of human measurement of prompt health effects is the controlled laboratory test. There are several major limitations inherent in laboratory tests. Ethical considerations demand that overt manifestation of health effects must be strictly limited to reversible effects in controlled human studies. Laboratory tests on humans must be strictly limited in duration of exposure. Finally, it is very difficult to cover the wide range of test conditions that would mimic the actual ambient combinations and variations of pollutants and other environmental conditions. Therefore, the results, of necessity, are highly inferential.

As an alternative to human tests, toxicologic laboratory tests offer the chance to manipulate experimental variables more freely, including use of chronic exposures or even tests involving successive generations. But the most notable difference is the opportunity for invasive

Table 4–2. Classification of air pollution studies by type of disease, type of study, and considered pollutant variables.

This table is developed from three general reviews and the studies referenced within these reviews are tabulated by disease and pollutant according to the following classifications:

1. *Severity*—indicates aggravation of symptoms in persons with existing disease.
2. *Prevalence*—indicates the number of persons with a given condition in a certain population.
3. *Incidence*—indicates the number of new cases of a condition found in a certain population over a defined time period.
4. *Laboratory/Experimental*—controlled studies of the relationship between air pollution and health.

The most direct indication of the transition probability of going from health to illness is given by the incidence rate. Experimental studies can also be designed to investigate transition probabilities but they are more generally designed to elucidate mechanisms of pollutant damage.

The table indicates that relatively few incidence studies exist. Most human morbidity studies are of the severity or prevalence type.

Several other observations on the table should be made. Because classification into one of the four study categories requires reinterpretations of the three authors' reviews, many of the studies may be misclassified in the table. Careful review of this table will indicate several cases where a single study is classified under more than one study category. This may be due to 1) the studies actually investigated more than one category; and 2) different interpretations of original authors' work given by different reviewers.

An *F*, *L* + *S*, or *S* following the study indicates where it was reviewed. An *F* stands for Ferris 1978, *L* + *S* for Lave and Seskin 1977, and *S* for Stern 1977.

Disease	Possible Cause	Aggravation of Existing Symptoms (severity)	No. Existing Cases (prevalence)	Rate of New Cases (incidence)	Laboratory/ Experimental
Asthma	Oxidants	Schoettlin and Landau [1961] L + S	Hausknecht [1962a] S Hausknecht [1959] S	Schoettlin and Landau [1961] S	Schoettlin and Landau [1961] S

Nitrogen dioxide			Orehek, et al. [1976] F
Silica combustion particles	Lewis, Gilkeson, and McCaldin [1962] L + S		
Smoke levels	Derrick [1970] L + S Smith and Paulus [1971] L + S		
Smoking	Phelps [1965] L + S Phelps and Koike [1962] L + S Phelps [1961] L + S Phelps, Sobel, and Fisher [1961] L + S		
Sulfates	Zeidburg, Prindle, and Landau [1961] L + S		Zeidburg, Prindle, and Landau [1961] L + S
Sulfur dioxide and/or suspended particles	Goldstein and Block [1974] F Cohen, et al. [1972] F Yoshida, et al. [1969] L + S Kenline [1966] L + S	Cohen, et al. [1972] F Sultz et al. [1970] L + S	
Unspecified pollution	Brown and Ipsen [1968] L + S Greenburg et al. [1965] L + S		Phelps [1965] L + S Phelps and Koike [1962] L + S Phelps, Sobel, and Fisher [1961] L + S

(Table 4–2. continued overleaf)

Table 4-2.   continued

Disease	Possible Cause	Aggravation of Existing Symptoms (severity)	No. Existing Cases (prevalence)	Rate of New Cases (incidence)	Laboratory/ Experimental
Bronchitis and Emphysema	Coal consumption	Lambert and Reid [1970] L + S			
	Cotton dust		Lammers, Schilling, and Walford [1964] L + S		
	Fog		Cornwall and Raffle [1961] L + S		
	Nitrogen dioxide		Pearlman, Finklea, Creason, Shy, Young, and Horton [1971] F		
			Shy et al. [1970a] F		
			Burrows, Kellogg, and Buskey [1968] L + S		
	Oxidants		Durham [1974] S		
	Smoking habits	Lambert and Reid [1970] L + S	Waller [1971] L + S		
			Lawther, Waller, and Henderson [1970] L + S		
			Chapman [1965] L + S		
			Lammers, Schilling, and Walford [1964] L + S		
			Pemberton [1961] L + S		
			Lawther [1958] L + S		
			Waller and Lawther [1955, 1957] L + S		

Smoke pollution		Lammers, Schilling, and Walford [1964] L + S
Sulfur dioxide and suspended particles	Lambert and Reid [1970] L + S	Waller [1971] L + S Carnow, et al. [1970] S Lawther, Waller, and Henderson [1970] F, L + S Burrows, Kellogg, and Buskey [1968] L + S Petrilli, Agnese, and Kanitz [1966] L + S Lammers, Schilling, and Walford [1964] L + S Pemberton [1961] L + S Lawther [1958] L + S Waller and Lawther [1955, 1957] L + S
Unspecified pollution	Motley, Smart, and Leftwich [1959] L + S	Ishikawa et al. [1969] S Cederlöf [1966] L + S Higgins [1966] L + S Holland and Reid [1965] L + S Fairbairn and Reid [1958] L + S Reid and Fairbairn [1958] L + S Reid [1956, 1958] L + S
Cancer Particulates		Selikoff, Hammond, and Seidman [1973] S Mancuso [1970] S Lee and Fraumeni [1969] S Committee on Biological Effects of Atmospheric Pollutants,

(Table 4–2. continued overleaf)

Table 4-2. continued

Disease	Possible Cause	Aggravation of Existing Symptoms (severity)	No. Existing Cases (prevalence)	Rate of New Cases (incidence)	Laboratory/ Experimental
			Selikoff, Hammond, and Churg [1968] S Selikoff, Churg, and Hammond [1965] S Selikoff and Hammond [1964] S	NAS [1972] S Wynder and Hoffman [1965] S	
	Smoking		Carnow and Meier [1973] S Pedersen et al. [1969] S Hitosugi [1968] S Stocks [1966] L + S, S Kreyberg [1956] L + S	Eastcott [1956]	
	Smoking and pollution levels	Higginson and Muir [1973] S Schneiderman and Levin [1972] S			
	Sulfur trioxide		Shy [1976] F Pike et al. [1975] F Hammond [1972] F Lee and Fraumeni [1969] S		
	Unspecified pollution		Buell, Dunn, and Breslow [1967] S Hagstrom, Sprague, and Landau [1967] S		

Effect	Pollutant			
Cardiovascular	Unspecified pollution	Stocks [1966] S Buck and Brown [1964] S The Cancer Registry of Norway [1964] S Auerbach et al. [1962] S Winkelstein et al. [1967] S Auerbach et al. [1961] S Levin et al. [1960] S World Health Organization [1960] S Hammond and Horn [1958] S Mancuso and Coulter [1958] L + S Haenszel, Marcus, and Zimmerer [1956] S Hoffman and Gilliam [1954] L + S	Sinnet and Whyte [1973] S	Ayres, Gianelli, and Mueller [1970] S, L + S Ayres et al. [1969] S
Irritations	Carbon monoxide		Cassell et al. [1968] L + S	
	Hydro-carbons		Cassell et al. [1968] L + S	
Eye and throat (e.g. chronic cough)	Oxidants	Hausknecht [1962b] S Hammer, Hasselblad and Portnoy [1974] S, L + S Richardson and Middleton [1965] S Renzetti and Gobran [1957] F	Hammer, Hasselblad, and Portnoy [1974] S, L + S Hammer, Hasselblad and Portnoy [1965] F Richardson and Middleton [1958] F Renzetti and Gobran [1957] F	Glasson and Tuesday [1970] S Heuss and Glasson [1968] S

(Table 4–2. continued overleaf)

Table 4-2. continued

Disease	Possible Cause	Aggravation of Existing Symptoms (severity)	No. Existing Cases (prevalence)	Rate of New Cases (incidence)	Laboratory/Experimental
	Sulfur dioxide		Cassell et al. [1968] L + S Skalpe [1964] L + S	Yoshii et al. [1969] L + S	
	Suspended particles		Cassell et al. [1968] L + S Winkelstein and Kantor [1969] L + S		
	Unspecified pollution		Kagawa and Toyama [1975] F HEW, [1969] S Loudon and Kilpatrick [1969] L + S Oshima et al. [1964] S Takahashi [1970] S	Oshima et al. [1964] L + S	
Pulmonary function	Nitrogen dioxide	Heuss, Nebel, and Colucci [1971] S Petr and Schmidt [1967] S	Speizer and Ferris [1973a, b] F Shy et al. [1970a, b] F, S Pearlman et al. [1971] S Vigdortschik et al. [1937] S		Von Nieding et al. [1973] F
	Oxidants	Kagawa, Toyama, and Nakaza [1976] F Rokaw and Massey [1962] F			Silverman et al. [1976] F Hackney et al. [1975c] F Hazucha and Bates [1975] F Bates and Hazucha [1973] F

Pollutant			
Ozone	Kagawa and Toyama [1975] S; Mueller and Hitchcock [1969] S	Kagawa and Toyama [1975] F	Goldsmith and Nadel [1969] F; Holland et al. [1968]; Johnson, Mohler, and Armstrong [1971]; Friberg [1950] S
Particulates	McDonald et al. [1970] S; McEwen et al. [1970] S; Stumphius and Meyer [1968] S; Lieben and Pistawka [1967] S; Newhouse and Thompson [1965] S; Wagner, Sleggs, and Marchand [1960] S		
Sulfur oxides	Petr and Schmidt [1967] S	Hausknecht [1962a] S	
Sulfur dioxide	Chapman et al. [1976] F; Ferris et al. [1976] F; Mandi et al. [1974] F; van der Lende et al. [1974] F		

(Table 4–2. continued overleaf)

Table 4-2.   continued

Disease	Possible Cause	Aggravation of Existing Symptoms (severity)	No. Existing Cases (prevalence)	Rate of New Cases (incidence)	Laboratory/ Experimental
	Sulfates	Emerson [1973] F; Ferris et al. [1973] F; Carnow et al. [1969] L + S; Ferris and Anderson [1964] F			Sackner et al. [1977] F; Sackner, Reinhart, and Ford [1977] F; Sackner, Dougherty, and Chapman [1976] F
	Unspecified pollution		Oshima et al. [1964] S, L + S		
Nonspecific respiratory disease	Carbon monoxide		Durham [1974] L + S		
	Coal consumption		Douglas and Waller [1966] L + S		
	Dustfall	Toyama [1964] L + S	Manzhenko [1966] L + S; Ferris and Anderson [1964] L + S; Prindle and Coauthors [1963] L + S		

Hydrocarbons		Durham [1974] L + S	
Nitrogen dioxide		Chapman et al. [1973] L + S Pearlman et al. [1971] L + S Shy et al. [1970a, b] F, L + S, S Rokaw and Massey [1962] L + S Schoettlin Landau [1962] L + S	
Oxidants	Linn et al. [1976] F Kagawa and Toyama [1975] F Ury and Hexter [1969] S Remmers and Balchum [1965] S Rokaw and Massey [1962] F Motley, Smart and Leftwich [1959] F	Durham [1974] F, L + S Cohen et al. [1972] F McMillan et al. [1969] L + S Rokaw and Massey [1962] L + S Schoettlin [1962] L + S, S California, State of [1955] S	
Ozone		Rokaw and Massey [1962] L + S	
Smoking habits		Ferris and Anderson [1962] L + S	
Soiling		Ferris and Anderson [1964] L + S	
Sulfates		Dohan [1961] L + S Dohan and Taylor [1960] L + S	
Sulfur dioxide	Ferris et al. [1976] F Fletcher et al. [1976] F	Durham [1974] L + S Lunn, Knowelden and Roe [1970] F	Angel et al. [1965] L + S Bell and Sullivan [1963] S

(Table 4-2. continued overleaf)

Table 4-2. continued

Disease	Possible Cause	Aggravation of Existing Symptoms (severity)	No. Existing Cases (prevalence)	Rate of New Cases (incidence)	Laboratory/ Experimental
		Ferris et al. [1973] F	Holland et al. [1969] L + S		
		Holland et al. [1965] L + S	Lunn, Knowelden and Handyside [1967] F, L + S		
		Toyama [1964] L + S	Douglas and Waller [1966] F		
			Manzhenko [1966] L + S		
			Spicer, Reinke and Kerr [1966] L + S		
			Holland et al. [1965] L + S		
			Holland and Stone [1965] L + S		
			Ferris and Anderson [1964] L + S		
			Reid et al. [1964] F		
			Prindle et al. [1963] L + S		
			Schoettlin Landau [1962] L + S		
	Suspended particles		Durham [1974] L + S		
			Silverman [1973] L + S		
			Spodnik et al. [1966] L + S		
			Holland et al. [1965] L + S		
			Holland and Stone [1965] L + S		
			Prindle et al. [1963] L + S		
	Tarry substances		Manzhenko [1966] L + S		

Unspecified pollution	Holland et al. [1969] L + S
	Meyer [1976] S
	Lave and Seskin [1972] S, L + S
	Lave and Seskin [1971] S
	Colley and Reid [1970] L + S
	Gregory [1970] L + S
	Heimann [1970] L + S
	Ishikawa et al. [1969] S, L + S
	Deane [1965] S
	Phelps [1965] S, L + S
	Spotnitz [1965] S, L + S
	Beard, Horton and McCaldin [1964] S
	Motley and Phelps [1964] S, L + S
	Yoshida, Oshima and Imai [1964] S
	Speizer and Ferris [1973a, b] L + S
	Ferris and Anderson [1962] L + S
	Phelps and Koike [1962] L + S
	Spicer and Kerr [1970] L + S
	Holland, Spicer, and Wilson [1961] L + S
	Huber et al. [1954] S

measurements that are not permitted in human subjects. The animal tests attempt to mimic and/or understand the underlying mechanism of human processes. However, applying the animal tests to predict human ef'ects is a challenging activity. In other cases, as in guinea pig bronchoconstriction, the detailed action may differ from that of human subjects, but the careful study of the effect still gives insight to the action of the pollutant on the mammalian system. A further useful aspect of animal toxicology is the opportunity to use matched animals, as in a purebred strain. The great reduction in biological variability increases the statistical power of the tests and leads to fewer false negative results (finding no effect where an effect actually exists).

Finally, epidemiologic studies examine large populations for the physiologic responses to air pollution exposures. The responses ascertained vary from a disease's specific mortality to measures of pulmonary dysfunction.

Epidemiologic investigations are often designed to confront the two major sources of uncertainty in relating air pollution with human health: (1) measurement of the pollution exposure or dosage that individuals receive and (2) the magnitude and nature of biological response at a given pollution exposure. The relative uncertainty of (2) increases at lower doses.

4.2 HEALTH EFFECTS OF SULFUR OXIDES AND PARTICULATES

At sufficiently high concentrations of certain air pollutants, life-threatening effects, even death, occur. Pulmonary edema (swelling), bronchitis, massive destruction of lung tissue, or asphyxiation have been observed for each of the pollutants described in Chapter 3 in certain specially aggravated situations.

> ... Concentrations of SO_x in the ambient air twice the current standards are associated with adverse health effects. A considerable body of evidence suggests that there may be discernible human health effects from exposure to concentrations approximating the current standards. ... Since the scientific basis for this judgement is incomplete, further scientific information will be required either to validate the present standards or to justify alteration of these standards (Rall 1974, p. 97).

Of all the air pollutants for which the Environmental Protection Agency (EPA) has set criteria, sulfur oxides have received the most attention and have been the most frequently reviewed, possibly for the historic and psychologic reasons given in Chapter 1.

The attention given to sulfur oxides is due, but only in part, to an early recognition of this pollutant and to the availability of methods for its detection. However, there are still many significant gaps in our understanding of the chemical and biological activity of sulfur pollution. Originally adverse effects were assigned to the gas sulfur dioxide (SO_2), but increasing attention is being paid sulfites and sulfates, which are particulates, as well as other non-sulfur-bearing particulates. This is because SO_2, the gas, is highly soluble and is absorbed by the mucous lining of the nose and upper respiratory tract. During nasal breathing it is unlikely that more than 10 to 20 percent of the inhaled SO_2 penetrates beyond the upper airways [Andersen et al. 1974; Frank et al. 1969; Speizer and Frank 1966]. However, switching from nose to mouth breathing during exercise will increase the dose of SO_2 to the lower respiratory tract. Even though direct effects on the lower respiratory system are unlikely to occur from SO_2 alone, concentrations (around 1 ppm) of this gas that are high compared with 1979 ambient air pollution levels have been shown to cause functional changes such as airway narrowing [Andersen et al. 1974; Lawther et al 1975; Frank et al. 1962]. Less is known about the respiratory penetration and biological effects of SO_2 dissolved in small aerosol droplets.

4.2.1   Animal Experiments with Sulfur Oxides

Animal toxicologic studies of the effects of sulfur oxides allow for controlled experimental designs with higher exposures and more definite effects than are ethically permissible with human subjects, but we know of no attempts to derive human responses at ambient pollution concentrations from animal experiments. An exposure of 72 hours to concentrations almost 100 times greater than the 24-hour SO_2 standard (0.14 ppm) are required to produce pulmonary edema and cell necrosis (death) in mice [Giddens and Fairchild 1972]. Even higher concentrations of SO_2 in several different animals have produced changes that are similar to changes observed in chronic bronchitis in human subjects [Chakrin and Saunders 1974;

Asmundsson, Kilburn, and McKenzie 1973]. Changes that appear to be neurally mediated in respiratory frequency in mice occur from short-term concentrations that are over 100 times the 24-hour SO_2 standard [Alarie 1973a]. This response subsides after several minutes. Amdur [1969, 1971, 1973; Amdur and Corn 1963; Amdur and Underhill 1968; Amdur et al. 1975] has extensively studied increased respiratory resistance to sulfur compounds in the guinea pig. She and her associates have found the constriction response to be dependent upon the chemical form of sulfur. Sulfuric acid (H_2SO_4) and other sulfate salts were found to elicit a greater response than SO_2 (see Table 4-3). This may arise from histamine release, contraction of muscle, decreased mucous secretions, or swelling.

Charles and Menzel [1975] have shown that histamines are released by the lungs of guinea pigs in the presence of several sulfate salts. The degree of histamine release to the various sulfate salts corresponds well with the in vivo bronchoconstriction results of Amdur.

Amdur et al. [1975] also found that H_2SO_4 particles of 0.1 micrometer (μm) were five times more irritating than 2.5 μm particles when normalized to the same total intake by weight. These responses are observed in the laboratory for SO_4 concentrations in the range found in urban areas. This finding is in contrast to that of Pattle and Collumbine [1956], who found in guinea pigs that 2.7 μm (mass-media diameter) particles of H_2SO_4 were more lethal at high concentration than 0.8 μg particles. It appears to us that the more recent work of Amdur is more reasonable and reliable. Other recent work by Sackner and Reinhardt [1977] and Sackner, Reinhardt and Ford [1977] on dogs and sheep indicates that a ten-minute exposure to H_2SO_4 mist at a concentration of 1,000 μg/m^3 produces no change in several measures of respiratory function. A fifteen-minute exposure to 7,000 μg/m^3 H_2SO_4 increased pulmonary resistance.

Schlesinger, Lippmann, and Albert [1978] have recently reported a decreased rate of bronchial clearance in donkeys in the absence of any change in respiratory mechanical function. This was observed after a one-hour exposure to submicrometer particles of H_2SO_4 in concentrations between 194 and 1,364 μg/m^3.

Charles and Menzel [1976] have shown that the cations that correspond with the sulfate anion determine the removal rate for sulfate. Eatough and Colucci [1975] have postulated that stable atmospheric sulfite species may be responsible for a portion of the observed

Table 4–3. Comparative toxicity of sulfates in guinea pigs.

Compound	No. Animals	Concentration mg/m³	mg/S/m³	Change in Flow-Resistance
Zinc ammonium sulfate[a]	10	1.1	0.24	+81
Zinc sulfate[a]	7	0.9	0.17	+40
Ammonium sulfate[a]	6	1.0	0.24	+29
Ferric sulfate[b]	15	1.0	0.24	+77
Ferrous sulfate[b]	20	1.0	0.21	+ 2
Manganous sulfate[b]	10	4.0	0.75	− 1
SO_2	71		0.35	+12.8

a. Data from Amdur and Corn [1963].
b. Data from Amdur and Underhill [1968].

health effects from sulfur oxides. Relatively few inhalation studies on the effects of sulfite species exist. Alarie, Wakisaka, and Oka [1973b] report a greater change in the respiratory rate of mice from sodium metabisulfite ($Na_2 S_2 O_5$) than from SO_2.

4.2.2 Chronic Exposures to Sulfur Oxides

Alarie and co-workers [1970, 1972, 1973c, 1975] found that exposure of animals to SO_2 alone for more than one year at concentrations up to 160 times the annual primary standard (0.03 ppm) failed to produce functional or histologic pulmonary damage. At the upper end of Alarie's exposure concentration, Lewis et al. [1973] found evidence in beagles that ventilatory distribution was altered after almost two years' exposure. Alarie et al. [1973c, 1975] report that, with exposure to $H_2 SO_4$ alone and in combination with SO_2 and fly ash, the $H_2 SO_4$ in concentrations of 1,000 $\mu g/m^3$ produced an increase in pulmonary flow resistance, an altered ventilation distribution, and damage to bronchi and bronchiole epithelium.

In addition to the work of Schlesinger, Lippmann and Albert [1978] noted earlier, other studies have indicated an effect of sulfur oxides on resistance to infection and mucociliary clearance rates. Short-term exposure to higher than ambient levels of SO_2 has been reported to increase susceptibility to infection in mice [Lebowitz and Fairchild 1973; Fairchild, Roan, and McCarroll 1972]. This effect may have resulted from direct injury to the affected tissues. Sackner and Reinhardt [1977] observed little change in tracheal mucous velocity in sheep following exposure to ammonium sulfate ((NH_4)$_2$ SO_4).

Ferin and Leach [1973] have shown that the rate of clearance of titanium dioxide (TiO_2) aerosol initially increased and finally after twenty-five days, decreased with exposure to SO_2 at 1 ppm. Ferin and Leach argue that this finding indicates that extended exposure to low concentrations of SO_2 may have a more significant effect upon the clearance mechanism than do short-term high-concentration exposures. Hirsch, Sweson, and Wannen [1975] exposed dogs to 1 ppm SO_2 for one year and found evidence that the mucous flow rate had diminished in the trachea. No difference was found between controls and experimental animals on other functional parameters. Zarkower [1972] has found that up to six months' exposure to

2 ppm SO_2 in mice produced complicated changes in antibody formation. Further experiments are necessary to determine the possible significance of this finding.

4.2.3   Other Effects of Sulfur Oxides on Animals

Amdur and Underhill [1968] have found that pulmonary flow resistance in guinea pigs to SO_2 in combination with various salts correlated with the solubility of SO_2 in those salts. This finding indicates that some interaction between SO_2 and other constituents may influence the health effect of the gas.

Cavendar et al. [1977] and Last and Cross [1978] produced conflicting evidence on the possible synergistic effects of H_2SO_4 and ozone (O_3). Cavender and his co-workers histologically examined rat and guinea pig lungs and found that O_3 alone (2–7 days' exposure) at 2 ppm was responsible for all observed damage to bronchioles and alveoli. Last and Cross exposed rats for three days to ozone concentrations of 0.4–0.5 ppm and to sulfuric acid at 1 $\mu g/m^3$. At these lower concentrations they found that the combination of O_3 and H_2SO_4 produced a high level of glycoprotein and nucleic acid secretion that was not observed for either O_3 or H_2SO_4 alone.

4.2.4   Sulfur Oxides in Human Experiments and Epidemiology

The controlled conditions necessary to determine human response to air pollutants come from dose-limited human experiments (Table 4–4). As in animal studies, airway constriction resulting in increased flow resistance is one of the first responses observed in high-concentration, short-duration exposures. For most people, concentrations of 14,300 $\mu g/m^3$ are required to produce sufficient smooth-muscle constriction to cause an increase in pulmonary resistance [American Thoracic Society 1978]. Persons with a sensitivity to SO_2 have been reported to respond with increased flow resistance at concentrations between 2,100 and 5,700 $\mu g/m^3$, which is still six to sixteen times greater than the 24-hour SO_2 standard [Weir and Bromberg 1973; Nadel et al. 1965; Frank, Amdur, and Whittenberger 1964]. Andersen et al. [1974] found a decrease in forced expiratory maneuvers and an increased nasal flow resistance in

V

Table 4–4. Pulmonary functional responses (ventilatory and mechanical) to SO_2 concentrations under 5 ppm in human volunteers.

Procedure	Results	Reference
15 healthy men, 20–28 yr old, exposed at rest to 1 ppm for 6 h	Nasal mucus flow rate not significantly changed; nasal airway caliber reduced ($p < 0.05$); decreased $FEF_{25\%-75\%}$; generally, changes progressively greater with time; no subjective complaints	Andersen et al. [1974]
4 normal subjects exposed at rest to 0.37 or 0.75 ppm for 2 h	No functional changes at 0.37 ppm; progressive reduction in ventilatory function at 0.75 ppm; recovery incomplete 30 min after end of exposure	Bates and Hazucha [1973]
9 healthy men, 18–27 yr old, exposed to 0.40 ppm during intermittent exercise for 2 h	No changes in pulmonary mechanics, maximal voluntary ventilation, or closing volumes; subjects could not distinguish subjectively between clean air and SO_2	Bedi et al. [1978]
10 healthy men, 25–34 yr old, exposed at rest and during hyperventilation to 2.1 ± 0.2 ppm for 30 min; mouth-breathing	One "possible" reactor; otherwise no changes in pulmonary mechanics or thoracic gas volume	Burton et al. [1969]
11 healthy men, 22–56 yr old, exposed at rest to 1 ppm for 10 min; mouth-breathing	One subject with history of occasional wheezing had increased flow resistance ($+ 7\%$, $p < 0.01$); one subject had decreased flow resistance (-23%, $p < 0.01$)	Frank et al. [1962]
40 healthy nonsmokers, 25 ± 5.7 yr old, and 40 subjects with mild asthma, 27 ± 9.2 yr old, exposed at rest to 0.5	Equivocal changes in healthy subjects, i.e., average ventilatory performance increased less during exposure to SO_2	Jaeger, unpublished cited in NAS [1978]

Table 4–4.   continued

Procedure	Results	Reference
ppm for 3 h; mouth-breathing	than during sham exposure; this pattern more pronounced in asthmatics; airway resistance did not change; one healthy 13 yr old boy experienced shortness of breath and increased airway resistance after exposure; two asthmatics required treatment for dyspnea the next night	
Healthy volunteers 18–47 yr old:	Effects were short-lived	Lawther et al. [1975]
a. *13* subjects exposed at rest to 1 ppm for 1 h	a. No changes in airway resistance or ventilatory function	
b. *12* subjects took 25 maximal breaths of 1 ppm	b. Specific airway resistance increased 12% (p < 0.001) above changes produced with clean air; no change in ventilatory function	
c. *14* subjects took 8 deep breaths of 1 ppm	c. Results inconclusive	
d. *17* subjects took 2–32 deep breaths of 3 ppm	d. Increase in airway resistance related to number of deep breaths of SO_2 taken	
15 healthy volunteers exposed at rest to 2.5 ppm either by nose or by mouth for 10 min	Decrease in specific airway conductance maximal after 5 min; averaged 18% for nose-breathing, 26% for mouth-breathing	Melville [1970]
Healthy men, 18–45 yr old, exposed at rest by mask to 1.34–80 ppm for 10 min or in chamber to 1–23 ppm for 60 min	Little change in flow resistance or clinical findings below 30 ppm by mask or 5 ppm in chamber	Sim and Pattle [1957]

(Table 4-4. continued overleaf)

Table 4–4.   continued

Procedure	Results	Reference
9 healthy volunteers, 20–40 yr old, exposed at rest to 0.5 and 1 ppm for 15 min, mouth-breathing	$MEF_{50\%VC}$ reduced at 1 ppm (averaged about 6%; $p < 0.02$); 1 ppm SO_2 mixed with saline aerosol produced no functional change	Snell and Lucksinger [1969]
3 healthy men exposed at rest to 2.0–2.5 ppm for 5 min; mouth-breathing	Flow resistance (interrupter method) increased 15–21%	Toyama [1964]
a. 4 groups of 3 healthy men, 21–28 yr old, exposed at rest in random sequence continuously for 120 h to 0, 0.3, 1, and 3 ppm	a. Minimal reversible decrease in airway conductance and in dynamic compliance at 3 ppm only; no dose-related symptoms	Weir and Bromberg [1975]
b. 7 male smokers 25–49 yr old, with early evidence of chronic obstructive pulmonary disease, exposed at rest continuously for 96 h in random sequence to 0, 0.3, 1, and 3 ppm	b. Intersubject and intrasubject variability in functional tests and subjective responses exceeded variance due to exposure to SO_2 : no evidence of greater susceptibility than in healthy subjects	

Source: Modified from National Academy of Sciences, Committee on Sulfur Oxides, Board on Toxicological and Environmental Health Hazards, *Sulfur Oxides.* Washington, D.C., 1978.

healthy subjects with 6-hour exposure to SO_2 at 1 ppm. The lowest level at which healthy subjects were found to respond functionally to SO_2 alone is reported in experiments of Bates and Hazucha [1973], who found a decrease in the maximal expiratory flow rate after one half-hour exposure to SO_2 at 0.75 ppm. At 0.37 ppm no measurable functional impairment was found after two hours of exposure.

Sackner, Reinhardt, and Ford [1977] found no functional respiratory impairment in human subjects exposed to up to 1,000 $\mu g/m^3$ of

H_2SO_4 mist, zinc ammonium sulfate $(Zn(NH_4)_2(SO_4)_2)$, ammonium sulfate $((NH_4)_2SO_4)$, and sodium sulfate (Na_2SO_4).

Hackney, Linn, and Bell [1978] exposed healthy, "sensitive," and asthmatic individuals for 2 to 4 hours on two to three consecutive days to 100 $\mu g/m^3$ $(NH_4)_2SO_4$, 85 $\mu g/m^3$ ammonium bisulfate (NH_4HSO_4), or 75 $\mu g/m^3$ H_2SO_4 aerosol. They found no evidence of adverse health effects.

Relatively few human studies have been conducted investigating changes in response to sulfur oxides that might be of immunologic significance. In one, Andersen et al. [1974] found that a 6-hour exposure of healthy men to 1 ppm SO_2 was associated with a decrease in nasal mucus flow. Recently, several investigators have explored the relationship between sulfur oxides, various other factors including additional pollutants, and human response under experimental conditions. McJilton, Frank, and Charlson [1973] found that increasing humidity increased functional respiratory response to SO_2. M.S. Morgan et al. [1977] reported on the effect of combined exposure to NaCl aerosol and SO_2 during exercise and rest. For SO_2 at 1 ppm and NaCl aerosol at 1.0 mg/m^3, Morgan found a much greater response, as measured by flow resistance, in exercising than in resting subjects. The response diminished in subsequent exercise periods and the resistance returned to normal following cessation of exercise even with continued SO_2 and sulfate exposure.

Provocative indication of synergism between SO_2 and O_3 comes from the studies of Bates and Hazucha [1973], Hazucha and Bates [1975], and Bell et al. [1977], who found that exposing Montreal residents to a mixture of SO_2 and O_3 at about 0.37 ppm produced a functional respiratory response that was greater than the sum of response to each gas individually. Bell et al. [1977] similarly exposed Los Angeles residents, with less conclusive results. Analysis of the chamber mixtures indicated that 100–200 $\mu g/m^3$ of sulfate aerosol was formed in the chamber used by Bates and Hazucha. However, we have already seen that this sulfate concentration range produced a functional response at a lower concentration than has been demonstrated for selected sulfate species. In contrast, recent experiments by Bedi et al. [1978] do not indicate this synergism of O_3 and SO_2.

Ferris and Speizer [1979] reviewed recent epidemiologic studies of the relationship between sulfur oxides and health and is recommended for a more detailed review. Buechley et al. [1973] and Buechley [1975] studied daily mortality rates over two time periods in New York. They were concerned with the effect of peak exposure on mortality. During the earlier, more polluted, time period they found that daily mortality was related with SO_2 concentration in excess of 300 $\mu g/m^3$. During the second period, they found the same relationship for mortality and SO_2 at 30 $\mu g/m^3$. Schimmel and Murawski [1975] and Schimmel and Jordan [1977] reanalyzed the same data set, using different analytical methods, and concluded that daily mortality was not related to SO_x concentration of less than 300 $\mu g/m^3$ but was still related to particulates. Goldstein and Lando-witz [1977] disagree because the statistical error of the method did not provide a test that was powerful enough to detect an effect at the hypothesized levels.

Van der Lende et al. [1974] studied one community in Holland during two time periods, three years apart. Smoke and SO_2 pollution levels were reduced during this time and the investigators found that lung functions improved. Fletcher et al. [1976] found a reduction in phlegm production in London subjects during an 8-year period when smoke levels fell to nearly a third of the initial level while SO_2 levels remained about constant. Ferris et al. [1973, 1976] conducted a series of prospective studies in a pulp mill town. During the first 6-year period, pulmonary function and respiratory symptoms improved as TSP levels dropped by about 25 percent (from 180 to 110 $\mu g/m^3$) and SO_2 (as measured by the lead-peroxide method) fell more than 30 percent. In the subsequent study period, during which TSP continued to decline (from 110 to 82 $\mu g/m^3$) while SO_2 rose above its original level, no changes in respiratory function or symptoms were observed. All levels were below primary standards. Sulfate levels were not isolated.

Lawther, Waller, and Henderson [1970], using a population with chronic bronchitis, found increased symptoms when both SO_2 and smoke values were in excess of 250 $\mu g/m^3$; but in a more recent time period during which smoke values decreased, SO_2 excursions above 250 $\mu g/m^3$ were unrelated with symptom aggravation. They inferred that particulates were more important than SO_2 in producing adverse health effects.

Cohen et al. [1972], studying asthmatics near a coal-burning power plant, found a weak association between frequency of asthmatic attacks and SO_2 in excess of 200 $\mu g/m^3$ when total suspended particulate (TSP) levels were also in excess of 150 $\mu g/m^3$. Stebbings et al. [1976] studied pulmonary function change in schoolchildren following a high SO_2 concentration episode. They found no change in pulmonary functioning that could be related to the increased SO_2 exposure. However, the study was started after the period of exposure.

Several investigators have conducted cross-sectional studies of air pollution and health effects utilizing only two monitoring points. Sawicki and Lawrence [1977] monitored high-pollution and low-pollution locations and found more evidence of respiratory disease in the high-pollution area. Chapman et al. [1976] studied pulmonary function in children in two communities, one with higher levels of TSP and SO_2 than the other. The major concentration difference was in particulates and the authors attribute lower pulmonary function in the higher-pollution region to particulate exposure. But social class was a question in this study. Hammer et al. [1976] studied four areas within New York City and found more childhood acute lower respiratory disease in the higher-pollution areas.

Mostardi and Leonard [1974] and Mostardi and Martell [1975] studied schoolchildren in two communities. The SO_2 levels were 100 $\mu g/m^3$ in the high-pollution urban community and about half that in the lower-pollution rural community. Decreased pulmonary function and decreased maximum oxygen consumption were found in the higher-pollution community.

With two-location, cross-sectional studies confounded associations are likely. Many alternative explanations can generally be postulated to explain observed relationships. However, the evidence is suggestive if several different studies demonstrate an association between health and pollution gradients.

Several cross-sectional studies have been conducted in more than two communities. Douglas and Waller [1966] imputed pollution exposure to the amount of coal burned and found an association during childhood between respiratory infection where SO_2 and smoke values were each around 120 $\mu g/m^3$. Lunn, Knowelden, and Roe [1970] investigated respiratory illness in children in several districts in Sheffield County, England. They found decreased pulmonary function and increased illness when annual SO_2 rose above 120 $\mu g/m^3$ and smoke above 100 $\mu g/m^3$.

Finklea et al. [1974] as part of the CHESS program of the U.S. Environmental Protection Agency, taking the data as a whole, found

Figure 4–1.   Asthma attack rate as a function of sulfate concentration at various locations in Riverhead, Queens and the Bronx from Roth, Viren, and Colucci [1977]. Finklea et al. [1974] averaged this to get the dotted line (their figure 5.4.6, page 581). The Lave and Seskin death rate correlation (multiplied by 4) is superimposed as a solid line.

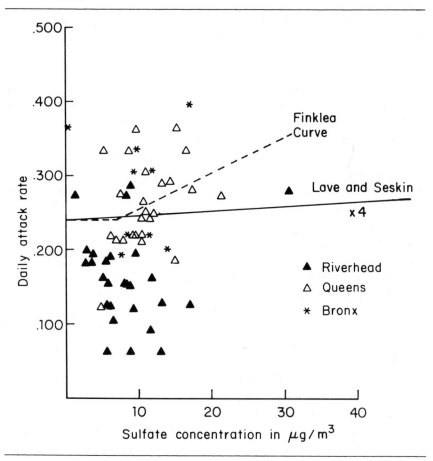

a positive relationship between various respiratory ailments with sulfate SO_2 and particulate concentrations. This led to a number of ad hominem charges [U.S. House 1976]. At the same time that detailed methodological criticisms have been made, reanalysis of some of the data sets has been performed by others [Roth, Viren, and Colucci 1977] who, looking at New York City alone, failed to find a significant correlation. This can be seen in Figure 4–1 from Roth et al. [1977]. Roth, Viren and Colucci, using the original data from

Finklea et al. [1974], plot the asthma attack rate as a function of sulfate concentration. There is no obvious correlation. Finklea et al. averaged these points and derived the dotted curve with a threshold at 7 μg/m^3. An optimist could say that the whole effect could be merely an urban–rural difference, with a larger effect in the Queens and Bronx than in Riverhead.

It is important to realize, however, that Roth, Viren and Colucci [1977] do not prove that there is no effect. As an indication of the size of effect that other data might suggest, we note from section 6.3 the claim by Lave and Seskin [1977] that an annual average sulfate concentration of 100 μg/m^3 increases the death rate by 5 percent. If we make a rough guess that the asthma attack rate is four times the death rate (or 4 X 5 percent = 20 percent) we derive the solid curve.

The data cannot distinguish clearly between the dashed line (threshold), solid line (no threshold), or horizontal line (no effect). However, an effect of the size suggested is not disproven.

This illustrates a psychological phenomenon among researchers. We will repeatedly be coming up against a contradiction that is apparent rather than real. One group says that the data are consistent with an air pollution-related effect; the other group says that an air pollution-related effect cannot be proved! The problem is that experiments are not sensitive enough to give us decisive answers to questions that the U.S. public is legitimately asking.

A summary of reported responses in human and animal subjects to various levels of SO$_2$ is presented in Table 4−5. Where the effects are not derived from human studies the test animal is identified.

4.2.5 Health Effects of Particulates

As discussed in Chapter 3, coal burning produces primary particles and the precursor gases for secondary particle formation. These particles are in the smaller sizes. Industrial emissions, wind-blown dusts, and pollen comprise additional sources of primary particles in the atmosphere.

Mass-related fine atmospheric aerosols pose special problems for health and property; current measurement techniques and regulatory standards may be inadequate or inappropriate for obviation of the damages [Comar 1978; Perera and Ahmed 1978]. Present regulations and measurement techniques are based upon the total mass concen-

Table 4–5. Health effects of SO_2.

Comments	Concentrations		Reported Effects (exposure time)
	$\mu g/m^3$	ppm	
		10	edema, necrosis, desquamation of respiratory epithelium (24 hr) most effects in nasomaxillary turbinates, Giddens and Fairchild [1972]
	5,214	2.0	change in antibody formation in mice, effect alone or with carbon particles, Zarkower [1972]
short-term concentration experienced adjacent to point sources	2,607	1.0	decrease human mucus flow rate ⎱ (1–3 hr) decrease cross-sectional nasal area ⎰ Anderson decrease $FEV_{1.0}$, $FEF_{25\%-75\%}$ et al. [1974] increased reactivity to acetylcholine, dogs (1 hr) Islam, Vastag, and Ulmer [1972] decreased clearance on 25th day post challenge, Ferin and Leach [1973] decreased clearance (1.5 hr. exposure 2x/day, 5 day/week, 1 yr)
		0.26	mean exposure (0.03–0.65 range) for 12.8% increase in guinea pig flow resistance, Amdur [1973]
EPA 24 hr. standard	365	0.14	
EPA annual standard	80	0.03	
		0	

tration of particles that may have little or nothing to do with the chemical behavior and biological activity of the atmospheric aerosols. Mass concentration of those particles that are inhaled could be preferable.

Some of the early work in industrial hygiene that was done during the 1950s showed that particles from 1 to 3 μm are deposited in the alveoli, which are the sites for gas exchange in the lung [Eisenbud 1952]. Since that time there has been a growing consensus that the fine particulates—those less than a few micrometers in diameter—rather than particles larger than 10 μm are responsible for causing human health effects. Particles greater than 10 μm are prevented from reaching the lungs by the filter of the nose [Hatch 1961]. Even though large particles are filtered out, 30 percent of particles with a diameter of 1 μm are deposited deep within the lungs (see Figure 4–2 [HEW 1969]).

The primary mechanism for filtration of larger particles ($> 3 \mu$m) in the lung is impaction onto nasal hairs and on the numerous branching of the tracheobronchial tree. These larger particles are then swept out of the pulmonary system by the combined forces of the mucous layer and the wavelike pumping motion of cilia. Smaller particles ($< 3 \mu$m) lack the inertia of larger particles and follow the gas streamlines deep into the lungs and below the terminus of the protective mucous and ciliated regions. These particles are referred to as "respirable" particles [Lippmann 1970a]. These small particles can then settle and diffuse onto the lung wall, where their half-lives are on the order of months. Even though it is primarily geometric (size) parameters that influence this deep penetration, the chemical and other physical properties of the small particles will determine their health effects.

Many toxic elements, including the heavy metals, have been shown to concentrate on these smaller particles because of their high surface-area-to-volume ratio [Fennelly 1975; Lee and von Lehnden 1973]. This may have special significance for coal combustion, where 90 percent of the mercury in coal has been reported to go up the exhaust stack [Billings and Matson 1972]. The surface-area characteristics and long atmospheric residence time of the smaller particles also lead to the absorption of toxic gases such as SO_2 [Fennelly 1975]. Therefore, the small particles that penetrate deep within the lungs will also concentrate toxic elements.

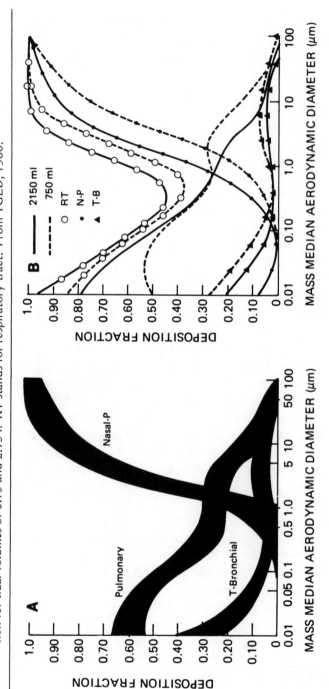

Figure 4-2. Particle deposition in the respiratory system. (A) The shaded areas show the deposition in each compartment when the geometric standard deviation (σ_g) of the particles varies from 1.2 to 4.5 for the indicated mass median aerodynamic diameter, with a tidal volume of 1.45 l; (B) Indicates order and direction of changes in deposition for tidal volumes of 0.75 and 2.15 l. RT stands for respiratory tract. From TGLD, 1966.

Existing laboratory and epidemiologic studies on particulates do not isolate the disease-causing mechanisms [HEW 1969]. In part because of the complicated relationships of particle deposition, laboratory results do not provide the quantitative relationships required to predict adverse health effects. Of the few available epidemiologic studies on particulates, only the CHESS study records the concentration of even one chemical species (water-soluble sulfates) [Finklea et al. 1974].

The significance of particle size is displayed for soluble H_2SO_4 particles in Figure 4-3. In this work from Amdur [1973] it can be seen that within a certain range, particle size is a more important determinant of change in respiratory resistance than is sulfur concentration; for a given concentration the flow resistance is proportional to size. In Figure 4-4 from the same work, we show that the breathing loss is proportional to H_2SO_4 concentration for a given particle size.

At sufficiently high concentrations, certain particulates have been shown to produce fibrosis of lung tissue. MacFarland et al. [1971] exposed cynomolgus monkeys for over one year to fly ash and sulfuric acid mist. At higher exposures they found fibrosis occurring with no associated functional change.

Some particles such as asbestos fibers penetrate the lung lining and lodge in the interstitial space or are carried by the lymphatic and circulatory systems to other body sites. In these cases, the manifest health effects may not be respiratory but rather carcinogenic or systemic. The toxic elements contained in insoluble aerosols may ultimately affect extrapulmonary tissue irrespective of where they enter the body.

Both chemically inert and chemically active aerosols have been shown to alter the mechanical behavior of the lungs [NAS 1979a]. The principle response of different aerosols as measured by animal and human controlled studies is to increase air flow resistance in the lungs. This indicates that lung airways are in some way altered. Several physical mechanisms are known to cause this symptom: reflex constriction of the trachea and bronchi, excessive secretions of mucus, edema, and local enzyme stimulation causing smooth muscle constriction. In addition to changing airway resistance or causing systemic changes, aerosols can cause harm by altering lung clearance, transporting dissolved gases that may damage lung tissue, and affecting the compliance (tissue elasticity) of the lung. In general the

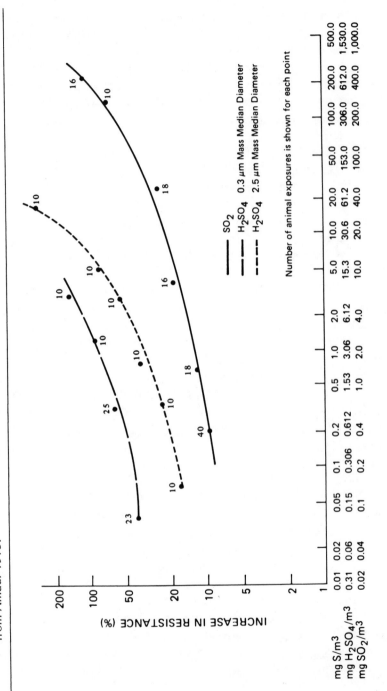

Figure 4–3. Pulmonary flow resistance in guinea pigs following one hour exposures to SO_2 or H_2SO_4 particles of different size. Small particle sulfuric acid mist (H_2SO_4) is seen to have the greatest effect. Adapted by NAS 1979a from Amdur 1973.

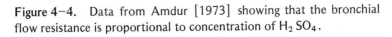

Figure 4−4. Data from Amdur [1973] showing that the bronchial flow resistance is proportional to concentration of H_2SO_4.

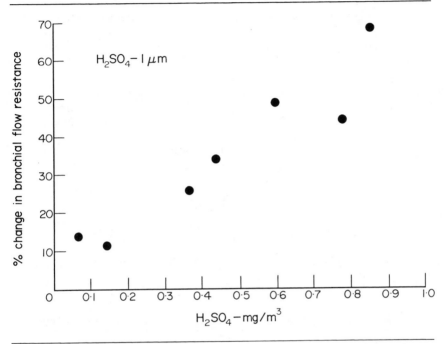

deeper the relatively insoluble particles are deposited in the lungs, the longer the time required for their clearance.

Since sulfur oxides and particulate pollution tend to vary together in the atmosphere, it has been difficult to separate their effects. A summary of the acute health effects of undifferentiated particulate matter and sulfur oxides is shown in Figure 4−5 and the following quotation from a report by the National Academy of Sciences [1979a, p. 139]:

> ... particulate atmospheric pollutants may be involved in chronic lung disease pathogenesis, as causal factors in chronic bronchitis, as predisposing factors to acute bacterial and viral bronchitis, especially in children and cigarette smokers, and as aggravating factors for acute bronchial asthma and terminal stages of oxygen deficiency (hypoxia) associated with chronic bronchitis and/or emphysema and its characteristic form of heart failure (cor pulmonale).

Figure 4–5.   Dose-response curve for the combined effect of SO₂ and TSP. Based on personal communication from Ferris as appeared in NAS [1975].

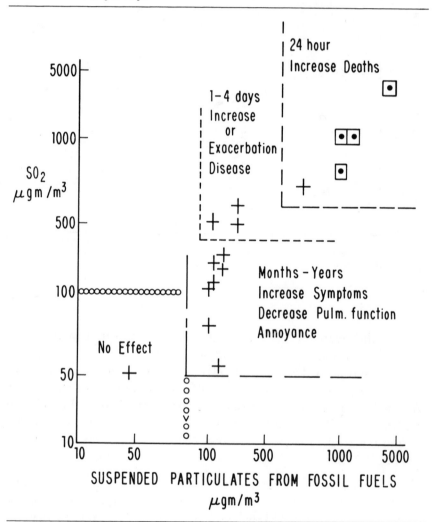

4.3   HEALTH EFFECTS OF NO_2

NO_2 is a powerful oxidizing agent, readily losing an oxygen atom to other molecules and often creating free radicals in the process somewhat in the manner of O_3. One of its major modes of biotransformation is preoxidation of lipid molecules [Menzel 1976]. NO_2 is also known to attack elastin and collagen molecules and to interfere in metabolic reactions.

The present assessment of health effects focuses on concentrations in the range of 3,000 $\mu g/m^3$ (or 15 ppm) to 100 $\mu g/m^3$ [EPA 1978a]. The observed threshold of cell sensitivity to NO_2 is less than 200 $\mu g/m^3$ for in vitro conditions. Experiments have shown that some human subjects can smell NO_2 at that level [Shalamberidze 1967]. At exposure levels ranging from 200 to 500 $\mu g/m^3$ inhalation of NO_2 has been found to impair dark adaptation [Shalamberidze 1967; Bondareva 1963]. It has not been ascertained whether this response represents an effect on nerve transmission or on the light receptors, which are known to be a heavily defended system, but this response indicates extraordinary sensitivity of internal processes.

Two in vitro experiments have also indicated short-term effects of NO_2 at that level: one on alveolar macrophages in culture [Voisin et al. 1977], and another on a culture of bronchial epithelial cells [Samuelsen et al. 1978], each a major component of the defense system in its part of the air-lung interface. Alveolar macrophages are scavenger cells responsible for defending the lungs from particulates. The exposure level to the tissue culture requires correction for absorption of NO_2 during the respiratory cycle. Measurements of overall absorption of NO_2 range from 50 to 90 percent. Following mathematical modeling considerations [Miller 1977], a nasal exposure several times higher might be required to attain an average alveolar or bronchial exposure of 200 $\mu g/m^3$. In addition, the macrophage preparations are probably not nourished as well in vitro as in vivo. Their removal from the alveoli probably places great stress on them even prior to comparing their metabolism and effectiveness with and without exposure. Nevertheless, this added stress on the macrophage may be a good model for a pulmonary crisis condition where normal nutrient is unavailable. The epithelial cell culture is unherently less viable and, therefore, suspect to being oversensitive to environment.

Exposure levels also produce different effects under different atmospheric and physiologic conditions. An atmospheric effect has been shown by Nakamura [1964] where soluble particles in air result in an increase in airway resistance in normal human subjects for a fixed level of NO_2 (5,600 $\mu g/m^3$). Evidently the hygroscopic effect coats particles with water, thus enhancing transport of NO_2 to lower airways. A physiologic condition of increased breathing frequency increases the effective concentration of NO_2 in the airways. Individuals with an increased rate of breathing, as occurs during exercise, will thus be more at risk to a fixed level of external exposure.

Epidemiologic evidence of children's increased respiratory symptoms with chronic exposures to NO_2 was obtained by Shy and his co-workers at concentrations of 150–280 $\mu g/m^3$ [Shy et al. 1970a, b, 1973a, b]. It has also been found that at an exposure level of 200 $\mu g/m^3$, 13 out of 29 slight to mild asthmatics experienced substantially more bronchoconstriction in a carbochol provocation test than without the NO_2 [Orehek et al. 1976]. For these 13 highly sensitive subjects the dose necessary to provoke a 100 percent increase in airway resistance was, on the average, cut in half by this concentration of NO_2.

At exposure to concentrations of 3,000 $\mu g/m^3$ for 15 minutes, von Nieding and Krekeler [1971] has observed increased airway resistance in patients with chronic obstructive lung disease. In 9 subjects tested for pharmacologic mechanisms [von Nieding 1978], the increased airway resistance was not reversed by either an agent (atropine) to inactivate the reflex or an agent (orciprenaline) to stimulate the constricted smooth muscles of the airways. The increased airway resistance was reversed by an antihistamine, signifying a local irritation of cells to an extent that could not be overridden by specific stimulation of sympathetics. Beil and Ulmer [1976] observed that normal subjects required concentrations of 5,000 $\mu g/m^3$ NO_2 to exhibit an increase in resistance and even higher concentrations to cause the provocation resistance to be increased significantly for the group.

An experiment that tends to confirm these effects for concentrations above 2,000 $\mu g/m^3$ was that of Rokaw et al. [1968]. Four normal subjects required 5,600 $\mu g/m^3$ exposure to produce an increased resistance at rest and an exposure of 2,800 $\mu g/m^3$ when exercising. Four subjects with chronic obstructive lung disease experienced increased resistance starting at $3,800 \, \mu g/m^3$ at rest and at $2,800 \, \mu g/m^3$ while exercising.

Thus it appears that the airway resistance of normal subjects is not necessarily highly sensitive to the irritant effects of NO_2, that bronchitics and others with chronic obstructive lung disease are probably more sensitive, and asthmatics are very sensitive, at least in the provocation test.

A number of changes in lung biochemistry have been detected in guinea pigs for one-week exposures to concentrations in the range $750-1,000 \, \mu g/m^3$. An exposure of 4 hours per day increased acid phosphates [Sherwin, Margolick, and Aguilar 1974]. For continuous exposures, there was an increase in lung protein content, most likely due to plasma leakage (edema) [Sherwin and Carlson 1973]. An 8 hour-per-day exposure caused a detectable rise of serum proteins, one of which was statistically significant and was thought to indicate generalized lung damage. There was also a rise in lung lysozyme possibly indicating damage to alveolar macrophages, and a decrease of a red blood cell peroxidase [Donovan et al. 1976; Menzel et al. 1977].

A number of morphologic changes of the lung at concentrations of somewhat less than $1,000 \, \mu g/m^3$ have been observed. In mice, 10 days of exposure produced bronchial damage through loss and shortening of cilia, and also produced alveolar edema [Hattori 1973; Hattori and Takemura 1974]. For 30-45 days of continuous exposure, the mice exhibited damage to tracheal mucosa and cilia. For longer periods there was both alveolar leakage and fibrosis [Blair, Henry, and Ehrlich 1969; Hattori et al. 1972; Nakajima et al. 1969]. In one experiment on rabbits [Buell 1970], an exposure of 470 $\mu g/m^3$ for 4 hrs/day for 5 days each week for a total of 24 days produced abnormal collagen that could lead to fibrotic lesion (scar tissue). After removal of all pollution for a week following the exposure, the collagen of exposed animals could not be distinguished from normal collagen.

The animal infectivity model has also shown marked effects at exposures as low as $1,000 \, \mu g/m^3$ over one week of continuous exposure [Gardner et al. 1977] or $3,000 \, \mu g/m^3$ for 1.5 hours (see Figure 4-6). Immediately after such exposure, animals were challenged with enough streptococci (or other organisms) to kill a substantial fraction of normal animals. The excess deaths caused by NO_2 are an indicator of decreased lung defenses. One useful interpretation

Figure 4–6.   Effect of NO$_2$ on mortality of mice exposed to *Streptococcus pyogenes*.

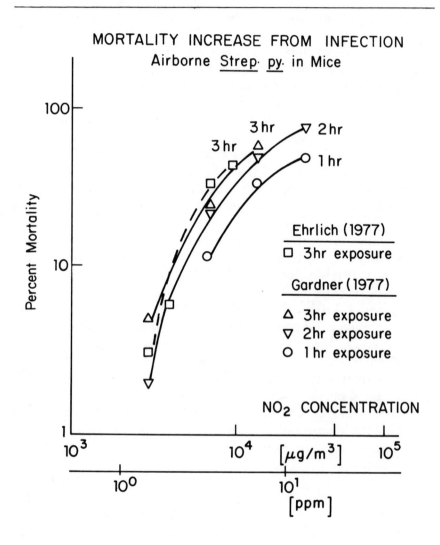

is that this massive challenge on a normal animal mimics the ordinary challenge to a sick animal. An independent determination of decreased bacterial effectiveness caused by 17 hours of NO_2 exposure of $2,000 \, \mu g/m^3$ has also been obtained [Goldstein, Eagle, and Hoeprich 1973].

This evidence indicates that there are several small mammals that are affected by $1,000 \, \mu g/m^3$ of continuous or repeated NO_2 exposure over the course of a week or more. Epithelial cells are damaged and these animals incur biochemical changes in a manner that might be expected of human tissues. Furthermore, the defenses of normal mice against massive bacterial infection are reduced at that concentration, perhaps mimicking the response of a severely impaired animal to a normal exposure of bacteria. *In vitro* preparations to macrophages are impaired at $200 \, \mu g/m^3$; however, this figure must be corrected upward to account for the effect in *in vivo* exposure where mechanisms of homeostasis are in effect. Given these adjustments, culture results can be considered consistent with the *in vivo* increases in bacterial susceptibility. Table 4−6 provides a summary of health effects at various ambient levels.

4.4   HEALTH EFFECTS OF O_3 AND OTHER PHOTOCHEMICAL OXIDANTS

Ozone and other photochemical oxidants are highly reactive compounds. When inhaled, they damage the respiratory tract and pulmonary parenchyma (lining). At moderate exposure, O_3 can alter lung structure and reduce the normal capacity of the lung to resist infections.

In combination with other pollutants in the photochemical oxidant mix, O_3 is associated with chest discomfort, eye irritation, and cough at concentrations that are typical of several urban areas.

4.4.1   Ozone, O_3

Several experiments have shown that exposures of small mammals to O_3 concentrations in the range of $160−200 \, \mu g/m^3$ (0.08−0.1 ppm) for 2.5 to 3 hours have reduced their resistance to infection. Coffin et al. [1968], Ehrlich et al. [1977], and Miller, Illing, and Gardner [1978] noted increased mortality in mice exposed to bacteria and O_3

Table 4–6.   Health effects of NO_2.

Comments	Concentrations	Reported Effects (exposure time)
	$\mu g/m^3$, ppm	
	2,000 — 1	infectivity, mouse (3 hr inter-polated, 17 hr)
	1,800 —	
	1,600 —	multiple biochemical changes, guinea pig (8 hr/day, 1 week)
	1,400 —	increase in protein uptake by lung, guinea pig (4 hr/day, 8 days)
	1,200 —	
Peak level in Los Angeles	1,000 — 0.5	infectivity, mouse (1 week)
		tracheal mucosa and cilia, autoimmune response, mouse (one mo.)
		cilia, Clara cell, and alveolar edema, mouse (10 days)
Peak levels in four cities	800 — 0.4	acid phosphates rise and serum pro-teins enter lung, guinea pig (1 wk.)
EPA lowest suggested 1 hour standard	600 — 0.3	detected in blood, monkey (9 minutes) / human dark adaptation impairment (immediate) / collagen, rabbit (20 hr/wk for 24 days)
Peak levels in many cities	400 — 0.2	human asthmatics provocation resistance, (1 hr)
WHO highest (320) lowest (190)	200 — 0.1	bronchial epithelial cells, alveolar macrophages (1–2 hr, in vitro) human odor perception (immediate) human dark adaptation impairment (5 and 25 min)
suggested 1 hour standard		
	0	

simultaneously. In addition, information on the chemical basis of the decrease of resistivity was obtained by Gardner et al. [1971], who noted that in rabbits a protective lung factor was partially inactivated. At 500 $\mu g/m^3$ Hurst [1970] noted in rabbits that the chemical activities in alveolar macrophages were reduced. Many other impaired lung defenses have been noted in the range of 800–1,000 $\mu g/m^3$ and above.

Physical alterations of alveolar cell populations, where type II cells replace type I cells, have been observed in rats at exposures of 750–1,000 $\mu g/m^3$ for 2 hours [Stephens et al. 1974]. Exposures for 3 hours per day for 16 days at 600 $\mu g/m^3$ caused changes in Clara cells and altered the surfaces of cilia and other bronchial cells [Sato et al. 1976]. In monkeys, rats, and mice, exposures of 400 $\mu g/m^3$ for 8 hours per day for 7 days produced lesions in respiratory bronchioles, excess growth (hyperplasia and hypertrophy) of the bronchial wall, damage in ciliated and Clara cells, and an increase of type II alveolar cells [Dungsworth 1976; Mellick et al. 1977; Schwartz 1976; Schwartz et al. 1976]. Continuous exposures of young rats to 400 $\mu g/m^3$ for 30 days produced an excessive increase in lung volume and individual alveolar dimension as well as a reduction of lung elasticity [Bartlett, Faulkner, and Cook 1974]. At exposures of 500 $\mu g/m^3$ and above for 4 to 6 hours, lungs of cats were found to have damaged airway cells, type I alveolar cells, and alveolar capillaries.

Pulmonary function tests showed increased breathing resistance at exposures as low as 500 $\mu g/m^3$ for 2 hours in cats [Watanabe, Frank, and Yokoyama 1973]. Increased respiration frequency and decreased tidal volume were also noted in guinea pigs at exposures of 750 $\mu g/m^3$ for 42 hours.

In some of the earlier laboratory tests on human subjects, resistance increases have been reported at concentrations as low as 200 $\mu g/m^3$ [Goldsmith and Nadel 1969]. These results have recently been confirmed by von Nieding in 11 healthy subjects in a 2-hour exposure [von Nieding et al. 1976]; this author also found that the difference between alveolar and arterial oxygen tension was increased at that concentration. Such results are contrary to those from a long series of experiments by Hackney et al. [1975a, b, c], who found effects in 5 of 7 of the normal subjects only on the second successive day of 2-hour exposure to 1,000 $\mu g/m^3$ and not on the first.

Another group of 7 subjects, 3 of whom had hyperreactive airways and 4 of whom were normal, did not exhibit any measured response to 500 $\mu g/m^3$ of O_3, even with other pollutants added. Still another group of 5 subjects did not exhibit any measured response to 700 $\mu g/m^3$. On the other hand, Mohler et al. [1978] found an alteration in gas mixing at 730 $\mu g/m^3$ but found no change by conventional tests. Hazucha, Parent, and Bates [1977] found changes in forced expiratory ability at concentrations as low as 500 $\mu g/m^3$. More recently Silverman [1978] has found that 6 or 18 asthmatics developed decrements of pulmonary function at 500 $\mu g/m^3$ for 2-hour exposures.

Because some of the discrepancy among the results of work in various laboratories was along national lines, it was of particular interest to note the results of studies by Hackney on the question of actual differences of characteristics of human populations. In one study, 6 men with respiratory hyperactivity were exposed to 1,000 $\mu g/m^3$ for 2 hours on 4 successive days [Hackney et al. 1977a]. One subject showed little measurable response. The other 5 subjects showed decrements in lung function of the first 3 days but a substantial reversal on the decrement on the fourth day. Recently, Hackney reported that 9 new arrivals in the Los Angeles area (highest oxidant levels) showed substantial decrements in pulmonary function for exposures of 800 $\mu g/m^3$ for 2 hours, whereas 6 local residents did not show measurable effects. In a further study in Hackney's laboratory, 4 southern Californians and 4 Canadians were exposed to 700 $\mu g/m^3$ for 2 hours with light intermittent exercise [Hackney et al. 1977b]. The Canadians exhibited discernible decrements in lung function, but the Californians did not. Thus a possible adaption or a selective migration effect is inferred. Any adaptation would mean an altered lung configuration caused by air pollution, which is an effect that should arouse considerable concern. Table 4—7 summarizes some health effects of O_3.

4.4.2 Human Epidemiology

Although a number of epidemiologic studies have been initiated and some results have been obtained, little that is definitive has come from studies of effects of long-term exposures of general popula-

Table 4–7.   Health effects of O_3.

Comments	Concentrations	Reported Effects (exposure time)
Peak levels in Los Angeles area	($\mu g/m^3$) (ppm)	
	1200	
	1100	
	1000 .5	resistance doubles, guinea pigs (2 hrs) multiple lung defenses, rabbits (3 hrs) macrophage membrane, rats (2 hrs) lowered rate of DNA synthesis, mice (6 hrs)
	900	
	800 .4	Canadian–American function differs (2 hrs) multiple lung defenses, mice (3 hrs, 4 hrs)
		human increased breathing rate and decrements of lung function (2 hrs)
	700	alveolar cell transformation, rats (2 hrs)
Peak levels in four cities	600 .3	welders pneumonia, human population symptoms bronchial cells (3 hrs/day for 16 days)
	500	human asthmatics, resistance (2 hrs) alveolar macrophages, rabbits (3 hrs) resistance cats (2 hrs) lymphocyte chromosome damage, hamsters (5 hrs)
Peak levels in many cities	400 .2	exercising children, chest pain and breathing difficulty multiple morphological and biochemical, rats (8 hrs/day for 7 days)
	300	
Proposed 1 hour standards Existing	200 .1	human resistance and oxygen tension (2 hrs) protective alveolar fluid, rabbits (2.5 hrs in vivo, 0.5 hr in vitro) infectivity, mouse (2.5–3 hrs)
	100	
	0	

tions. A summary of the most positive of these results is quoted from a recent Air Quality Criteria document [EPA 1978b] :

> Several studies suggest a tenuous association between long-term oxidant exposures and chronic respiratory morbidity. Deane, Goldsmith, and Tuma [1965] have observed higher rates of persistent cough and phlegm in Los Angeles telephone workers aged 50 through 59 years than in comparable workers in San Francisco. Hausknecht [1962a] has reported that a disproportionate number of chronic respiratory disease patients, randomly selected from the whole state of California, lived in the Los Angeles area. Linn et al. [1976] has observed higher rates of non-persistent cough and phlegm in women living in Los Angeles than in San Francisco, a finding which would reflect an effect of oxidant on chronic or acute respiratory morbidity. However, the findings of these studies are not clear with respect to the existence of an association between oxidant exposure and chronic respiratory disease. Nor do the findings allow confident inference as to the level or duration of oxidant exposure necessary to promote increased rates of chronic respiratory illness.

The concentrations of ozone vary with time, and high concentrations occur for a few minutes to hours. Because of these high exposures, it is important to study acute human health effects [EPA 1978b].

Eye irritation is consistently related to high concentrations of photochemical oxidants. O_3, by itself, is not an eye irritant at concentrations presently prevailing. Other photochemical oxidants, or reaction products of O_3 and other substances, probably account for the irritation, although the relative importance of the constituents is unknown.

Hammer, Hasselblad and Portnoy [1974] report for the United States that both the number of persons reporting eye irritation and the severity of the irritation increase as the concentration of oxidants increases from $200 \mu g/m^3$ (0.1 ppm) to $294 \mu g/m^3$ (0.15 ppm) (hourly average). This compares with the standard, newly adopted in the United States, of 0.12 ppm as an hourly maximum. Makino and Mizoguchi [1975] have observed eye irritation in Japan at much lower levels than Hammer reports.

Longer term health effects resulting from transient irritation have not yet been demonstrated. However, there are observations of functional lung impairment. In a study of Japanese schoolchildren, Kagawa and Toyama [1975] found that O_3 was responsible for short-term impairment of lung function. Their measurements indicated that O_3 was associated with increased resistance of airways, which suggests that the function of large airways was altered. The degree of pulmonary impairment depends on level of physical exercise [Bates et al. 1970, 1972; Lebowitz et al. 1974; Folinsbee, Silverman, and Shephard 1975]. Wayne, Wehrle, and Carroll [1967] found that hourly oxidant concentrations above 200 $\mu g/m^3$ (0.1 ppm) reduced the performance of high school cross-country runners. Folinsbee, Silverman, and Shepard [1975] suggest that the impairment of performance may be due to oxygen uptake reduced because of the restricted lung function.

Several investigators have studied the effect of photo-oxidants in those with existing disease. Schoettlin and Landau [1961] found that when oxidant levels exceeded 0.25 ppm, the rate of asthma attacks increased. Several studies have found that filtration of air to remove oxidants helped persons with chronic respiratory disease [Motley, Smart, and Leftwich 1959; Remmers and Balchum 1965; Ury and Hexter 1969]. Hammer, Hasselblad and Portnoy [1974] reported that when hourly oxidant concentrations exceed 588 $\mu g/m^3$ (0.3 ppm), the frequency of cough, chest discomfort, and headache increases.

Equivocal results have been obtained by Brant and Hill [1965] and by Sterling et al. [1966], and Sterling, Phair and Pollack [1967] relating short-term oxidant levels and hospital admissions. Although positive associations were observed, the effect of other pollutants could not be ruled out.

5 STUDIES AND MODELS OF HUMAN HEALTH EFFECTS

The inhabitants groped in a world of unreality, in which everything was shadowy and assumed fantastic shape. Judgment of the reasoning faculties was betrayed by the false witness of misted perception. The fog alone was materially in evidence, and it blinded, choked and chilled them. The pervasive phenomenon penetrated their thought. There wasn't anything else to engage it. Nothing is more contagious than hysteria. (Description of the Meuse Valley air pollution episode, *Louisville Courier—Journal* [1931]).

5.1 AIR POLLUTION EPISODES

Unusual meteorologic conditions have from time to time provided a natural laboratory in which illness or mortality more obviously follow directly from air pollution. The first well-described air pollution episode occurred when air pollutants were trapped in the Meuse Valley of Belgium in December 1930 [Firket 1931, 1936]. Mortality and morbidity (illness) increased markedly during the days of high air pollution and several days thereafter. Less well-documented air pollution episodes have occurred at earlier times. Mortality and meteorology records have been used to indicate periods of excess mortality associated with "fogg" in London during the early part of this century and the later half of the nineteenth century [Goldsmith and Friberg 1977]. The quality of historic mortality data from

161

London permits investigators to make such retrospective associations. It is likely that other episodes went unrecorded in rapidly industrializing centers from the mid–1800s on.

Large, well-identified disasters are always perceived more readily than a large number of small incidents. It is these pollution episodes that have had the major effect on stimulating interest in regulatory action and pollution-control programs. The large health effects during these unusual meteorologic conditions foreshadowed the potential consequence, even during normal meteorologic conditions, of unbridled growth in combustion. The obvious impact of episodes led to public recognition that air pollution affects human health.

Although the particular combination of coal burning and adverse meteorology that led to the London fogs of the last 100 years were unique to that setting, certain characteristics were common to all the air pollution episodes. Meteorologic conditions were atypical. The Meuse Valley in Belgium; Donora, Pennsylvania, in the valley of the Monongahela River; and London in the valley of the Thames were all susceptible to episodes during temperature inversions that effectively trapped emitted pollutants in the interior of the valley. Normal conditions would transport much of the pollution out of the valley and dilute it with cleaner air from less industrialized areas. Most episodes were also associated with the onset of colder weather during November and December. In addition, fog or moisture was associated with many of the episodes. The fog may have permitted chemical transformation of pollution constituents in droplets or permitted respiratory deposition of pollutants dissolved in fine droplets.

The mix of industrial sources was complicated in areas where episodes occurred, making it difficult to determine the pollution factor or factors responsible for the associated health effects. Adding to the problem of identification, important pollutants were often determined by imprecise methods or even unrecorded.

Illness and death have been associated with the major episodes. In most cases mortality occurred in elderly individuals with existing heart or lung disease. Illness and respiratory symptoms, however, affected people of all ages. The excess in deaths was recognized only after the episodes were over, which made it impossible to take preventive action (e.g., reduction in emission).

The episode in the Meuse Valley of Belgium began the first week of December, 1930. Death of cattle and of about 60 persons was recorded during the week of the episode. Several hundred people became severely ill from the increased concentration of pollutants from several different industrial sources [Firket 1931, 1936]. Firket estimated that SO_2 concentrations were in the range of 25,000–100,000 $\mu g/m^3$ (25-100 mg/m^3). Autopsy results of the victims showed only general irritation and congestion of the mucosa of the trachea and the large bronchi in the upper respiratory tract.

The air became so hazy during the episode in Donora, Pennsylvania in October 1948 that one could not see across the street. In the population of 14,000, 20 deaths were attributed to the episode and about one-half the population became ill. Approximately 10 percent of the population was reported severely ill [Schrenk et al. 1949]. Those who died had existing cardiovascular and respiratory diseases. Donora also had a complex mix of pollution sources that included a steel mill and a sulfuric acid plant. SO_2 concentrations were estimated to be 1,400–5,500 $\mu g/m^3$ (1.4–5.5 mg/m^3). Particulate pollution was also very high. Autopsy findings showed only nonspecific irritation of the respiratory tract. Ciocco and Thompson [1961] performed a ten-year follow-up of those exposed to the Donora episode. They found that those who were affected during the episode had a subsequently higher mortality rate than those who had not been affected. This would imply that the episode either induced damage or disease in those affected or acted like a test that detected subclinical disease in existence at the time of the episode.

From December 5 through 9, 1952, air pollution and fog increased in London during a temperature inversion as air was trapped in the whole Thames River Valley. When the episode was over, between 3,500 and 4,000 excess deaths had occurred. Many more people became severely ill, and hospital admissions increased for heart and lung diseases [Ministry of Public Health 1954]. Air pollution measurements indicated a smoke concentration (somewhat different than the U.S. measure of total suspended particulates) of 4.5 mg/m^3 and a SO_x concentration of 3.8 mg/m^3 (as measured by the hydrogen peroxide method).

Autopsy findings were similar to those in Meuse and Donora. Attributable deaths included heart disease, bronchopneumonia, and chronic bronchitis. Death rates rose for most causes and remained high for several weeks following cessation of the episode. Because death rates rose for all causes, the mortality is attributed in many

Figure 5−1.   Deaths registered in greater London associated with air pollution episode of December 5−8, 1952. Broken line indicates average deaths. (From J. R. Goldsmith and L. T. Friberg, "Effects on Human Health," in A. Stern (ed), *Air Pollution*, New York: Academic Press, 1977.)

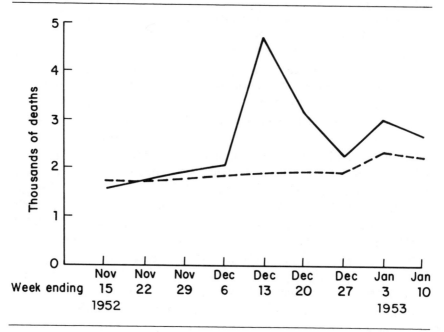

cases to those who were already severely ill. Figure 5−1 shows the number of deaths in each week and the average death rate for the five preceding years.

Stimulated by these episodes, Goldsmith and Friberg [1977] have presented data on other episodes in London that took place from 1873 to 1948. Later episodes in London were recorded by Gore and Shaddick [1958], Martin and Bradley [1960], and Scott [1963].

Table 5−1 presents a summary of London episodes during the 1950s. There is a marked difference between the episode of December 1952 and that of January 1959. The concentrations of both soot (particulates) and SO_2, (and probably sulfates also), were less in 1959 than in 1952 by factors of 3 and 5, but the numbers of excess deaths were less by a factor of 20. In a later episode in 1962, the concentrations of particulate were low, but the SO_2 levels were the

Table 5—1. Survey of selected acute air-pollution episodes in greater London.

	Dec. 1952	Jan. 1956	Dec. 1962	Dec. 1957	Dec. 1956	Jan. 1955	Jan. 1959
Duration of the cumulation period in days	5	5	5	5	10	11	5
Number of days with maximum pollution	2	2	1	1	5	1×3^{a}	1
SO₂ level preceding episode	500	300	400	300	300	300	300
SO₂ maximum	4,000	1,500	3,300	1,600	1,100	1,200	800
SO₂ increase per day	1,200	500	1,000	325	400	450	250
Soot level preceding episode	400	500	200	400	400	500	400
Soot maximum	4,000	3,250	2,000	2,300	1,200	1,750	1,200
Soot increase per day	1,200	1,300	600	500	400	600	400
Number of days with excess deaths	3,900	1,000	850	800	400	240	200
Number of days with excess mortality	18	10	13	10	6	6	6
Daily mortality expected under normal circumstances	300	330	310	300	270	320	325
Average daily mortality in the period (as a percent of normal)	170	130	120	125	125	112	110
Excess mortality (as a percent of normal)	70	30	20	25	25	12	10

Remark: The SO₂ and soot concentrations mentioned are average values over 24 hours expressed in $\mu g/m^{3}$.

a. Maximum pollution values of one day's duration occurred three times.

Source: HEW, 1970a, p. 125.

same as in 1952 while the death rate was lower. Possible explanations include (1) only particulates contribute to the death rate; (2) sulfur and particulates contribute in a synergistic (multiplicative) way; (3) there is a threshold for adverse effects, and in 1962 pollutant levels were close to that threshold; (4) radio and TV warnings to the elderly and bronchitics in 1962 encouraged people to stay indoors and avoid exertion. These might have been effective.

A slightly different type of air pollution episode was recorded in New York City during November 18–22, 1953. Rather than a temperature inversion occurring in a valley, as were the earlier episodes, the entire air mass over much of the eastern United States became stagnant and accumulated pollutants. SO_2 and particulate concentrations were elevated. For the period of November 15–24, 244 deaths per day were reported in contrast to the average death rate of 224 deaths per day for the same ten-day period during the three years preceding and succeeding the episode [Greenburg et al. 1962; Greenburg and Field 1962].

Improved health and pollution monitoring permitted the detection of an unusual worldwide episode that occurred during December 1962. In London, 700 excess deaths were attributed to this episode, which was characterized by increased levels of SO_2 and particulates [Scott 1963]. High pollution levels and excess deaths were also detected in the eastern United States [McCarroll and Bradley 1966]. Increased mortality was associated with elevated pollution during the same time in Rotterdam, the Netherlands and Hamburg, Germany. High pollution levels without concommitant health summaries were reported in Paris, France and Prague, Czechoslovakia [Goldsmith and Friberg 1977]. In Osaka, Japan, Watanabe [1965] attributed 60 excess deaths to particulate levels in excess of 1,000 $\mu g/m^3$ (1 mg/m^3) and SO_2 in excess of 285 $\mu g/m^3$ (0.3 mg/m^3). This large, nearly global episode appears to have been caused by very unusual meteorologic conditions.

There are many methodologic difficulties in relating mortality and morbidity to even high pollution levels. The relationship was starkly obvious and generally recognized as causal in the case of the 1952 London episode, where death and illness rates rose sharply with increasing pollution levels and returned to normal shortly following the episode. However, in the more frequent episodes of lower pollution and lower death rates the association is more difficult to make. Even where the association of air pollution to health exists, one should ask if the weather conditions that caused the increased pollution level could not be the cause, as well, of ill health. An illustration of this problem comes from a reference to Leonard, Crowley, and Belton [1950] from the 1970 Criteria Document for sulfur oxides [HEW 1970a]. Figure 5-2, from Leonard, Crowley, and Belton [1950], is an 11-year time series giving deaths from respiratory disease and the concentration of suspended particulates in Ireland. If one were to look only at the years 1938-1940 it might be suggested that particulate air pollution causes respiratory deaths. However, reduced pollution levels from 1941 to 1947, following a switch from coal to peat as a residential fuel, were not matched by an equal decline in respiratory mortality. Coal was again burned in 1948 and the death rate did not increase. This example shows that a correlation that may appear obviously causal at one epoch may not be so obvious at another; or else the magnitude of the correlation changes for some unidentified reason.

Some problems are mitigated in part by studying large populations. A World Health Organization report suggests that populations in the millions are necessary to ensure detection of acute mortality [WHO 1962]. Insufficiency of air pollution measurements create further problems. An associated difficulty is identifying the source or sources responsible for the health effects when the pollution emitted into the stagnant air is complex. Despite the accumulated evidence from air pollution episodes, such problems make it impossible to determine analytically defensible associations between the excess deaths that are known to have occurred and the specific constituent or constituents in the air pollution mix that are causing those deaths. Furthermore, the nonspecific autopsy findings from those who died during these episodes do not provide very useful insight into the cause of death. Recognition of these limitations suggests the difficulty of relating much lower levels of air pollution to health outcome.

Lawther [1963], in a summary of the London incidents, estimated that when SO_2 concentrations exceed $710 \, \mu g/m^3$ (0.71 mg/m^3) with smoke levels above $750 \, \mu g/m^3$ (0.75 mg/m^3) excess mortality is observed for conditions in the elderly and in those with chronic obstructive lung disease or heart disease.

It is generally agreed that the U.S. and world air quality standards and emission regulations will prevent such high exposures in the future. However, study of such high exposures is useful for the light

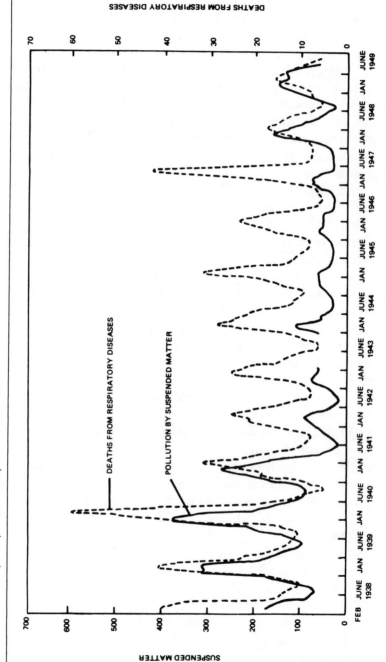

Figure 5–2.   Time series of air pollution levels and death rates in Dublin, Ireland, 1938–1949. (From A.G. Leonard, D. Crowley, and J. Belton, "Atmospheric pollution in Dublin during the years 1944–1950," *Royal Dublin Soc. Sci. Proc.* 25 (1950): 166–167.)

they may throw on effects of more widespread exposure at lower levels.

5.2 DOSE-RESPONSE RELATIONSHIPS

The effects on health discussed so far in this chapter have one feature in common with all early environmental problems. They involved massive doses of the poisonous pollutant to relatively few people, many of whom died as a result.

For many years, it was the prevailing view that in small doses pollutants and poisons were harmless and that they become harmful only after a threshold dose level or concentration is exceeded. When it became clear that some people died at dose levels where others remained unaffected, the idea became prevalent that there is a distribution of individual sensitivities, and if one could know in advance one's own sensitivity, one could avoid problems merely by refraining from exposures at greater than that critical level of sensitivity. We would expect that if the sensitivities are genetically determined, natural selection would lead to thresholds of sensitivity above the natural background levels. This natural selection would primarily lead to sensitivities to those ailments that cause morbidity and death before the age of procreation.

This viewpoint was modified by a consideration of cancer incidence, particularly of radiation-induced cancers. A completely different theoretical idea arose [Armitage and Doll 1954] that it is a random process that determines whether or not a person's cells are modified enough to become cancerous, but that people need not have different inherent sensitivities. For tobacco cigarette smoking, as noted in the Introduction, the incidence of cancer increases roughly proportional to the number of cigarettes smoked. Cancer is considered to be mainly induced environmentally, but with genetic predeterminants [Higginson, 1960, 1969, 1975]. The Armitage-Doll theory does not distinguish between environmental and genetic effects but considers that both genetic and environmental effects contribute, particularly in multistage cancer theories.

We are therefore led to consider in Figure 5—3 several possible relationships between health response and concentration of a pollutant. Curve A has an approximate threshold below which no effects

Figure 5−3.   Possible alternate dose-response relationships discussed in the text.

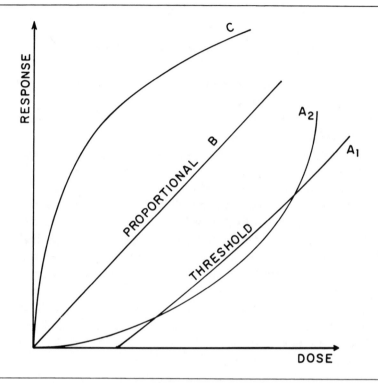

are formed; curve B shows a relation in which the incidence is proportional to dose in accordance with this theory; and in curve C low doses are relatively more important in producing effects. For convenience we consider that curve A also approximates a relation in which health effects vary as the square of the dose at low levels, leading to an approximate threshold level.

The difference among these three dose-response relationships has important policy implications. Understanding the difference is vital for our attempts to reduce pollution-related diseases and other health problems. Our knowledge of the effects of a pollutant usually comes from only a small number of exposed persons—perhaps 1,000. If one person in this group dies each year, after 10 years we have a 1 percent effect, which may be just statistically significant if there is no background death rate from this cause. If the same exposure is given to 200 million people, we expect 200,000 deaths a year so that although the effect can only just be seen in the experimental sample, it is huge when applied to the population as a whole.

If there is a threshold, we can perhaps reduce the exposure a factor of ten to eliminate the effect; if there is no threshold, reducing the exposure a factor of ten only reduces the death rate, in our example, by the same factor to 20,000 per year. This argument is brought out most clearly by the discussion of carcinogenesis induced by saccharin [Federal Register 1977].

We do not wish to go into detail about cancer theories, but we do want to emphasize that in models that assume linearity, the parameter that matters is the concentration of the substance averaged over a long time—an appreciable fraction of a lifetime.

The noncancer hazard to health of air pollutants cannot be assessed as easily as the cancer hazard because the theory is even less reliable. Cancer is considered to be the action of a pollutant on an individual cell, and although the cells can be repaired, there may be only a constant fraction repaired, a process which maintains linearity. On the other hand, it is possible that even if, for example, bronchial restriction is proportional to exposure [Amdur 1973], no adverse health effect appears for small restrictions. These and other considerations suggest that health effects follow an S-shaped curve, of which the bottom end is like curve A_2 in Figure 5–3.

However, if we ignore these indications and appeal directly to the experiments of Figures 1–2 and 1–3, we have indications of a linear behavior. Indeed, for example, we would expect some bronchitis with no air pollution. The graphical plot in Figure 1–3 has a suppressed zero and the linear curve is smoother and therefore more plausible. Stokinger [1972] summarized the changing perceptions:

> The premise on which the concept of thresholds rests is that, although all chemical substances produce a response ... at some concentration, if experienced for a sufficient period of time, it is equally true that a concentration exists for all substances from which no response of any kind may be expected no matter how long the exposure. ...
>
> Of late, two factors have done much to throw doubt on the threshold concept: (1) the development of increasingly sensitive indicators of response, and (2) a general concern that highly injurious agents at high dosages may still be injurious at any concentration, however small, either per se or as a factor in diseases of multiple causation.

5.3   CARCINOGENIC EFFECT
OF AIR POLLUTANTS

Ever since Percival Pott showed 200 years ago that chimney sweeps died of cancer of the scrotum, the combustion products of coal have been logical candidates for carcinogens. In Tables 3-2 and 3-9 we showed a list of some of these products in the ambient air.

A pulmonary disease (mala metallorum) was described in 1531 by Paracelsus as occurring in miners in Schneeberg, Germany. Harting and Hesse [1879] identified it as cancer. As early as the 1920s it was suggested that radioactivity was a cause of the disease in this limited area. More recently, an increase of cancer has been identified among gas workers [Armitage and Doll 1954; Lawther, Commins, and Waller 1965] where radioactivity is a less likely cause.

In the early twentieth century when lung cancer rates rose throughout the industrialized world, air pollution from coal was an obvious possible cause. However, we now know that cigarette smoking is the major cause of lung cancer in the general population [SGR 1979].

The investigations of air pollution therefore address three questions:

1.   Does air pollution per se cause lung cancer?
2.   Cancer is believed to have a multiple etiology. Does air pollution cause cancer in association with smoking or with occupation?
3.   What are the factors in air pollution of etiologic importance?

There are several general references addressing these questions [IARC 1977; Higginson and Jenson 1977; Pike et al. 1975; Doll 1978; Carnow 1978; Natusch 1978; Friberg and Cederlof, 1978]. In this book we will attempt to derive the best presently available numbers on the cancer rates for discussion of the policy implications.

5.3.1   Etiologic Factors of Cancer Caused by Air Pollution

Among the carcinogenic air pollutants only benzo(α)pyrene (BαP) has been systematically measured. It is normally used as a surrogate for other polycyclic aromatic hydrocarbons (PAH), and these are generally considered the most probable carcinogenic agents, although heavy elements are not ruled out. For epidemiologic studies, it is usual to assume—because it is the best that can be done—that all burning processes produce these PAH in roughly equal proportions, and therefore BαP may be an adequate surrogate. However, the concentrations of BαP vary by a factor of 10,000 from efficient coal combustion to inefficient (hand-stoked) coal or wood combustion (Table 3–9) so that BαP concentrations are not proportional to heavy element or SO$_2$ concentrations. It is also known that BαP is not proportional to other PAH. To the extent that BαP is a bad surrogate, the epidemiology will tend to underestimate the air pollution effect (see Section 6.4.1).

The role of heavy elements in lung cancer is unclear. High concentrations of arsenic, nickel, and chromium have been accused of causing lung cancer, but of these only arsenic has been accused of causing cancer outside the workplace [Blot and Fraumeni 1975], and then only near smelters.

5.3.2 Epidemiology of Cancer Caused by Air Pollution

There are two major approaches to discussion of the human cancer risk. In the epidemiologic approaches, in Table 5–2, lung cancer rates are compared for rural areas with little air pollution and urban areas with considerable air pollution. In all of these there has been a clear difference between lung cancer rates in urban and rural populations.

As an example of such data we present in Table 5–3 data from a Liverpool–North Wales population by Stocks and Campbell [1959]; a small effect was found among nonsmokers that was not significant in Doll [1978]. The evidence from other studies is also not conclusive. There are obvious difficulties in comparing the smoking populations of rural and urban populations. Doll [1978], by comparing the rates only for British physicians (where occupational differences do not occur) and by numbers of tobacco cigarettes smoked, has probably corrected for these confounding factors.

Most of these studies seem to agree that there may be a small increase of lung cancer among nonsmokers caused by air pollution;

Table 5-2. Epidemiologic studies of cancer related to air pollution.

Study	Location	Effect Observed
Stocks [1958]; Stocks and Campbell [1959]	N. Wales	Yes
Haenszel, Loveland, and Sirken [1962]	U.S.	Yes
Hammond et al. [1976]	U.S.	No
Doll and Peto [1977]	U.K.	Yes
Hitosugi [1968]	Japan	Yes
Dean [1959]	S. Africa	Yes
Golledge and Wicken [1964]	Teesside, England	Yes
Dean [1966]	Ireland	No
Cederlöf et al. [1975]	Sweden	Yes
Wicken [1966]	Ireland	No
Higgins [1974, 1976]	U.K.	Yes
Carnow and Meier [1973]	U.S. (regression)	Yes
Lave and Seskin [1977]	U.S. (regression)	Yes

Table 5-3. Age-standardized lung-cancer mortality (per 100,000 per year) for men aged 35-74 years by amount of cigarettes smoked in Liverpool and rural North Wales.

Packs/day (approx.)	Mortality Rates	
	Rural Area	Liverpool
Non-smokers	22	50
½	69	168
1	147	248
1½	232	389
2	344	327

Source: P. Stocks, "Report on cancer in North Wales and Liverpool Region," in British Empire Campaign, Supplement to Part II of 1957 Annual Report, London.

this is at most half the total incidence among nonsmokers, which is already small. The increase of lung cancer among smokers caused by air pollution is four times greater than the increase among nonsmokers, and is statistically significant. A possible reason for the apparent synergism between smoking and air pollution comes from the work of Cohen, Arai, and Brain [1979], who showed that smoking inhibits the action of cilia in long-term dust clearance from the lungs.

To illustrate the way in which these data can be used to give a crude estimate of the risk, and the total cancer mortality from air pollution, we follow here the argument of Pike et al. [1975] in analyzing the data of Stocks [1958] (Table 5–3). The BαP concentrations were estimated to be 77 ng/m^3 in Liverpool compared to 7 ng/m^3 in North Wales. Then the probability of death per ng/m^3 B α P becomes:

$$\frac{50-22}{10^5} \times \frac{1}{(77-7)\ \text{ng/m}^3} = \frac{0.4}{10^5} \frac{\text{deaths}}{\text{people} \times \text{ng/m}^3} \qquad \text{for nonsmokers}$$

$$\frac{248-147}{10^5} \times \frac{1}{(77-7)\ \text{ng/m}^3} = \frac{1.4}{10^5} \frac{\text{deaths}}{\text{people} \times \text{ng/m}^3} \qquad \begin{array}{l}\text{for smokers} \\ \text{(1 pack/day)}\end{array}$$

Carnow and Meier [1973] use a regression analysis in which they analyze the cancer mortality simultaneously in terms of air pollution and where cigarette smoking habits were considered by a measure of cigarette sales per person.

The results suggest that an increase in urban pollution associated with an average BαP concentration of 1 ng/m^3 corresponds (in the United States) to an increase of 5 percent in the lung cancer death rate of 40 cancers/10^5 persons. This is about

$$\frac{5}{100} \times \frac{40}{10^5} = \frac{2}{10^5} \frac{\text{cancers}}{\text{persons}}$$

which is twice the size of the effect suggested by Stocks. Carnow's result is taken almost verbatim by NAS [1972]. However, Carnow and Meier point out that they relate 1960 deaths to 1968 BαP exposures, and overall B α P exposures have been falling (see Figure 3-18). We therefore reduce the number a factor of 2 to 1 cancer/10^5 persons.

It should be pointed out that these numbers are to be taken in the context of a need for policy decisions and are not to be taken as a proof that either air pollution or BαP has been proved to be causally related to the health effects as seen in the regression equation. Far less, however, should they be taken as disproof of an air pollution-related effect of this magnitude [Carnow and Meier 1973; Friberg and Cederlof 1978].

5.3.3 Cancer Caused by Air Pollution — Estimates from Occupational Studies

The second approach is to take data from highly exposed workers, such as the gas workers studied by Doll et al. [1965], and assume (1) that cancer mortality is proportional to exposure even at low exposures, and (2) that BαP is a good indicator of the relative carcinogenic effect of the air pollution in the workplace and in the city.

Doll et al. found a lung cancer rate of $360/10^5$ among gas workers. With a proper adjustment for the age of the population, he expects $200/10^5$ among the rest of the population of the same age—a mixture of smokers and nonsmokers. This is a risk ratio of $360/200 = 1.8$ with an exposure level [Lawther, Commins, and Waller 1965] of $2,000$ ng/m^3 for each working day (22 percent of the year) or roughly 440 ng/m^3 average. The gas workers were not exposed all their lives; guessing a factor of 2 to correct for this (Pike et al. [1975] did not make this correction) we might estimate that the same risk ratio would be obtained by continuous lifetime exposure to about 200 ng/m^3. This then suggests for this age group a mortality rate from continuous exposure to pollutants, using BαP as a surrogate

$$\frac{360-200}{10^5} \div 200 \text{ ng/m}^3 = \frac{0.8}{10^5} \frac{\text{deaths}}{\text{people} \times \text{ng/m}^3 \text{ (B}\alpha\text{P)}}$$

The fact that this number agrees with the epidemiologic evidence ($0.4/10^5$ for nonsmokers and $1.4/10^5$ for smokers) makes it more plausible.

In another study Hammond et al. [1976] found that roofers and waterproofers working with pitch and asphalt had 60 percent more deaths from lung cancer than expected from others in the same trade union. The exposure of the roofers to BαP was estimated by samples of dust collected on masks worn during jobs; each worker inhaled 17 μg BαP per 7-hour day or 700 ng/m^3 in a breathing volume of 24 m^3/day. This corresponds to about a 200 ng/m^3 lifetime average. If we combine this with the lung cancer risk from this trade union, we find the final risk of $0.2/10^5$ deaths/ng/m^3 averaged over all ages.

High risks have also been found for Japanese gas workers [Kawai, Amamoto, and Harada 1967] and U.S. coke oven workers [Lloyd 1971; Redmond, Strobino, and Cypress 1976], but these exposure levels are even more approximate.

5.3.4   Animal Studies of Carcinogens

Carcinogenic substances have been administered to animals by inhalation, intratracheal instillation, ingestion, or implantation into the airways. The more recent protocols have involved regular administration by inhalation or ingestion for a lifetime. In addition, BαP and dibenzanthracene have been found to be mutagenic in mutagenesis tests.

Chu and Malmgren [1965] fed BαP to Chinese hamsters and from their data Meselson and Russell [1977] show that for 0.1 mg/kg body weight ingested daily, 50 percent developed tumors.

Peacock and Spence [1967] exposed a strain of mice especially sensitive to pulmonary edema to 500 ppm in air of SO$_2$ and found that they had an incidence of tumors exceeding that of controls; this experiment is not yet generally accepted as evidence of carcinogenicity. Laskin, Kuschner, and Drew [1970] reported that when rats were exposed to BαP at 10 mg/m^3, 1 hour per day, 5 days per week for 98 weeks, no cancer was observable. But when the rats were also exposed to levels of SO$_2$ of 3.5 ppm or 10 mg/m^3, 1 hour per day, 2 out of 21 developed squamous cell carcinomas. Of these also exposed to SO$_2$ at a level of 30 mg/m^3 for 5 hours per day, 5 days per week, 5 out of 21 developed carcinomas.

This increase of *average* sulfur dioxide concentration of a factor of 15 increased the number of tumors only by a factor of 2.5 with a large error. Nonetheless, the fact that no tumors were observed in the absence of sulfur dioxide suggests that sulfur dioxide increases the carcinogenic potency of benzo(α)pyrene, even though sulfur dioxide may not be carcinogenic by itself.

If we look at the benzo(α)pyrene concentrations, we find the average concentration in air over the 98 weeks was

$$10 \ \frac{mg}{m^3} \ \times \ \frac{5 \ days}{7 \ days} \ \times \ \frac{1 \ hour}{24 \ hours} \ = \ 0.3 \ mg/m^3$$

We take only the number when sulfur dioxide was present to get a risk estimate.

To find out the amount of B α P inhaled by the rats, we note that a typical 250 gram rat breathes about 0.25 m³ (= 250 litres) of air per day. This air will then contain 0.25 m³ $\times$ 0.3 mg/m³ of B α P or 0.3 mg/kg body weight.

At 98 weeks 10 percent of the rats got cancer. This gives a carcinogenic potency derived in the usual way

$$\frac{1}{0.3 \text{ mg/kg body wt} \times \text{day}} \times \frac{1}{10} = 0.33 \quad \frac{\text{kg body wt} \times \text{day}}{\text{mg intake}}$$

This potency is 20 times less than the potency suggested by the ingestion data of Chu & Malmgren [1965]. A small part could be due to the different lengths of the experiment. Since cancers occur roughly as the fourth power of age, more rats would die in the 120 week experiment of Chu & Malmgren than in the 98 weeks of exposure by Laskin, Kuschner & Drew. This would result a $(120/98)^4$ or approximately factor of 2 difference in the number of cancers induced. Also, one experiment was by inhalation and the other by ingestion, and different species of rats were used.

The same specie and protocol have been used for studies of the carcinogenicity of vinyl chloride by both ingestion and inhalation by Maltoni [1977]. There is agreement.

Returning to the bottom of page 177 we found that 10 percent of rats developed cancer with an average B α P concentration for 98 weeks of 0.3 mg/m³. If the concentration were 1 ng/m³ we multiply the fraction by 0.3 $\times$ 10^{-6}; if the rats had been exposed for a full life of 120 weeks we multiply by 2, so the cancer risk becomes 0.3 $\times$ 10^{-6} $\times$ 2 $\times$ (10%) = 6 $\times$ 10^{-8}. Assuming that this is the same lifetime risk for *people* exposed to this concentration, the *yearly* risk for people is 6 $\times$ 10^{-8}/72 years or approximately 10^{-9} per 1 ng/m³ average continuous exposure to B α P.

We must here note that this risk is only for B α P and SO_2 whereas for air pollution many hydrocarbons are included, of which at least one, dibenzanthracene, is as potent as B α P and according to subcutancous injection data may be thirty times more sensitive [Bryan and Shimkin 1942]. The risk from the animal data should therefore be expected to be smaller than the risk from real polluted air.

Petering and Shih [1975] and Shapiro, DiFate, and Welcher [1974] indicate that bisulfite may have a weak mutagenic effect, and therefore probably a weak carcinogenic effect. Chrisp, Fisher, and Lammert [1977] show that respirable fly ash from coal is mutagenic. It would be pleasing to have a consistent set of animal data on all the carcinogens listed in Table 3-9 and a mixture therefore with lifetime inhalation and ingestion tests on the same strain of animals and modern protocols. This is a long task but it is an important one.

5.3.5 Risk Summary and Mortality Estimates from Cancer

Table 5-4 summarizes estimates of cancer risks from 1 ng/m^3 average exposure to B α P. Risk in this table is interpreted as the number of expected cancers resulting from a 1 ng/m^3 average B α P exposure to 100,000 people per year. These estimates are all crude and cannot be taken as proof of causality. The risk, in fact, could be zero. It is unlikely that the risks are actually zero because they are derived from data at or near ambient levels. For public policy purposes radiation risks from environmental causes have been extrapolated from much higher exposures assuming linearity.

Measurements from the National Air Surveillance network in 1969 showed that many American cities had concentrations of B α P as

Table 5-4. Crude lung-cancer risk estimates for exposure to ambient air pollution, BαP as an indicator (per 100,000 = 10^5 persons at risk and per ng/m^3 BαP).

Animal Data (BαP *only*)	inhalation	0.0001
	ingestion	0.002
Epidemiology (direct)	Stocks [1958]	1.0
	Carnow and Meier [1973]	1.0
Epidemiology (high levels)	Gas workers [Doll et al. 1965]	0.2
	Roofers (Hammond) [1976]	0.2
Value taken in text for discussion		0.5
Value used in CAG [1978]		0.12

Risk of cancers other than lung cancer may equal lung cancer risk.
Risk of heart disease may be about equal to risk of all cancers and additional.

high as 3.5 ng/m^3 out of doors [NAS 1972], giving a yearly lung cancer risk of 1.7 × 10^{-5} *. Indoors the concentration may be less, with a lower risk of, say, 10^{-5}. Taking this risk and applying it to the 50 million persons within these cities leads to a death rate from lung cancer of 500 per year. We then make the plausible assumption that air pollution, like cigarette smoking, leads to as many cancers in other sites as in the lung, and also leads to heart problems and genetic problems; and the overall cancer rate in the United States might then become 1,000 per year. These are each 1 percent of the overall effect caused by air pollutants estimated in the next chapter. This percentage is small but it has its interest. Whereas some argue that there is a threshold for the adverse health effects of sulfates and other particulates, fewer scientists argue that there is a threshold for carcinogens, and those that do tend to admit that for pollutants that are inhaled such as BαP the level may already be sufficiently high as to exceed the threshold, particularly for cigarette smokers.

We must also be aware that sulfates and other particulates, although not carcinogenic themselves, may by their irritant action or otherwise act as cocarcinogens.

Doll [1978] implicitly accepts the quantitative value given here while stressing the smallness of these numbers:

> . . . by far the most important cause of lung cancer is cigarette smoking . . . the combined effect for all such atmospheric agents cannot be responsible for more than about five cases of lung cancer per 100,000 persons in European populations.

This 5/10^5 is consistent with European BαP concentrations of 3 to 5 ng/m^3.

The suggestion that air pollution causes cancer more among cigarette smokers than nonsmokers may have either of two origins. The air pollution and cigarette smoking can be synergistic in their causation of cancer and the cancer effect linear with either dose. Then the cancer effect will be proportional to the product of the air pollution dose and the cigarette dose. Alternatively, the two effects can be additive, but proportionality with dose is not a correct assumption (curve A$_1$ or A$_2$ of Figure 5-3). Then air pollution by itself would produce a small effect, but the pollution caused by cigarette smoke would bring the dose far to the right of the curve, well above

---

*A risk of 1.7 × 10^{-5} per person is the same as 1.7 cases per 10^5 persons.

any threshold, and the differential effect of the air pollution could be much bigger. There is a slight preference in Stocks [1958] data of Table 5−3 for this second alternative.

5.3.6 Attribution of Cancer Risk to Various Causes

Although it seems probable that a cancer rate caused by air pollution of about the amount discussed in Section 5.3.5 is correct, it is quite another matter to assign it to a specific cause.

The amounts of BαP emitted by fossil fuels of different types was listed in Table 3−10. The crucial feature is that there is a difference of 10,000 in emission of BαP per unit energy between the best coal-burning procedure and a hand-operated furnace. It is clear that as coal has been burned in more centralized burners (electric generating stations and large factories) the emissions of polycyclic aromatic hydrocarbons have been reduced, and they have been emitted at higher elevations so that the molecules have a greater chance of being broken up by ultraviolet light before reaching the ground. The concentrations of BαP in London have dropped by a factor of more than ten (Section 3.4) [Lawther and Waller 1978]. It seems likely, therefore, that whereas in 1930 the BαP concentrations in urban areas were largely due to coal burning, in the 1970s BαP is due to a combination of coal and wood burning in the few remaining residential hand-stoked burners and to automobile exhausts. Studies in Leningrad by Zabezinsky [1968] showed that B α P concentrations were by that time higher in the center of the city, because of automobiles, than were B α P concentrations in an industrial area.

The large factor of 10,000 between good and bad burners suggests that the concentrations will be dominated not by the many automobiles (or other burners) that are well designed but by the few that are badly designed or by the few days in which they are badly adjusted. The implications for public policy is that the cancer hazard will be dominated by the large number of small burners that regulatory authorities can barely control. Any trend back to coal or wood burning in small city houses, or even apartment or business buildings, would cause a major increase in BαP concentrations and in the cancer hazard. The recent increase in use of wood stoves for residential heating may sharply increase BαP exposure inside homes using wood stoves and within surrounding residential communities.

The studies reviewed by Shabad [1977] confirm that the adjustment of the automobile engine is important, that a rarefying mixture leads to less BαP, that a diesel engine can reduce BαP tenfold, and that a platinum-aluminum catalytic element can reduce BαP a factor of three. These findings are consistent with the fact that all these steps lead to more complete combustion.

It is also interesting that the age-adjusted lung-cancer rates in London have been falling recently; this is correlated with the fall in BαP concentration [Lawther and Waller 1978] and suggests an attribution.

5.3.7   Uncertainty of Cancer Estimates

The cancer estimate we have made here, based on polycyclic aromatic hydrocarbons, is crude because (1) we have very bad monitoring data for polycyclic aromatic hydrocarbons, and even BαP as an indicator is monitored in only a few cities; (2) we include all sources of BαP and the biggest are likely to be industrial (in such places as Altoona, Pennsylvania) or automobiles (Los Angeles); (3) we do not know the atmospheric chemistry of BαP and the other hydrocarbons, and we do not know the extent of long-range transport; and (4) we do not know the dose-response relationship, although it is becoming accepted public policy to assume linearity. The estimates are also unsatisfying because of lack of agreement with animal data, even though there is a probable reason—use of BαP as a sole pollutant—for the disagreement.

We proceed no further with considerations of BαP in our discussions in Chapters 7 and 8 on mitigation because the problems seem to be overshadowed by those of the sulfates and other respirable particulates.

5.4   HEALTH EFFECTS OF RADIOACTIVITY RELEASES

Coal contains many impurities and among them are radioactive elements—uranium and its daughters, and thorium and its daughters. The range of concentrations of radium and thorium in ores is a factor of ten [Swanson et al. 1976] but the average is about 1 ppm uranium and 3 ppm thorium. If we were to use the uranium and thorium in the coal in a nuclear reactor, ten times more energy could be gained than by burning the coal! This at once suggests that the problems of radioactivity release, when added up over all the coal burning in the United States, could be significant and possibly comparable to effects from nuclear power [Eisenbud and Petrow 1964; Terrill, Harward, and Leggett 1967; Martin, Harward, and Oakley 1969; Hull 1971, 1974; Lave and Freeburg 1973; Wilson and Jones 1974; Bedrosian, Easterly, and Cummings 1970; Jaworowski et al. 1073; McBride et al. 1978; and Travis et al. 1979].

We outline here calculations of McBride et al. [1978] that applies to advanced power plants with good particulate suppression. It is assumed that the uranium and thorium are in the ash, and that 1 percent is released to the atmosphere in particulate form. Then for one year of operation of a typical 1,000 MWe power plant at 80 percent capacity factor the releases are 23 kg uranium and 46 kg of thorium and associated products in equilibrium. The releases in curies of radioactivity are shown in Table 5−5.

The dose to the public depends a little upon the exact site, although not so much as for a nuclear power plant because the radioactivity does not decay fast. The doses can be calculated using the dispersion calculations of Section 2.3.

The radioactive materials can be inhaled or passed along the food chain and ingested. The vast amount of work on intake of radionuclides through food chains can then be used to estimate dose. Because these radioactive elements last for long periods of time, the dose will continue for some time after ingestion and therefore the dose commitment over a 50-year period is an appropriate number to consider. Radium is a well-known "bone-seeker"−that is, it is deposited in the bone, where it remains. It is therefore no surprise to find that radium 226 and 224 are the major contributors to the whole body and to specific organ doses. The deposited radionuclides were assumed to build up for 50 years in estimating surface exposure as shown in Tables 5−6 and 5−7 and as compared with doses from nuclear plants calculated in a similar way. It is clear that they are comparable.

This model generating plant consumes 2,000,000 tons of coal per year. If all U.S. coal were burned with as good a particulate suppression system, with as tall a stack, and in a similar area, then we would have 300 times as much, or about 6,000 man-rems commitment of

Table 5-5.   Estimated annual airborne radioactive materials released from a model 1,000 MWe coal-fired power plant (source term).

Isotope[a]	Release (Ci/year per radionuclide)
Uranium–238 chain ^{238}U, ^{234}Th, ^{234}Pa, ^{234}U, ^{230}Th, ^{226}Ra, ^{218}Po, ^{214}Pb, ^{214}Bi, ^{214}Po, ^{210}Pb, ^{210}Bi, ^{210}Po	8×10^{-3}
Uranium–235 chain ^{235}U, ^{231}Th, ^{231}Pa, ^{227}Ac, ^{227}Th, ^{223}Ra, ^{211}Pb, ^{211}Bi	3.5×10^{-4}
Thorium–232 chain ^{232}Th, ^{228}Ra, ^{228}Ac, ^{228}Th, ^{224}Ra, ^{212}Pb, ^{212}Bi	5×10^{-3}
Radon ^{220}Rn	0.4
^{222}Rn	0.8

Assumptions: (i) the coal contains 1 ppm U and 2 ppm Th, (ii) ash release is 1 percent, (iii) ^{220}Rn is produced from ^{232}Th in the combustion gases at the rate of 1.38×10^{-9} (Ci/sec per gram of Th, (iv) the annual release of natural U is 2.32×10^{4} gram and of ^{232}Th is 4.64×10^{4} gram, (v) 15 seconds is required for the gases to travel from the combustion chamber to the top of the stack.

a. Except for ^{222}Rn, radionuclides with half-lives less than several minutes are omitted.

Source: J.P. McBride et al., "Radiological impact of airborne affluents of coal and nuclear plants," *Science* 202 (December 8, 1978): 1045.

dose over a 50-year period for each year of coal burning, or about one cancer case per year from power plant operation.

These calculations of McBride et al. [1978] omit radon emissions from waste heaps. Radon is now widely recognized to be the major source of radiation in homes [Moeller and Underhill 1976; Cohen 1979; Hurwitz 1979] and in the environment generally [Travis et al. 1979]. Radon gas is continuously emitted from uranium over a billion-year time span. When in its original ore in the ground, the radon gas does not usually reach the surface of the earth before radioactive decay has changed it to a nonradioactive material. However, once the coal is brought to the surface, and if the waste remains near the surface, the radon gas can spread. A portion is inhaled by

Table 5-6.   Population-dose commitments from the airborne releases of model 1,000 MWe power plants (88.5 km radius).

	Population-dose Commitment (man-rem/yr)					
	Coal-fired Plant[1] Stack Height (m)					
Organ	50	100	200	300	BWR[2]	PWR[2]
Whole body	23	21	19	18	13	13
Bone	249	225	192	180	21	20
Lungs	34	29	23	21	8	9
Thyroid	23	21	19	18	37	12
Kidneys	55	50	43	41	8	9
Liver	32	29	26	25	9	10
Spleen	37	34	31	29	8	8

The population-dose commitments are for a midwestern site. The ingestion components of the dose commitment are based on the assumption that all food is grown and consumed at the reference location.

1. A plume rise due to buoyancy of hot stack emissions was assumed. The dose commitments are for an ash release of 1 percent and coal containing 1 ppm U and 2 ppm Th.

2. Source terms for the nuclear plants are from NRC. The release height was assumed to be 20 m with no plume rise.

Source: J.P. McBride et al., "Radiological impact of airborne affluents of coal and nuclear plants," Science 202 (December 8, 1978): 1045.

people and the decay products remain in the body, but the most important dose is by ingestion of deposited radioactivity.

This effect remains until the uranium is adequately covered again or until a billion years, whichever is sooner. If the uranium is not covered for half a million years, for example, the radioactivity deposited will build up 10,000 times as much as calculated for Tables 5-6 and 5-7 and give doses of about 10,000 cancer cases per year of operation—a large number. Because the wastes are widely dispersed, it is impracticable to artificially cover them all—unlike the analagous problem for nuclear power, where the mine tailings remain concentrated and can easily be covered. However, it is reasonable to assume natural covering, or runoff to the sea, within this period of 50 years.

These effects are small compared to the effect of other airborne carcinogens in the coal described in Section 5.3, but the comparison with nuclear power is of psychological interest. They are also much smaller than the dose from radioactive elements (primarily polonium

210 and lead 210) absorbed with cigarette smoke. This can reach 10 picocuries in the lungs with a maximum 25-year accumulated dose of 200 rem [Little, Radford, and Holtzman 1967]. The total population dose may be as high as 10^7 man-rems/yr.

5.5 PROSPECTIVE AIR POLLUTION STUDIES

In 1978 two air pollution epidemiologic studies were in progress in the United States. An acute respiratory disease study, conducted by the EPA Health Effects Research Laboratory, focused on the effects of NO_2 and O_3. The second study, by the Harvard School of Public Health, investigated long-term chronic changes in lung functions and

Table 5–7. Population-dose commitments from the airborne releases of model 1,000 MWe power plants as a function of food intake.

| Plant Type and Organ | Population-dose Commitment (rem/yr) if Percentage of Food Grown and Consumed in Area is | | | | |
	0	10	30	50	100
Coal-fired plant[1]					
Whole body	1.2	3.2	7.2	11.1	21
Bone	31	50	89	128	225
Boiling-water reactor[2]					
Whole body	4.3	5.2	6.9	8.7	13
Bone	5.7	7.1	10	13	21
Pressurized-water reactor[2]					
Whole body	3.1	4.1	6.1	8.1	13
Bone	4.9	6.4	9.4	12.5	20

Midwestern site, 88.5 km radius.

1. Population-dose commitments are for coal containing 1 ppm U and 2 ppm Th. The releases are from a 100–m stack with a plume rise due to buoyancy of the hot stack emissions.

2. Source terms for the nuclear plants are from NRC. The release height was assumed to 20 m with no plume rise.

Source: J.P. McBride et al., "Radiological impact of airborne affluents of coal and nuclear plants," *Science* 202 (December 8, 1978): 1045.

illnesses resulting from exposure to SO_2, sulfates, and particulate matter.

The first objective in the EPA study was to develop better methods to characterize human exposures to air pollutants. The second objective was to measure respiratory diseases in acute response to NO_2 alone, and to NO_2 in combination with other pollutants, especially O_3. This study included selected populations from the Los Angeles basin. Hence, pollutants other than NO_2 and O_3 will be considered. In 1979, a single station in each of the study areas was being used to estimate exposures for each study participant. In order to define individual pollution exposures more accurately than in previous studies, the daily movement of subjects was recorded in conjunction with personal monitoring. Monitoring was also performed using fixed-location monitors.

The other major U.S. epidemiologic study underway, at the Harvard School of Public Health, examines long-term changes in lung function in both children and adult populations. This study concentrates on SO_2, suspended respirable sulfates, and respirable particulates; NO_2, TSP, O_3, total sulfates, and meteorological paameters are also measured. Portable spirometers (used to measure lung function) are brought to homes in order to ensure a high percentage of measurements of subjects. Approximately 10,000 adults are involved in the study, from six different communities. In addition, about 12,000 children are re-examined every year in these six communities.

The air monitoring associated with this study has involved a three-stage approach to assessing exposures. Fixed-location central-site monitors continuously measure the gaseous pollutants. In addition, a network of satellite monitors is located throughout the communities to measure pollutant levels inside and outside homes. To further quantify human exposures, personal monitoring has been done in three of the communities. The goal is to develop an exposure model such that every individual in the health survey will have an exposure index assigned to him or her on the basis of community levels, home characteristics, personal characteristics, and commuting habits. The air pollution data, along with social and demographic data, will finally be merged with the health effects data to assess the effects of changes in air pollution on human health.

In order to make this assessment possible, six cities were carefully chosen. These six cities initially had very different air pollution levels. Further, it was expected that the air pollution levels would change in time. Both Portage, Wisconsin, and Topeka, Kansas, have very clean air. However, new coal-fired power plants being built close to these areas are expected to result in some small degradation in this air quality.

Both Watertown, Massachusetts, and Kingston/Harriman, Tennessee, have air pollution levels that are currently lower than the upper limit set by federal standards. However, the quality of the air in these two communities has not remained constant. Kingston/Harriman, Tennessee, which is near a TVA coal-fired power plant, has experienced very high acute exposures in the past. However, with the building of two 330–meter (1,000-foot) stacks, the air quality has been dramatically improved. In Watertown, Massachusetts, the air quality levels have shown some degradation as the strategy for sulfur-clean fuels has been relaxed in the Greater Boston area.

Both Steubenville, Ohio, and St. Louis, Missouri, have had levels that exceed the federal standards for both SO_2 and TSP. In both these communities, however, there has been tremendous improvement in the quality of the air over the last three years. The levels have improved such that there are only occasional violations of the 24-hour standards in each of these communities, and the number of alert days in Steubenville, Ohio, has been reduced from about 50 per year to fewer than 20 per year.

The study design allows for evaluation of air-pollution health effects in both comparisons of lung function in different cities (cross-sectional comparisons) as well as the change in lung function with time (time-series or longitudinal studies). Because of the inherent difficulties of comparing different cohorts of populations on account of social, economic, and ethnic differences, greater reliance in the six-city study may be placed on the change with time.

Each individual serves as his or her own control for changes in lung functions over the six years of study. The rate of growth in lung function for children or rate of decline for adults is related to changes in air pollution exposures. It is of fundamental interest that the lung function seems to show a permanent change with air pollution, even though there is no more obvious health effect. This contrasts with a naive view that people recover completely from effects of pollution when the pollutant is removed. With an anticipated 15 percent per year dropout rate, a simple statistical prediction shows that a cohort of about 665 per city should be adequate to detect a

5 percent change in lung function. In 1979 it was too early to describe any health effects from ambient exposures to air pollution. However, a statistically significant decrement in lung function has been reported in children who are living in homes with gas stoves, particularly those without adequate ventilation [Speizer et al. 1979]. This is presumably related in these homes to NO_2 levels that often exceed the criteria for ambient external air. The nitrogen oxides are formed by the combination of nitrogen and oxygen in the flame, a process that will be enhanced and catalyzed by carbon deposits present under most cooking utensils. Also indicated is an increased prevalence of respiratory disease in children living in homes where adults smoke.

5.6 MODELS FOR HEALTH EFFECTS OF LOW LEVELS OF AIR POLLUTION

We would like always to derive health effects at present ambient levels from data. However, in this chapter we have seen that health effects of air pollution are easily observable only at higher levels. We are, therefore, led to discuss three alternative logical models of air pollution and ask whether the data are consistent with the models. These models may be useful as a guide to our thoughts about the reliability and consistency of the epidemiologic evidence discussed in the next chapter, and as a guide to the mitigation methods we discuss in later chapters.

1. In the first model, the direct biological effect of air pollutants is assumed to be primarily due to respirable particulates that pass the filters in the nose and settle in the lung. There they have an irritant effect; it is uncertain whether this irritant effect is due simply to the fact that they are particles or to their specific chemical form. An assumption in this often used model is that the major effect is due to the sulfate particulates. If they are due to the chemical form, it is not sure whether the anion or cation is important.

The large particulates are trapped by particulate suppression devices, but even if they were not they would not penetrate the nose. Therefore these particles are not expected to directly damage the lungs. But in addition to those that start out as particulates, the gases SO_2 and NO leave the stack and form particulates in the atmosphere. These small particulates remain in the air for days.

This model has been most explicitly worked out by Hamilton and his collaborators [Hamilton 1979], who note that although some of the load of fine particulates—and hence the health hazard—in the past was due to sulfur conversion, recently this fraction has increased and is likely to increase further as particulate suppression improves and total sulfur emissions increase in the absence of more stringent controls.

2. The second model again assumes that particulates cause the health hazard, but that some other component (e.g., metallic fraction) causes the hazard. The argument is that sulfur oxides and nitrogen oxides do not produce permanent damage even though they can, and do, cause acute (prompt) health effects at high concentrations. On the other hand, metals such as manganese (Mn), iron (Fe), cadmium (Cd), and lead (Pb) accumulate in the body, and the build-up can cause adverse health effects. However, the amount of these substances that is inhaled, even in the most polluted cities, is far less than the amount ingested in the daily diet, so it is unclear why the build-up should be related only to air pollution.

3. The third model is that only large concentrations of sulfur oxides or particulates are significant. In this more optimistic model the low levels prevalent in 1980 air pollution have no health effect whatever; health effects exist only if the concentrations are allowed to rise in air pollution episodes or by relaxation of regulations.

6 EXTRAPOLATION OF EFFECTS AT LOWER EXPOSURES

Chapter 6 makes a logical break with the arguments of Chapters 4 and 5 and discusses retrospective studies of human populations. An examination of some older studies reveals that they were too insensitive to show small risks of widespread low levels of pollution. The effects we will consider can, if the pessimists are correct, contribute 3 to 10 percent of the total U.S. mortality rate. If we were to evaluate this effect through detailed observation of changes in the health of individuals in a prospective or laboratory study, we would need to study, say, 100,000 persons exposed to polluted air, and 100,000 persons exposed to clean air. This sample size would be enough to measure a 2 percent effect with a statistical error of the difference of about 0.5 percent.

Although prospective studies with 10,000 people are being done, and will give very valuable information, studies with 100,000 people are not being done, and we discuss instead retrospective studies where total mortality rates are related to air pollution concentrations. The statistical error due to the population size is small, but the retrospective nature of the study leaves us with very difficult problems of discussing the causal nature of the effect.

Following the discussion of some smaller scale studies we introduce a set of data from large statistical studies relating death rate with air pollution and socioeconomic variables. These studies are

191

fraught with difficulties of methodology and interpretation. All studies in the United States seem to show a higher death rate in geographical areas of high pollution than in areas of low pollution. Whether this is due to air pollution directly, or to some other variable, is the important question at issue. We will consider the most obvious of these confounding variables: occupation, migration, and cigarette smoking. We come to no firm conclusion but discuss a suggested procedure for approaching differences of expert opinions and deriving suitable numbers for policy analysis. Finally, we have a list of suggestions for further research.

6.1   A FEW SMALL-POPULATION STUDIES

In Section 1.2 we mentioned the Oslo study of Lindberg [1968] and the Japanese study of Nishiwaki et al. [1971]. In the Nishiwaki study the emphasis was on bronchitis and chronic obstructive lung diseases generally.

Increased prevalence of obstructive lung diseases were observed in the Netherlands [van der Lende et al. 1972], in Japan [Oshima et al. 1964; Oshima, Imai, and Kawagishi 1972], and in Poland [Sawicki 1972]. The question arose, however, whether the effects observed in these studies were due to air pollution or to socioeconomic status. Three other studies failed to show an effect: Comstock et al. [1973] in New York City and Washington, D.C.; Hrubec et al. [1973] in twins from the U.S. Veterans Registry; and Colley, Douglas, and Reid [1973] in the United Kingdom. However, the first was too insensitive to find even the known strong correlation between pulmonary symptoms and smoking, and the last two used crude measures of air pollution—by postal zip-code area and domestic coal utilization—which made them very insensitive to small risks caused by air pollution.

6.2   PROBLEMS WITH EPIDEMIOLOGY

Hill [1965] listed several items to be considered in the evaluation of causality.

1. The strength of the association. As we will see, the predicted effect is smaller than many nonrandom variations of the death rate,

but is larger than some others. The strength of the association can therefore be categorized as moderate (tobacco cigarette smoking gives a strong association).

2. The consistency of the results. Several people have analyzed different data sets in slightly different ways, and all find an association of roughly the same magnitude.

3. The specificity of the results. The association with air pollution is not specific to any particular air pollution variable; nor is it specific to body sites or disease types. But Hill stresses that specificity is not a necessary condition.

4. Temporality. Time-series studies indicate that exposure is simultaneous with or precedes observed health effects, satisfying temporality.

5. Existence of a biological gradient. The health effects increase with increasing air pollution.

6. Biological plausibility. Opinions of experts vary widely, from those who believe it plausible that air pollution at almost any level is dangerous until proved safe to those who argue that there must be a biological threshold. We discuss mainly the plausible assumption that the effect is due to sulfate particulates and is proportional to dose (exposure).

7. Coherence. Unfortunately, there has been no good attempt to make a coherent picture of air pollution-related mortality relating laboratory experiments, air pollution episodes, and low-level studies. However, animal models have verified biologic damage from air pollution exposure.

8. Experimentation. No serious attempt has been made to relate experimentally chronic health effects to long-term exposure to air pollutants, but by analogy we turn to higher level short-term experimentation as discussed in Chapter 4.

9. Analogy. The analogy is often made with carcinogens (e.g., Section 5.3) where the probability of tumor incidence is often believed to be proportional to dose (exposure), and the probability accumulates with time. Tobacco smoking presents another analogy.

The situation, therefore, is far from clear, and causality is far from proved. It is not disproved either, and for purposes of public policy an interesting and important dilemma remains.

6.3 REGRESSIONS OF CROSS-SECTIONAL STUDIES

We will consider here only the major recent work, that of Winkelstein et al. [1967], Lave and Seskin [1977], Mendelsohn and Orcutt [1978]; Koshal and Koshal [1973]; Gregor [1976]; Lipfert [1978]; Thibodeau, Reed, and Bishop [1979]; Bozzo [1977, 1978], and Hamilton [1979]. Winkelstein et al. [1967, 1968], studied the relationship of air pollution to mortality in Buffalo, New York. Lave and Seskin's [1970, 1973, 1977] studies compared air pollution and mortality using Standard Metropolitan Statistical Areas (SMSAs) distributed throughout the United States. They used the census data for the decennial years of 1960 and 1970, and correlated them with air quality data for the nearest available year. Lipfert [1978] restricted some of his analyses to the major cities and hoped thereby to avoid some of the spread caused by variation of socioeconomic variables between city and suburb. Bozzo [1978] correlated all U.S. death statistics with air pollution concentration calculated from emissions, using Meyer's meteorologic model (Section 2.3). More recently [Hamilton 1979], the same group compared the death rate with measured air quality data.

We describe first the regressions of Lave and Seskin, which have been in the literature for nine years. These authors correlated the 1971 deaths in metropolitan areas (114 SMSAs) with pollutant variables, by a regression analysis. In one of these regressions they relate

M = mortality rate per 10,000 population

$\bar{P}$ = average level of total suspended particulates in $\mu g/m^3$

$\bar{S}$ = average level of sulfate particulates in $\mu g/m^3$

S_{min} = minimum level of sulfate particulates in $\mu g/m^3$

P_{min} = minimum level of total suspended particulates in $\mu g/m^3$

D = population density in persons per square mile

NW = proportion of nonwhites in the population

(> 65) = proportion of people aged over 65 years in the population.

They find:

$$M = 20 + 0.04\bar{P} + 0.7S_{min} + 0.001D + 40\,(NW) + 700\,(> 65),$$

$$(6-1)$$

which becomes if we insert the values of the variables averaged over all SMSAs for 1971:

$$M = 20 + 4.5 + 3.3 + 0.7 + 5 + 58$$

$$= 92 \text{ deaths}/10,000 \text{ people} \tag{6-2}$$

(We have rounded off the numbers and simplified the notation.)

It is important to consider the socioeconomic factors (population density, proportion of nonwhites in the population, and proportion of elderly in the population) very carefully. The effect of elderly or nonwhites on mortality rates may not be merely an expression of different death rates among whites and nonwhites, but may be a surrogate for other factors such as social stress, income level, or disease propagation.

With this caveat in mind, the equation, with the numbers for 1971 below, tells us some important facts. First, air pollution in 1971 describes (although it may not be the cause of) 8 percent of the U.S. mortality in metropolitan areas. Second, population density per se was not an important factor; the racial characteristics are not as important as air pollution, but the age distribution of the population is very important.

In Figure 1–5 we plotted the data of Winkelstein et al. [1967] in a "three-dimensional" histogram, and this graphically illustrated the problems of the correlations among these variables. In the regression equation here, there are five variables and it is not possible to display these correlations so clearly, although it is possible to display them two at a time. When we consider the relevance of these other variables, we must therefore extend our "obvious" geometric instincts to the more complex algebraic expressions.

Lave and Seskin found that when they used minimum sulfate in their equation the coefficient was more significant. This seems counterintuitive and in this book we have concentrated on average pollutant levels. In order to be clear and simple we transform the variables in the previous equation to these units. We use a distribution of air pollution levels that is lognormal with a standard geometric deviation of 1.6 (obtained by a rough fitting of published data). We then find the average sulfate levels are equal to 2.6 times the minimum value ($\overline{S} = 2.6\, S_{min}$). Then the coefficient of the S term goes down by this factor of 2.6. Our "adjusted" equation becomes

$$M = 20 + 0.04\overline{P} + 0.27\overline{S} + 0.001D + 40\,(NW) + 700\,(>65) \tag{6-3}$$

Lave and Seskin have made attempts to separate out the various air pollution variables. They find that either sulfate or TSP concentrations describe the mortality well. This is physically plausible, because their concentrations tend to vary together, and sulfates are a subset of particulates.

Using TSP or sulfates alone, Lave and Seskin find [Lave and Seskin 1971, Equation 3−1):

$$M = 22.3 + 0.01P_{min} + 0.001D + 32 (NW) + 682 (> 65) + 13 (< \$3000)$$

$$(6-4)$$

or [Lave and Seskin 1971, Equation 3−2):

$$M = 22.6 + 0.85S_{min} + 0.001D + 33 (NW) + 652 (> 65) + 6 (< \$3000) \quad (6-5)$$

where ($< \$3000$) is the fraction of households with income of less than \$3000.

For purposes of comparison we estimate

$$\bar{P} = 2.6P_{min} ; \text{ and } \bar{S} = 2.6S_{min} .$$

We define the coefficients a_p and a_s as the coefficient of air pollution-related mortality assuming average TSP or average sulfates, respectively, as the sole air pollution variables. Then

$$a_p = 0.038 \, (\mu g/m^3)^{-1} \, (\text{deaths} \times 10^{-4})yr^{-1} \qquad (6-6)$$

$$a_s = 0.33 \, (\mu g/m^3)^{-1} \, (\text{deaths} \times 10^{-4})yr^{-1} \qquad (6-7)$$

With these adjustments we list the coefficients from a number of air pollution studies in Table 6−1. Sulfate levels are about 15 percent of TSP levels averaged over the population areas of the United States. Therefore we expect that a regression on sulfates alone will give us a coefficient (a_s) equal to six times the coefficient for a regression on TSP alone (a_p) in the same units. If the health effect of particulates are entirely due to the sulfate components, Lave and Seskin find $a_s/a_p = 8.7$, which is surprisingly close.

We collect in Table 6−1 several other air pollution regressions. In many of these TSP was measured; in a few, sulfates; and in others, many other variables as shown. These data cannot be called consistent, and imagination and thought are needed to reconcile them.

The negative association with sulfates of Schwing and McDonald [1976] is derived from the two variables of minimum and average readings; it could be masked by the positive association that they

Table 6-1.   Coefficients of mean SO_x and mean TSP in the yearly mortality equations (adjusted to mean where necessary).

Note	Author	$a_s (SO_x)$ Deaths $(\mu g/m^3)^{-1}$ $(\times 10^{-5})$	$a_p (TSP)$ Deaths $(\mu g/m^3)^{-1}$ $(\times 10^{-6})$
1.	Lave and Seskin [1977]	3.3	3.8
2.	Mendelsohn and Orcutt [1978]	10.2	−0.28
3.	Schwing and McDonald [1976]	−7	x
4.	Lipfert [1978] (regression I.3)	1.9	6.5
5.	Smith [1976]	x	11
6.	Gregor [1976]	x	26
7.	Winkelstein et al. [1967]	x	70
8.	Bozzo [1978]	3.56	x
9.	Used later in text	3.5	x

Variables included in addition to TSP and SO_x: (1) population density, % nonwhite, % > 65, % poverty; (2) $\overline{SO_2}$, $\overline{NO_3}$, $\overline{NO_2}$, CO, $\overline{O_3}$ specific for age, sex, race, region, % divorced; (3) SO_2 emission, radiation, cigarettes, NO_2, NO_{3min}, $\overline{NO_3}$, HC emission; (4) > 65, % nonwhite, % poverty, birth rate, % housing before 1950, Mn, BαP, cigarette smoking (corrected from geometric to arithmetic mean); (5) population density, % > 65, income; (6) SO_2, population density, > high school, no. days with precipitation, no. days < 32° F; (7) income; (8) income, specific for age and whites; (x) not included in regression.

measured with SO_2 emission. If *all* sulfur levels (concentration *and* emission) were increased by 1 percent, then Schwing and McDonald predict a 0.08×10^{-5} increase in mortality with an average sulfate concentration of 12 $\mu g/m^3$. This could be interpreted as an "effective" coefficient $a_s = 0.7 \times 10^{-5}$ deaths $(\mu g/m^3)^{-1}$ which is small but positive.

Lipfert finds always a positive relationship with manganese and with particulates, but not always with sulfate.

There is a definite difference between the regressions of Lave and Seskin and of Bozzo et al. on the one hand, and of Lipfert on the other. The former find a strong correlation with sulfates and attribute the major health hazard to sulfates and other fine particulates. Lipfert finds some negative, though insignificant, correlations with sulfates. It is important to understand this difference in detail.

There is a difference between the sizes of the regions studied. Lave and Seskin took SMSAs; and Bozzo et al. took all the white populations by counties. Lipfert took only the big cities, and suggests that when we include the rest of the SMSAs the sulfate coefficient increases.

Lipfert argues that by comparing only cities he gets a more reliable result. However, because sulfate measurement nets are inadequate, and sulfates are a larger proportion of TSP and are more acidic in the suburban and rural areas, he might merely be blinding himself to an interesting and important effect.

Lipfert's view has some support from the critics of EPA's CHESS study. Finklea et al. [1974] had, by combining the data, found a positive relationship of asthma and other ailments with sulfate levels. However, this diminished when cities were examined separately. The situation is, in our view, unsatisfactory, for no specific explanation seems to exist for the difference. There seems to be agreement that some air pollutants cause health problems at low levels, but no agreement on which ones.

The question of existence of a threshold is even harder to determine. Both Lave and Seskin, and Lipfert, used a method of dummy variables to see whether there is a threshold. Their results were not definitive.

Therefore, many persons regard it as prudent to assume that no threshold exists, or that it is lower than East Coast ambient levels. By analogy with radiation, they argue that definite effects of radiation on people have been found at only ten times the ambient levels, yet the prudent policy is to assume no threshold for radiation-induced health effects. As we proceed further in this book, we follow some logical consequences of this view.

6.4 RANDOM AND CONFOUNDING VARIABLES

6.4.1 Definitions

We do not know the "correct" causative variables to describe mortality. Some of the variables chosen may be nearly correct; others are merely surrogates for some mixture of causative agents; and some true causes are probably unknown.

Factors that are uncorrelated with an air pollution concentration, but that decrease the precision of the measurement, are called random variables. If these variables cannot be accounted for, they reduce the significance of the coefficient, but cannot produce a spurious effect. Examples of random variables might be measurement error and monitor siting. If, for example, we have random errors in the air quality samples, we would underestimate the coefficient of air pollution-related mortality. In Section 6.4.3 we see an example of this in the case of tobacco cigarette smoking; an inadequate measure of smoking leads to a coefficient smaller than we know is correct from other data.

On the other hand, Lipfert [1978] criticized Lave and Seskin's air quality sampling for the 1960 regressions because the "minimum" data was obtained in different ways in different cities, and this could lead to a systematic error. This did not occur for Lave and Seskin's 1969 data, because by then EPA had a more complete data set.

To the extent we consider a buildup of adverse health effects over time, with a buildup of small pollution-related bodily malfunctions, Lave and Seskin may also have been observing adverse health effects caused by higher concentrations in the past. This could also result in an overestimate of the health effect coefficient. If, however, the unknown causative variables that are not included in the regression are correlated with air pollution, they can produce a spurious result on the regression equation, which is a noncausal effect. These effects might either increase or decrease the coefficient of air pollution-related mortality calculated from the data.

There are various possible confounding variables that ideally will be included in the regression equations. There are two different types of problems that can arise from these variables. First, if a confounding variable varies strictly proportionally with the air pollution variable of interest, it is logically impossible to separate the effect of the confounding variable from that of the air pollution variable from statistical studies. If it is correlated, but not completely so, separation is possible but becomes more difficult with increasing degree of correlation.

Second, we do not always have available death statistics in a form where they can be related to those variables. Death statistics do not include smoking habits, occupational history, or migration. Surrogate variables such as income can be used in the regression equation to replace a missing confounding variable such as health and nutrition. A surrogate variable is good if it varies proportionally with the real confounding variable. Then the air pollution effect is unlikely to be overestimated. A poor surrogate variable will effectively be a combination of no variable at all and a correct confounding variable; the statistical error of the coefficient of air pollution-related mortality will not be reduced much and a spurious effect can be present.

If we had chosen the density of coal burning as a surrogate, we would obviously "capture" a portion of the air pollution-related coefficient because air pollution is certainly correlated with coal burning. On the other hand, coal burning is also an indicator of other industrial activity. There are no hard-and-fast rules to describe the best variables to put into the regression equation. There is no substitute for judgment and common sense.

The obvious confounding variables are occupation, smoking, migration of healthy people to pleasant areas, and the "urban factor." We discuss them in turn below.

Finally, if other causes of death are known reliably, they can be included in a mechanistic way. This is discussed in Section 6.4.3 on tobacco cigarette smoking. We now proceed to discuss whether these confounding variables could plausibly account for the observed correlation between air pollution and mortality.

6.4.2 Potential Confounding Factors — Occupational Exposures

We first make a tentative assumption that occupational exposures are causing the deaths that the investigators spuriously attributed to air pollution. Because in the United States women make up a relatively small percentage (approximately 20 percent) of the labor force in heavy industry, in which most occupational exposures occur, we would expect the relationship between air pollution and mortality to be smaller for women than for men. Thus, if we find equivalent relationships between air pollution and mortality for men and women, we might legitimately discount the hypothesis that occupational exposures are confounding the relationships.

On the other hand, if we find a stronger relationship between air pollution and mortality for men than for women, we might hypothesize (1) that all of the effect is due to occupational exposures; (2) that air pollution and occupational exposures are both causes of premature death that are acting alternatively to produce the observed coefficient; (3) that air pollution and occupational exposures act synergistically and the mortality is a product of both exposures; (4)

that because women smoke less than men, a synergy between air pollution and smoking may account for the difference; or (5) that women are genetically more resistant to the effects of air pollution. On assumption (2), if community air pollution levels and severity or extent of the occupational exposures are linearly related, we might say

$$a_{men} = a_{occupation} + a_{pollution} \qquad (6-8)$$

$$a_{women} = 0.2 a_{occupation} + a_{pollution} \qquad (6-9)$$

whence

$$a_{occupation} = \frac{a_{men} - a_{women}}{1.2} \qquad (6-10)$$

or

$$a_{pollution} = 1.25 a_{women} - 0.25 a_{men} \qquad (6-11)$$

We develop this question with the full understanding that it is at best a crude approximation of the truth. It is intended simply to help us evaluate the possibility that occupational exposures are confounding our results.

We find from equations from Gregor [1976] that in all age groups the coefficients of apparent air pollution-induced mortality are lower for women than for men. A similar result is found by Finch and Morris [1977] in their analysis of Winkelstein's data. However, Lave and Seskin [1977, pp. 140–144] report mixed evidence for particulates and sulfates (Table 6–2). At first sight these data suggest that a part of the health effect is actually related to occupation. However, in Section 6.4.3 we point out that the data could be a consequence of the fact that more men smoke cigarettes than women, and also of synergism with cigarette smoking. Smoking itself is somewhat related to occupation, in that it is practiced more by blue-collar than by white-collar workers.

One might argue that data from large cities would not be likely to identify industrial exposures per se as having a significant effect on community mortality, although industry could be creating significant air pollution, because the percentage of the city population employed as operatives in heavy industry would be low. But in order to follow this line of reasoning to its logical conclusion we would need

Table 6-2. Ratio of mortality experience for different age groups. Adopted from OTA, 1979.

I: Data from Gregor [1976]

Age Group (years)	Ratio of Coefficients of TSP a_p (women)/a_p (men)
< 45	0.35
45–64	0.60
> 65	0.93
Expected for all occupational exposure	0.20

II: Data from Finch and Morris [1977]

Age Group (years)	Ratio of Coefficients of TSP a_p (women)/a_p (men)
50–69	0.58

III: Data from Lave and Seskin [1977]

Age Group (years)	Ratio of Coefficients of TSP a_p (women)/a_p (men)			Ratio of Coefficients of SO_4 a_s(women)/a_s (men)		
	white	nonwhite	all races	white	nonwhite	all races
overall	0.91	0.66	0.88	1.95	1.28	1.87
< 14	2.86	0.23	2.54	0.74	1.13	0.79
15–44	1.42	1.00	1.39	−6.25	0.17	−5.46
45–64	1.10	0.90	1.07	1.84	1.72	1.83
> 65	0.42	0.75	0.46	1.47	0.79	1.39

IV: Data from Mendelsohn and Orcutt [1978]

Age Group (years)	Ratio of Coefficients $a_{(women)}/a_{(men)}$	
	TSP	SO_4
< 1	1.38	2.82
1–4	2.00	0.28
5–17	−0.33	−2.00
18–24	0.15	0.68
25–44	0.33	0.42
45–64	0.15	0.48
> 65	0.76	0.69

data on the age-specific mortality for each industrial group and the percentage employment in each group by county, SMSA, or city.

Finally, data from the recent work of Mendelsohn and Orcutt [1978], which are based on data from 404 county groups, also demonstrates relatively consistent differences in the coefficients for men and women (see Table 6–2). An alternative explanation for these results is that women are less susceptible to the effects of air pollution than men. Although we see no good reason for an intrinsic difference, there might be a synergism with smoking as discussed below.

In Table 6–3 we report the results of further regressions of Lave and Seskin [1977] that include variables reflecting occupational mix and potentially accounting for industrial exposures. These new variables include the percentage of the work force employed in each of the following categories: agriculture, construction, manufacturing (durables and nondurables), transportation, trade, finance, education, white collar, and public administration. The additional set of variables also accounts for the percentages of unemployed, of males in the labor force, and of the work force using public transit. Table 6–3 summarizes the effect of the addition of these variables on the estimated coefficients of air pollution-induced mortality.

Because all variables have been simultaneously added, including the percentages of males in the labor force and of workers using pub-

Table 6–3. Accounting for occupational exposure [Data from Lave and Seskin 1977].

Mortality Variate	Pollutant	Ratio of Coefficient/Corrected for Occupation/Uncorrected
Unadjusted total mortality	min SO_4	0.37
	mean TSP	0.71
Race-adjusted infant mortality	min SO_4	0.93
Total cancers	mean TSP	0.93
	min SO_4	0.52
Total cardiovascular disease	min TSP	0.06
	min SO_4	0.43
Total respiratory disease	max TSP	0.24
	min TSP	0.77

lic transportation,[1] the changes in the mortality coefficient for air pollution are difficult to interpret. However, they do indicate the possibility that proper consideration of occupational exposures could further reduce calculated effects caused by air pollution. But it seems unlikely that the calculated effect of air pollution would be reduced to zero. Lave and Seskin [1977] argue that an alternative interpretation is that these occupational variables might also be surrogates for unmeasured industrial emissions of community air pollutants, and that the cause of the decrease in the coefficients of measured air pollutants is open to debate. One last insight may be provided by Lipfert [1977], who adds manganese and BαP to the list of explanatory variables in a study of 136 cities; his coefficient of TSP-induced mortality is smaller than the others. Lipfert argues that atmospheric concentrations of manganese may be a surrogate for occupational exposures; but it could also be a surrogate for other air pollution.

In summary, although it is possible that the apparent relationship between air pollution and mortality is confounded by occupational exposures, we are of the opinion that a more plausible interpretation is that only a portion of the calculated coefficients of air pollution-induced mortality may be explained by occupational exposures.

6.4.3   Tobacco Smoking as a Confounding Variable

The crude Japanese data shown in Figures 1−2 show that a health-related air pollution effect, if real, is about as bad in the most polluted cities as smoking a pack of 20 tobacco cigarettes a day.

As a thought experiment, if we wish to explain the difference in mortality entirely by difference in cigarette smoking, the cigarette smoking must vary with particulate levels and must vary widely. The particulate levels in the two extreme cities in the Lave and Seskin studies, Phoenix, Arizona and Augusta, Georgia, were 250 $\mu g/m^3$

---

1. The percentage of workers using public transportation may be acting as a surrogate for urbanization, or all of those factors associated with living in the Northeast. To quote Viren [1978] : "In 1970 the mean transit use of 243 SMSAs was 12 percent with the rank ordering of the top five SMSAs being New York (48%), Jersey City, N.J. (35.6%), Chicago (23.2%), Philadelphia (20.6%), and Boston (20%). Less than 25 SMSAs had usage exceeding the mean, which indicates that the true median usage was well below 12 percent and that the distribution was very skewed due to the large metropolitan SMSAs which are densely populated and have a historical pattern of public transit usage. . . . "

and 66 μg/m^3, respectively. According to the correlation, this would lead to a difference, M, in mortality

$$\Delta M = a_p (247-66)$$

$$= 7.1 \times 10^{-6} \times 181$$

$$= 1.3 \times 10^{-3} / \text{person-year} \qquad (6-12)$$

In the United States, nearly 700 billion cigarettes have been made every year, and lead to 350,000 deaths per year [SGR 1979] or there are 2 million cigarettes consumed per death. Then the difference in mortality between Phoenix and Augusta could be explained by a difference of smoking (1.3 $\times$ 10^{-3} deaths per person-year) $\times$ (2 million cigarettes per death) = 2.6 $\times$ 10^3 cigarettes per person-year, or about 7 cigarettes per person-days, or 14 cigarettes per smoker-day.

It is common knowledge that cigarette smoking started in the countryside later than in the towns, and many people smoke a pack of 20 cigarettes a day, but it seems unlikely that the smoking difference between Phoenix and Augusta is this large.

If we knew the smoking habits of all Americans we could include tobacco smoking either in a regression equation or as a mechanistic correction. Unfortunately, the death statistics do not include smoking habits, so that a direct correction is not possible.

Nonetheless some investigators have attempted to include crude measures of smoking habits. In a study of mortality in Allegheny County, Pennsylvania census tracts, Gregor [1976] included as a variable in a regression equation the estimated annual average per capita expenditure for cigarettes in each census tract based on national average expenditures of people of the same age, sex, race, and income class. He was unable to produce a "statistically significant" coefficient for the smoking variable, and deleted it from his published results. His method for estimating exposure may not have captured the variability in smoking behavior. We know that smoking habits influence mortality rates, and his result is that differences in community smoking habits do not make major differences in mortality rates. We compare the ratios of Gregor's male and female age-specific data (uncorrected for smoking) from Table 6–4, with the pattern of excess mortality among male and female smokers from Hammond [1966]; we note a striking similarity. Hammond defined a mortality ratio r_1 equal to the mortality among male smokers

Table 6–4. Maie/Female mortality ratios. Adopted from OTA 1979.

| Age (years) | Hammond [1966] | | | Gregor [1976] | |
	$r_1 = \dfrac{male\ smoker}{male\ nonsmoker}$	$r_2 = \dfrac{female\ smoker}{female\ nonsmoker}$	$r_3 = \dfrac{r_2}{r_1}$	Age (years)	$\dfrac{a_p\ (women)}{a_p\ (men)}$
35–44	1.89	1.13	0.60	< 45	0.35
45–54	2.28	1.26	0.55	45–64	0.60
55–64	1.83	1.20	0.66		
65–74	1.51	1.17	0.77	> 65	0.93
75–84	1.23	0.99	0.80		

divided by the mortality among male nonsmokers; r_2 the same ratio for females.

Hammond [1966] also defined the ratio $r_3 = r_2/r_1$, which he found to be less than 1, and by considering smokers only we have taken account of the fact that fewer women smoked cigarettes than men 20 years ago. The ratio a_p for women to a_p for men is also less than one, which is consistent with an interpretation that the male/female ratio of this coefficient might in part have the same origin, such as difference in sensitivity, as the male/female coefficient of smoking.

Schwing and McDonald [1976] included a smoking parameter for each SMSA in their data. The parameter was the estimated per capita packs per year, based on 1960 state cigarette sales statistics. Within a state all SMSAs were assumed to be characterized by identical community smoking habits. Using this surrogate for smoking exposure, Schwing and McDonald found the following diseases to be statistically significant, positively related with smoking: arteriosclerotic heart and coronary artery disease, respiratory cancers, influenza and pneumonia, cirrhosis of the liver, cancer of the large intestine, and emphysema. This evidence is qualitatively consistent with our present understanding of the effects of smoking. Quantitatively, however, the smoking variable explained only 2.5 percent of white male mortality whereas smoking is generally believed to cause 15 percent [SGR 1979]. Clearly, this surrogate has described only part of the effect of smoking, and leaves open the suspicion that the air pollution is spuriously describing another part. Unfortunately, Schwing and McDonald did not include a measure of particulate matter, nor did they report the change in their own regression coefficients that accompanied entrance of the smoking variable.

Lipfert [1978] included cigarette smoking in his regressions by using sales data, adjusted by a tax gradient between states to give a consumption estimate. He found a fairly consistent coefficient. This was 0.01 deaths per 1,000 population per year per pack-per-year smoked. This amounts to 10^{-5} deaths per pack, or 5×10^{-7} per cigarette. With 7×10^{11} cigarettes smoked per year, this equals the 350,000 deaths per year found by the Surgeon General [SGR 1979].

This agreement with the Surgeon General suggests that the cigarette-smoking variable in Lipfert's regression neither "captures" some of the air pollution effect, nor is it confounding the results.

But even this statement is not as definite as we would like. Lipfert's regression for lung cancer alone gave a smaller coefficient for smoking-induced lung cancer than we know to be the case (see Section 5.3).

In conclusion, it would seem that the argument that differences in community smoking habits are confounding the results is unlikely, particularly for the regression of Lipfert. But cigarette smoking may be affecting the data although not confounding them. Cigarette smokers may be more sensitive to air pollution than nonsmokers just as the cancer data suggest in Section 5.3. (The work of Cohen, Arai, and Brain [1979] showing that smoking inhibits long-term dust clearance in the lung was mentioned earlier.) This could either be a synergism in which the health effect is proportional to the product of the air pollution variable and the number of cigarettes smoked, or the health effect could be related to the sum of these two variable (appropriately normalized) but with a threshold or effective threshold such as shown in curve A_1 or A_2 of Figure 5−3. Thus air pollution by itself might not be enough to cause bronchial damage, but once the breathing is hindered by cigarette smoke, the air pollution can produce a large effect. The data of Figure 1−2, curve C suggest that this could be the case. This was also suggested by Finklea et al. [1974].

Because men smoke more than women, and because blue-collar workers smoke more than white-collar workers, such a synergism would explain the apparently greater sensitivity of men to the effects of air pollution. Before the associations between air pollution and increased mortality are interpreted causally, this possibility deserves further investigation. It would be particularly useful to run separate correlations for smokers and nonsmokers.

6.4.4   Migration as a Confounding Variable

The principal effects of migration are that (1) it can produce areas where the age distribution of the population differs considerably from the U.S. average, and (2) persons who migrate may have a different health status than those left behind. Florida and Arizona have a larger proportion of older persons than does Alaska, and therefore they have a higher mortality. Lave and Seskin [1977] attempted to correct the first problem by using the statistic available to them, the fraction of the population over 65 years of age as an additional variable in the regression equation.

We can test further whether this is adequate by examining the age-specific air pollution-mortality correlations. These are available for a limited set of the data. In Table 6−5, we show their data on the age-specific correlations; we weight these by the average U.S. population distribution for 1966 to arrive at a weighted-average coefficient. This is almost exactly the coefficient obtained in the regression equation by using the fraction of persons over 65 years of age as a societal variable, thereby validating the method. Bozzo et al. [1978] also studied the age-specific mortality and are in agreement with the above. When enough data are available, it is desirable to calculate the age-specific mortality and then this problem is avoided. There remains, however, the second possibility that these persons who migrate into an area are more healthy than the average. This could happen if the healthy tend to be wealthy and wise enough to move to less polluted areas. If we had detailed statistics for migration, we could redo the regressions only for those persons who have remained in the area for 30 years. However, these data do not exist.

Lave and Seskin [1977] used two methods to address this question. First they put an extra variable into the regression equation to

Table 6−5. Comparison of age-specific and crude mortalities. Adopted from OTA 1979.

	Regression Coefficient, a	
	Minimum SO$_4$	*Average TSP*
U.S. population weighted sum of age-specific coefficients	0.865	0.473
Total mortality, not age standardized	0.825	0.465
Based on following age-specific coefficients:		
Age Group (years)	*Minimum SO$_4$*	*Average TSP*
< 14	0.077	0.222
15−44	−0.041	0.124
45−64	0.927	0.676
> 65	6.373	2.137

describe migration, for which overall statistics are available. This variable is the ratio of the 1960 to the 1950 population in the area. This, as shown in Table 6−6, reduced the magnitude of the correlation for sulfates, and increased the correlation for particulates, but both within the statistical error. Second they broke the data into two parts: those SMSAs where more than 25 percent of the population had migrated from somewhere else, and those SMSAs where fewer than 25 percent had migrated, with results as shown in the second part of Table 6−6. This procedure reduces the magnitude of both the sulfate coefficient and the particulate coefficient by a factor of two. This suggests that there is an effect of migration, but not one that explains all of the air pollution correlation.

6.4.5 Potential Confounding Factors—
The Urban and Other Geographic Factors

It has long been known that death rates are higher in cities than in rural areas, and no one knows exactly why. This was discussed already for cancer in Section 5.3.

Various obvious possibilities suggest themselves: occupation, air pollution, income level, level of medical care, disease propagation, exercise, stress of commuting. No one of these seems to describe the

Table 6−6. Effects of differential migration. Adopted from OTA 1979.

| Pollutant | Coefficient a | | $a_{regional\ dummies}$ |
	Basic Regressions	Dummy Variables	$a_{basic\ regressions}$
min SO_4	0.631	0.376	0.60
TSP	0.452	0.559	1.24

| | Coefficients | | |
Pollutant	a_f Full Data Set	a_h (areas with high in-migration)	a_e (low in-migration)
min SO_4	.631	.305	.330
mean TSP	.452	.293	−.021

whole variation of death rates, which is over 20 percent. There are inadequate data to include all of these in a regression analysis. Lave and Seskin [1977] analyzed only metropolitan areas, including suburbia, so the urban–rural difference can be only a small confounding factor. In addition, Lave and Seskin attempted to describe all these factors except air pollution collectively by a population density variable. In equation 6–2 we see that less than 1 percent of the death rate in these SMSAs is described by this variable, but that air pollution describes 8 percent of the death rate.

Other factors in Lave and Seskin's regression that contribute to the urban factor are the fraction of nonwhites, and those older than 65 years, in the population. To some extent these variables might be surrogates for this urban factor. This extent, however, cannot be large because as noted in Section 6.3 the air pollution factors for the average population are the same as the weighted average of the age-specific air-pollution effect.

Viren [1978] has argued that there are geographically related serial factors that may produce spurious associations between air pollution and mortality:

... cities in the Southwestern and Pacific regions of the country show age-adjusted ... mortality rates well below those of the urban Northeast and Middle Atlantic regions. ... Sulfates, SO_2, NO_2, and ozone all tend to vary consistently across the country. Sulfates and SO_2 have traditionally been higher in concentration in the East rather than the West. NO_2 and ozone (oxidants) generally have higher concentrations in the West rather than the East. ...

This leads immediately to the correlation observed. But, Viren notes [1978]

... the percent of workers using mass transit to go to work ("transit") follows an increasing West-to-East gradient. Age of housing follows an increasing West-to-East gradient, as does average age (of the population), birth rate, and many other socio-economic and demographic variables. ... the results (of cross-sectional studies) may be artificial due to inadequate control for very broad geographic covariation of numerous social, biological, and physical variables.

For Viren's explanation to be convincing, some socioeconomic or physical factor or factors must be suggested that cause premature mortality and that systematically vary northeast to southwest. In addition to Viren's list, it would be reasonable to argue that factors such as exercise in the glorious west, climate, and urbanization may have an east-to-west gradient and can influence mortality. Because direct measures of many such variables are not available, most of the cross-sectional studies attempt to control for these variables using socioeconomic surrogates. Viren's argument would be supported if some unaccounted-for causal factor or factors varies or vary with air pollution and not, or only poorly, with the socioeconomic surrogates.

The air-pollution coefficient would be overestimated if the causal factor increased west to east, the direction of increasing air pollution; but the air-pollution coefficient would be underestimated if the factor increased east to west.

One method of investigating whether there is an undiscovered confounding geographic factor is by looking for an effect of air pollution within a single region of the country. The broad geographic patterns postulated by Viren would probably not significantly influence mortality within a region. In Table 6−7 we present the regional coefficients for minimum sulfates by Viren.

Although the wide variation in the signs and magnitudes of the sulfate coefficients would initially appear to offer support for Viren's

Table 6−7. Variation of sulfate coefficient with geographic region. Adopted from OTA 1979.

Census Region	No. SMSAs	Estimated Sulfate Coefficient, a_s
		$\left(\dfrac{\text{deaths}/100,000/\text{yr}}{\mu g/m^3 \text{ minimum sulfate}} \right)$
New England	11	−0.38
East North Central	23	+3.84
Middle Atlantic	13	−4.95
South Atlantic	25	30.89
East South Central	8	−4.44
West South Central	14	13.28
West North Central	9	58.16
Mountain	5	−4.37
Pacific	9	−8.03
Simple average		$\bar{a}_s = 9.33$

Source: J.R. Viren, *Cross−Sectional Estimates of Mortality Due to Fossil Fuel Pollutants: A Case for Spurious Association.* Washington, D.C.: U.S. Department of Energy, for the Division of Policy Analysis, 1978.

argument, individual coefficients estimated from such small samples are statistically unstable. Although the regional units are not commensurate, a simple average of their coefficients may be instructive. In this case, the average coefficient (a_s) is 9.33×10^{-5}. This coefficient of minimum sulfates is converted to arithmetic mean sulfate levels by the method of Larsen [1969]. Using a standard geometric deviation of 1.6, we find

$$\Delta M = 3.26 \times 10^{-5} \, S_{\text{average}}$$

This estimate is very close to Lave and Seskin's estimate (Table 6–1) of 3.3×10^{-5} using the full data set. Thus, evidence that intuitively appeared to support the hypothesis of geographic confounding is consistent with no confounding.

Finally, Lipfert [1978], Mendelsohn and Orcutt [1978], and Lave and Seskin [1977] have used various methods to explore this problem. Lipfert [1978] compared results disaggregated into four census zones (North Central, Northeast, South, and West) with aggregated results. The total air pollution effect was only slightly reduced.

Lave and Seskin's similar results were reported in Section 6.4.4 on differential migration. Although the magnitude (and the sign, in two out of fourteen cases) of the calculated coefficients shift when Mendelsohn and Orcutt introduce regional dummy variables, there is no dramatic or consistent shift of the coefficients toward zero.

The validity of arguments of geographic confounding are dependent upon the assumption that there exists a set of unmeasured, or poorly measured, causal factors that vary geographically with air pollution so as to overestimate the air pollution effect. No consistent set of such causal factors has been suggested.

6.4.6   Potential Confounding Factors—Summary

We have discussed the evidence that one of the major factors (occupation, smoking, migration, or geography) can confound the results of air pollution studies. It is also possible that one or more of these factors is synergistic with air pollution and that, for example, an air pollution effect is large for smokers and smaller for nonsmokers, just as the lung cancer effect for tobacco smokers seems to be four times that for nonsmokers (Section 5.5). This could explain the larger apparent air pollution coefficient for men than women, because only recently have women smoked almost as much as men.

It is also possible that each of these four confounding factors acts independently, each explaining a part of the correlation.

Thus it is possible for reasonable men to disagree on the meaning of these results. There may be a large effect of air pollution on health but it has not been proved to everyone's satisfaction, nor can it be disproved to anyone's satisfaction.

6.5 TIME-SERIES STUDIES

In addition to the cross-sectional studies described in the previous sections, there have been more sophisticated time-series studies than the Oslo series described in Chapter 1. Several of these are described in the work of Lave and Seskin [1977].

There is no direct way of relating the time-series studies to the cross-section studies without a model by which air pollution operates. Conversely a comparison of the two types of study can tell us a great deal about the way in which air pollution causes mortality, if in fact it does. Unfortunately there has been very little work on models of air pollution-related mortality and little on the comparison of data. This work is breaking new ground.

We outline several possibilities here.

1. Suppose that air pollution is causally and instantly related to mortality, without a time delay, and that this relationship is proportional. We then expect the relationship of mortality to air pollution to be the same whether derived from comparison pollution concentrations at different places or at different times. Then the proportional relation implies that each little piece of pollution has its irritant effect on the probability of death.

We note here that Lave and Seskin's correlations, derived from a time-series study of the weekly mortality rate in Chicago as a function of air pollution, are 20 percent larger than those derived from the cross-sectional geographic variation of mortality in agreement with this model; but the time-series data of Schimmel, and the episodes give smaller effects (Figure 6−1).

2. If low levels of air pollution have no effect on health, and if the threshold at which effects begin is different for different people, we might expect that when the air pollution varies with time, the more sensitive people die first; we would expect, then, a reduction in mortality as soon as the air pollution levels fall again because the

Figure 6-1. Graph, modified from Lipfert [1978], showing the total excess deaths as a function of smoke concentrations. This shows a definite dose-response relationship with no obvious threshold above the ambient New England levels.

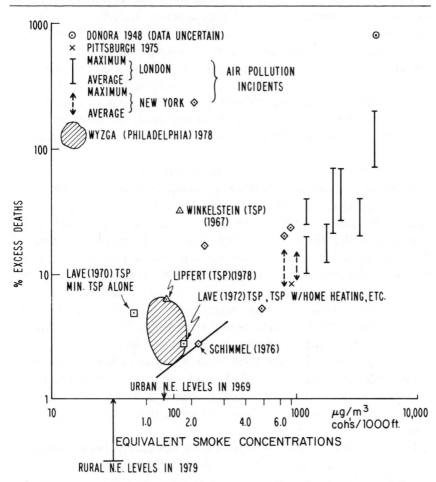

sensitive people, who were on the verge of death, have died already. This effect is known colloquially as harvesting. As studies are made with air pollution averaged over larger times, the correlation will then go down. And for the cross-sectional series where the averaging time is one year, the effect would readily vanish. The fact that the correlation from the cross-sectional series is 80 percent of that in the time series suggests that harvesting is at most 20 percent of the effect.

3. If there is a long time delay between the pollution insult and death — as in cancer — we would expect the cross-sectional studies to give a larger correlation than the time series. Indeed, a time-series study on cancer is almost impossible. There is a twenty-year time delay between the rise in cigarette smoking and the rise in lung cancer. A time-series study that merely compares cancer rates between summer and winter months could not possibly show an effect.

Data on air pollution episodes (Figure 5−1) suggest that the health effect observed lags behind the pollution insult by three weeks or less, although some effects at nearly one year were observed at Donora.

These episodes should be able to be directly related to the time-series studies, which can be regarded as repeated small episodes. This was done in Chapter 8 of Wilson and Jones [1974]. In Figure 6−1, taken from Lipfert [1978], we show a better attempt to relate the episodes to the time-series result. Also on the graph are some of the cross-sectional series studies we discussed earlier. A clear dose-response relationship is seen, with a larger effect seen from the cross-sectional series as suggested in comment 3, in contradistinction to evidence from episodes. This suggests either two different causes for the acute (prompt) effects and the chronic (long-term) effects, or that the cross-sectional series are spuriously high.

It is interesting to relate these models to our earlier discussion of the expected preconceptions of physical scientists and biologists. The biologist, with his threshold, might believe in harvesting as soon as the pollutant exceeds that threshold more than might the physical scientist.

Unfortunately, study of this subject has barely begun although some discussion is contained in Lipfert [1978], Wyzga [1978], Lave and Seskin [1977], and Schimmel and Greenberg [1973]. Any model to describe the correlations between air pollution and mortality must describe both sets — the cross-sectional studies and the time-series studies.

6.6 SUBJECTIVE COEFFICIENT OF AIR POLLUTION-INDUCED MORTALITY

Because various estimates of a, the coefficient of air pollution-induced mortality, are not random samples from the true distribution of a, averaging a among several different authors, as if they were independent measures, is not useful. Morgan et al. [1978a, b] have suggested that a subjective assessment may be useful in this instance. In Figure 6−2 we show their comparison between the classical (statistical) approach and the subjective approach; the dashed curve is their combination of all statistical estimates of a. By adding an understanding of the physical and biological processes involved and subjective assessment of the relative quality of the several estimates they produced the subjective distribution shown as the solid curve. Although they believed that it is very unlikely that sulfate air pollution reduces the mortality rate, they were not prepared to reject this possibility out of hand.

Accordingly, they placed 1 percent of the probability to the left of the origin. They believed that there was a finite possibility that sulfate air pollution has no effect on mortality, and they placed 14

Figure 6−2. Uncertainty in the slope of a linear damage function expressed as probability density functions. The dashed line is a student's 't' distribution using four regression studies as independent observations. The solid curve is a subjective distribution developed by Morgan, et al. [1978a].

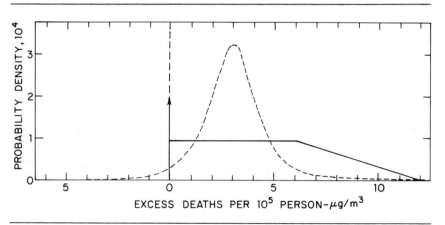

percent of the total probability in a delta function at the origin. Their assessment of all available evidence led them to believe that for positive slopes the classical estimate of uncertainty is too narrow, and so they increased it.

It is likely that a curve of this type can be derived from a "Delphi Method" survey of those who have studied the field in detail, where everyone would be asked his opinion and some sort of average taken. In that sense, therefore, it might be regarded as the best available indication of the magnitude of the effect of air pollution on health.

6.7 SULFATES AS A SURROGATE

The epidemiologic surveys of this chapter do not tell us which of the pollutants cause the mortality, even if we accept that one of them does. For purposes of discussion of mitigation methods and incentives for pollution reduction, this is important to know. In this section we argue that for the next decade sulfates and fine particles are a good surrogate, but perhaps their use understates the health hazard.

We consider the three possibilities of Section 5.6: first, that sulfates are irritant solely because of their particulate nature, and that the difference between sulfuric acid ($H_2 SO_4$) and other sulfates noted by Amdur's experiments [1968, 1973, 1975] with guinea pigs; second, it is the heavy element cation that is important; and third, only high concentrations are important.

In each case we will assume that the particulates are irritant not in direct proportion to their weight, but that smaller particulates are more irritant as suggested by Amdur's experiments.

To complete the analysis in the subsequent sections and chapters we neglect the effect of nitrogen oxides. We also neglect polycyclic aromatic hydrocarbons (PAH) because we have already discussed their carcinogenic effect in Section 3.1.9.

We suggest the continued use of sulfates as a surrogate, with periodic re-examination in cases where the sulfur emissions are reduced in greatly different proportions than the particulates, as well as continued examination of the role of heavy elements. In particular, if the indications of Lipfert that sulfate is not the hazard are confirmed, then respirable particulates of TSP must be used.

There are other possibilities. One is that particulates and SO_2 have a synergistic effect in such a way that the health hazard is proportional to the product of the concentrations of particulates and SO_2 or of some other combination of pollutants.

6.8 ANALYSIS

Anyone who believes that there is no adverse effect of low-level air pollution will presumably be willing to continue to allow such pollution. Those who believe that it causes a large hazard will want to prevent major increases in fossil fuel burning unless controls are increased to reduce emissions.

Suppose that the action is to demand a pollution charge—of perhaps $1,000,000 per life shortened by air pollution—then if we average the actions demanded by these various citizens, we average the charge levied. If further we assume proportionality of health hazard with concentration of pollutant we can average the opinions of the health hazard we discussed in Section 6.3. Stated in another way, it does not matter to an individual whether the probability of his dying from air pollution is half because it is equally likely that the optimist and the pessimist are right, or whether it is because there is an intrinsic uncertainty in the physical or biological process.

For the purposes, therefore, of much of the discussion in the subsequent chapters, we take from Section 6.6 a health hazard of 3.5 deaths per 100,000 persons per μg of SO_x exposure ($a = 3.5 \times 10^{-5}$ (μg/m^3)). This figure includes all health hazards; SO_4 is used as a surrogate for all other air pollutants.

In Section 2.3.4 we found a population-weighted exposure ($\int c\,P(c)\,dc$) over the United States of 1,520 (million persons) $\times$ (μg/m^3) for 22 million tons of SO_2 emitted per year [Meyers et al. 1978]. Multiplying these two figures (1,520 $\times$ 10 and 3.5/10 together, we find a central estimate of 53,000 persons per year dying due to air pollution. If each person dies, say, 17 years earlier than otherwise, there is a total of 17 $\times$ 53,000 or 900,000 years of life lost per year. This arises from emission of 22 million tons of SO_x, or 11 million tons of elemental sulfur, and leads to 4,800 lives per million tons of elemental sulfur, or 2.15 lives per million pounds of elemental sulfur emitted.

We carry these suggested numbers to the next chapters as a guide to the numbers and magnitudes involved.

6.9 RECOMMENDATIONS FOR THE FUTURE

1. The measurement of the air pollution variables is inadequate. There have been major criticisms of the sulfate measurements used by Lave and Seskin; that should not mean that we ignore Lave and Seskin's work but that we should insist on better and more complete pollutant measurements in the future, together with computer programs such as we discussed in Chapter 2 for interpolation.

2. The apparent discrepancies between Lave and Seskin and Bozzo et al. on the one hand, and Lipfert on the other, must be resolved.

3. Many of the problems of dealing with confounding variables can be avoided if we run separate regressions for subgroups of the population. We noted how the problems of relating the mortality to the age distribution of the population can be avoided by calculating the age-specific correlation between mortality and air pollution. The number of persons involved in the study should be sufficiently great that the population sample could be divided into smaller groups.

4. Because cigarette smoking is such an important factor in U.S. mortality we would like to calculate the air pollution correlations for nonsmokers, light smokers, and heavy smokers separately. For this purpose it would be nice to have data on present and past smoking habits. This information could be collated with census data or recorded on death certificates, but the reliability would be open to question.

5. We would like occupational history to be included in the correlation; it is not included with death statistics so that we cannot easily include corrections for occupation. Even radiation records, which are the best kept of occupational exposure records, are kept primarily to protect an employer from charges of negligence, and they do not include medical x-rays and are not included with death records. Migration data could also be included so that death rates could be compared for those who have always lived in one area and those who have moved once or more.

6. One consequence of U.S. national policy is that the rapid expansion of coal burning could lead to large, nearly uniform con-

centration of secondary transformation products of sulfates, nitrates, and fine particulates over large geographic regions. If this occurs while a variation of pollutant concentration within a region is reduced, there is less chance of finding a positive coefficient associating air pollution to health even if a causal relation really exists. Further, it would be impossible to reject the hypothesis that any variation in mortality that remained across large areas was confounded by other unexplained factors that also vary with geography. There should therefore be a sense of urgency in carrying out this research.

7 OPTIONS FOR POLLUTION REDUCTION

In this chapter we are concerned principally with hardware and non-hardware alternatives for reducing the exposure of people to particulates and to sulfur compounds. The hardware options are (1) the use of tall stacks, which disperse the flue gases but which do not reduce the total mass of pollutants; (2) various devices for removing particulates, of which the electrostatic precipitator, the traditional mechanism, is being improved, but the fabric filter appears to be the best choice for the immediate future; and (3) methods of removing sulfur from flue gases, of which methods wet scrubbing is virtually the only choice open at present, but fluidized-bed combustion seems to be the most promising new technology. If high removal efficiencies for sulfur can be obtained, the use of treated fabric filters may turn out to be the optimum near-future approach. The removal of nitrogen oxides from flue gases will also be reviewed.

We shall also mention the prospects for producing clean-burning fuels from coal, oil, and other fuel minerals. If these clean fuels can be produced economically, their use would not only virtually eliminate particulate and sulfur emissions from traditional generating stations, but would permit new, highly efficient, and dispersed systems such as fuel cells and high-temperature gas turbines to be employed.

Two nonhardware options for pollution reduction will also be discussed. The first can be applied immediately to existing plants:

intermittent control, whereby the peak allowable station emissions are varied in inverse proportion to the pollution levels in the area affected by the discharge, or on signal of certain meteorologic predictions. The second strategy can be used only for new plants: to site them in areas where the emissions will normally affect very few people, for example, the U.S. East Coast, on an offshore island.

7.1 THE USE OF TALL CHIMNEY STACKS

Till more effectual methods can take place, it would be of great service, to oblige all those trades which make use of large fires, to carry their chimneys much higher into the air than they are at present; this expedient would frequently help to convey the smoke way above the buildings and in a great measure disperse it into distant parts without falling on the houses below. [Evelyn 1661]

When the health hazard was thought to be entirely due to SO_2, and when a threshold was believed to exist below which no health hazard is present, all that seemed necessary was to reduce the ground-level concentration below the threshold, which if our regulatory bodies have been well advised, is the air-quality standard. This reduction could be accomplished by discharging the flue gases through tall stacks to allow for adequate dispersion in a mass of cleaner air.

In Section 2.3 various dispersion formulas for calculating the concentration of a pollutant at ground level from a stack of height h were discussed. With the inclusion of experimental data and with some approximations, we arrived at the following relation between the maximum ground-level concentrations C_{max} and stack height h:

$$C_{max} \propto 1/h^{1.8}$$

Therefore if the air quality standard corresponds to a physical threshold for primary emissions, all that is necessary to meet them is to have an adequate stack height. The advantage of a high stack is somewhat greater than these formulas suggest. A plume from a stack 1,000 feet high can penetrate an inversion layer and avoid being trapped in stagnant air beneath.

This advantage of tall stacks has been the basis of antipollution technology in Great Britain and has been particularly stressed in many articles by engineers from the Central Electricity Generating

Board (CEGB) [Clarke, Lucas, and Ross 1970; Ross and Franken-berg 1971].

It is probable that if sulfur dioxide (SO_2) were the only problem, tall stacks would be enough to solve it. But we have seen that SO_2 converts to sulfate (Section 3.1) and that sulfates have deleterious effects (Section 4.2 and Chapter 5). Nitrates (Section 3.2) are also present in addition to various heavy elements (Sections 2.2 and 5.4). The health effects of low levels of sulfates and particulates that accumulate from many plants form the scientific issue that makes the discussion important.

Tall stacks can completely solve the pollution problem for a small island where the pollution blows out to sea. This is probably the case, for instance, for Puerto Rico or New Zealand. However, Britain is on the borders of the continent of Europe and contributes sulfur to the continent, as well as receiving some in return.

The cost of tall stacks is moderate. Ross and Frankenberg [1971] quoted a cost in Great Britain of £650,000 for the 650 foot (200 m) Eggborough chimney (2,000 MW). This is $2.2 million at 1971 prices or $1.90/kw in 1979 dollars. The 1,200 foot (366 m) chimney at Mitchell cost $2.20 per kw in 1971 dollars or $4.4/kw in 1979. This is still a small fraction of the $600/kw of a typical coal-fired power station completed in 1978 in the United States.

7.2 FLUE GAS DESULFURIZATION

In this section we review the three principal alternative methods of removing sulfur from flue gases. It has long been realized that burning coal cleanly is a good idea as the following quotation suggests: "A method of charring sea-coal, so as to divest it of its smoke, and yet leave it serviceable for many purposes, should be made the object of a very strict enquiry and premiums should be given to those who were successful in it" [Evelyn 1661]. In the next section we shall discuss coal conversion processes.

7.2.1 Alkali Scrubbing

A typical scrubber is illustrated in Figure 7-1. A slurry of an alkali with water is sprayed down a tower up which is ducted the hot

Figure 7—1.   Diagram of flue gas scrubber.

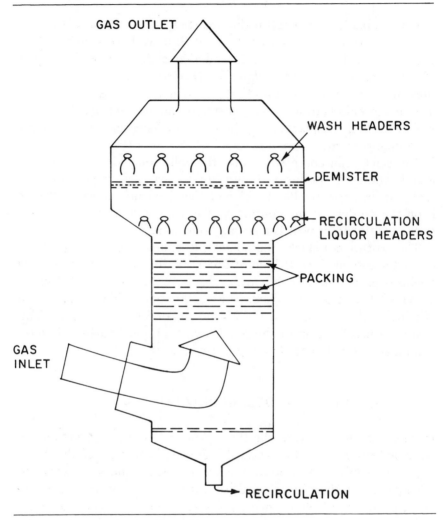

flue gases. Heat and mass are exchanged in this countercurrent contact. The slurry is collected at the bottom of the tower and is usually pumped up to be injected a second time. After this second passage, the slurry is passed to a thickening tank in which some of the water is removed, and the resulting sludge is disposed of in a pond or landfill.

The flue gas emerges from the scrubber in a cold, saturated state. A heat exchanger fed with extraction steam is normally fitted to raise the temperature by approximately $25°C$ ($45°F$) to give more buoyancy to the gas; to reduce condensation, corrosion, and deposition on downstream induced-draft fans; and to decrease the visibility of the plume.

These are the common essentials of flue-gas scrubbing. There are many alternative designs or operational conditions. The alkali used may be calcium oxide (lime or quicklime, CaO) or calcium carbonate (marble, limestone, chalk, calcite, aragonite, $CaCO_3$), or sodium carbonate (Na_2CO_3), sodium sulfate (Na_2SO_4), sodium hydroxide ($NaOH$), ammonium sulfate (($NH_4)_2 SO_4$), or ammonium hydroxide ($NH_4 OH$), with lime or limestone. The liquor may be saturated or unsaturated with respect to calcium dihydrate, or gypsum, $CaSO_4 \cdot 2H_2O$. The coal, besides having a high or a low sulfur content, may have a high or low chlorine content. Chlorine contributes to the acidity of the slurries and of the sludge and damages equipment, and also retards the rate of oxidation of the calcium sulfite to calcium sulfate.

Scrubber operation may be steady, or it may be set to follow load variations. The water may or may not be recycled. By 1979, scrubbers have not yet operated in all parts of this multidimensional matrix, and we do not know whether or not existing scrubbers are at near optimum conditions. However, scrubbers constitute the most readily available technology for flue gas desulfurization, albeit with a high first cost and operating cost (Table 7−1).

The operating costs include the energy costs of pumping the flue gases through the added pressure drop, of pumping the sludge, and of reheating the stack gas after scrubbing. There are also environmental costs. A large area has to be set aside for receiving the spent sludge, which has a low strength and appears at present to condemn the land for other uses for the foreseeable future. A possibly more expensive alternative is to fix the sludge in a cement-like solid, which, besides allowing future use of the site, reduces the threat to the ground water.

7.2.2 Regenerative Scrubbing

The alkali scrubbers just described are sometimes referred to as "throwaway" systems. An alternative approach is the regenerative

Table 7-1.   Ten clean ways to burn coal.

Method	Sulfur Removed (%)
Direct firing, pulverized low-sulfur coal	none
Direct firing, pulverized high-sulfur coal with wet alkali scrubbing and on-site sludge disposal	80-90
Direct firing, pulverized high-sulfur coal with regenerative scrubbing (using hydrogen produced on site) and reduction of gas to elemental sulfur	90-95
Direct firing, pulverized coal in a fluidized bed at atmospheric pressure with dry limestone added to the bed	80-90
Direct firing, pulverized solvent-refined coal	70-90
Liquid firing, petroleum-type fuel	90-95
Low-Btu gas firing (100-150 Btu/scf) after gasification in moving-bed, dry ash Lurgi gasifier and Selexol process for acid gas removal	> 95
Medium-Btu gas firing ($\sim$ 300 Btu/scf) after gasification in slagging moving-bed gasifier and Selexol process for acid gas removal	> 95
Low-Btu gas firing (100-150 Btu/scf) after gasification in atmospheric, two-stage entrained gasifier and Stretford process for acid gas removal	> 95
Medium-Btu gas firing ($\sim$ 300 Btu/scf) after gasification in pressurized, two-stage entrained gasifier and Selexol process for acid gas removal	> 95

Source: Combustion 48 (19), (April 1977): 23-25, "Ten ways to clean coal and their busbar costs," a review of an article in the *EPRI Journal*, November 1976.

type of system in which the sulfur is recovered. Regeneration of sulfur into three forms is being investigated: elemental sulfur; sulfuric acid (H_2SO_4), and gypsum. Only one-tenth as much elemental sulfur, by weight, is produced per ton of coal of given specifications as that of the wet sludge that would be produced by the alkali-scrubbing method; or one-third the weight if H_2SO_4 is produced, or 60 percent of the weight of wet sludge if gypsum is the product.

The technology required for the regenerative scrubbers is much more complex than for the comparatively simple alkali systems, and none has yet been thoroughly demonstrated in a full-scale commercial coal-fired power plant (see Table 7-1). The costs of electricity produced using high-sulfur coal are reckoned to be 10 percent higher at the power station (busbar cost) if regenerative scrubbing rather than alkali scrubbing is used. However, these estimates could be proved inaccurate by improvements in technology, or by changes in the costs of treating and disposing of the solid wastes from the throwaway process.

7.2.3 Dry Absorption

One form of regenerable scrubbing is dry absorption using activated charcoal derived from coal itself. This method was patented in Britain in 1879 but has been under intensive development only recently. An advanced process based on special activated-coke absorbent is that developed by Bergbau Forschung (BF) in Essen, West Germany. The gases to be cleaned pass up a tower down which is fed a moving bed of activated coke. SO_2 is absorbed, oxidized to H_2SO_4 when combined with the water vapor that is present and the coke emerges from the base of the absorber saturated with H_2SO_4. Particulates are also filtered out.

The coke is then regenerated by heating in an inert atmosphere using a moving sand bed at 595–650°C (1,100–1,200°F). The gases are reduced back to SO_2 and may be fed to a recovery process, such as Foster Wheeler's RESOX producing elemental sulfur (Figure 7-2).

Dry absorption would have two major environmental advantages over wet scrubbing, if it can be proved to be fully feasible. A potentially useful product of low volume, sulfur, is produced instead of a large volume of noxious sludge. Energy is saved because no reheating of the flue gases is needed.

A spray-dry process will be mentioned later in connection with fabric filtration.

7.2.4 Fluidized-Bed Combustion

A fluidized bed is a collection of particulate solids, sand for instance, confined in the lower part of a tall vessel or tube that rests on some form of porous surface through which a fluid is blown with a veloc-

Figure 7–2.   RESOX sulfur recovery process.

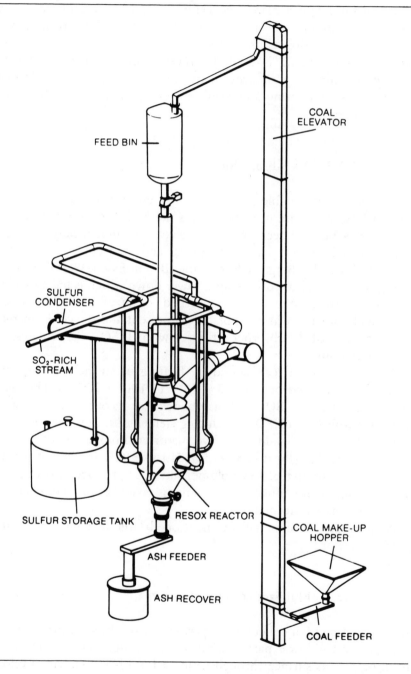

Figure 7–3.   A fluidized-bed combustor (schematic).

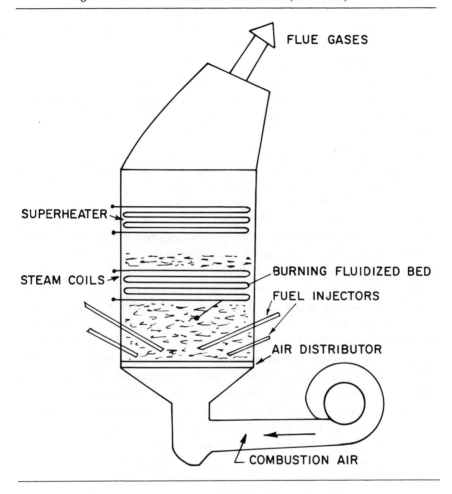

ity such that the solid particles have mobility (Figure 7–3). The upward fluid velocity must still be low enough, however, that a well-defined upper limit of the bed exists. It has been found possible to burn poor solid fuels, such as lignite, very efficiently when they are fluidized with air or oxygen. When an alkali, such as limestone, is added to the bed, a large proportion, for example over 90 percent, of the sulfur can be absorbed. These solids can be regenerated and the sulfur recovered for conversion into the usual possible by-products. A problem presently being investigated with regeneration is excessive attrition of the limestone, which would limit the number of regenerative cycles possible. Fluidized-bed steam generators are likely to have a first cost that is 20 percent less than conventional units, and to have higher availability because of their modular construction [Bryers 1972] (Figure 7-4).

Fluidized-bed combustion as a means of eliminating sulfur from flue gases would have large potential advantages in economy of investment and in energy savings if certain developments pay off. The savings in investment come about through combining the fluidized-bed burner with high-pressure combustion. This was pioneered in Switzerland and sold as the Velox boiler: a compressor feeds the combustion air to the steam generator, the whole of which, including the combustion chamber, is built as a pressure vessel. After leaving the combustion chamber, the gases transfer heat to the steam-raising surfaces, and then expand through a gas turbine whose only function is to power the compressor. This system does not necessarily increase efficiency, but it greatly reduces the size of the steam generator, and has been used on U.S. naval vessels.

However, it is possible to take more power out from the gas expander and less through the steam turbine. The limit to the development of gas turbine power from coal comes when the turbine inlet temperature is high enough for certain ash constituents to be sticky and for corrosion of and deposition on the turbine surfaces to occur. This limit will be removed if coal gasifiers are developed to work reliably on a large scale. By adding steam and reducing the air supply, a shift reaction can be induced, with the consequent production of a small volume of high-pressure fuel gas. This can be cleaned regeneratively and fed directly to a gas turbine, which would be able to achieve far higher thermal efficiencies than steam cycles.

7.3 REMOVAL OF PARTICULATES
FROM FLUE GASES

Only two or three decades ago, air pollution was almost synonymous, in the minds of most people, with soot particles from smoke stacks. The quantity emitted was so large that the mass, settling on and around the cities and turning white buildings black within a few years, came to dominate thinking. When typical coal is burned in an uncontrolled furnace the fly ash that is discharged from the stack can amount to one-tenth of the coal by mass.

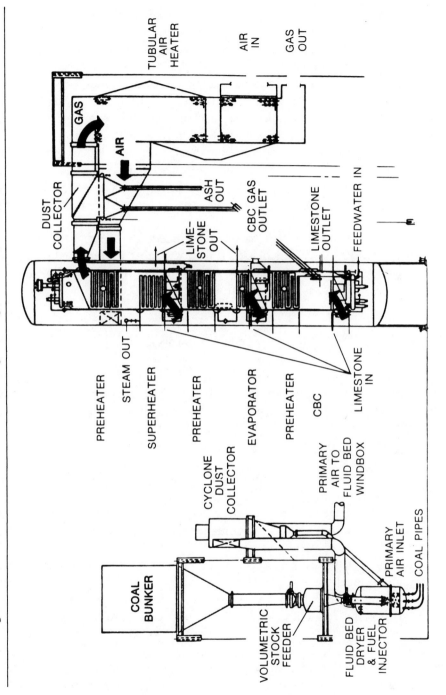

Figure 7–4. Fluidized-bed sulfur removal-steam generating plant. Adapted from (Bryers 1972).

The earliest control system used to remove particulates from flue gas was a settling chamber, simply a section of the exhaust duct with an enlarged cross-section so that the larger particles would respond to the steady gravitational forces rather than the now reduced aerodynamic entrainment forces. Later, the body forces were increased to well above the gravitational level by turning or swirling the flow to throw out the particulates in cyclones. Only the largest were removed by these means, and the improvement in the ambient-air quality was more aesthetic than health-related. Up to 80 percent, by mass, of the fly ash would be removed in cyclone air cleaners. As stated in Chapters 2–5, the smaller particles, generally below 2 micrometers (μm) in diameter, pass through the trachea into the lungs.

The next technology to be introduced, the electrostatic precipitator (ESP), described below, is now the most widely used means for cleaning stack gases. It may be supplanted by the fabric filter, or baghouse, for reasons to be explained in Section 7.3.2.

7.3.1   Electrostatic Precipitators

The electrostatic forces that are used (instead of the acceleration forces of settling chambers and cyclones) are induced in the entrained particles by passing the flue gases into parallel passages between vertical plates or within vertical tubes. A high voltage on a central wire or plate charges the particles, the larger of which reach the plates and adhere there through electrostatic forces. Periodically the plates or tubes are rapped while closed off from the gas flow, and the collected dust is removed.

Modern electrostatic precipitators have refinements such as multiple fields in series to trap the smaller particles. The attractive features of precipitators are that the energy consumption and the imposed pressure drops are small, while the efficiency on a total-mass basis is high, typically 99 percent. The gas stream remains in the dry state, minimizing downstream corrosion and plume opacity, and also minimizing any tendency of the plume to sink, all problems arising from the use of wet scrubbers. Precipitators are also relatively compact. The first cost, which formerly seemed large when it was virtually the only pollution-control system added to an electricity-generating station, now is small compared with sulfur-removal systems ($43/kw

Figure 7–5.   Effect of gas temperature and coal type on back-corona tendency.

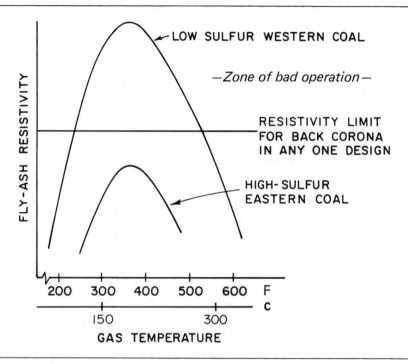

versus \$135/kw for wet alkaline scrubbers, both under optimum conditions in new power plants).

Despite these many advantages, electrostatic precipitators have problems. One is that the small fraction of the particles by mass that pass through are a large proportion of the total respirable particulates, and it is upon these that public health attention is principally focused today (Section 4.2.5). Another problem is that the high resistivity of the fly ash from low-sulfur (generally western U.S.) coal at the temperatures at which precipitators are generally operated (about 300°F or 150°C) can cause a blocking effect known as back corona. The variation of resistivity with temperature and with type of coal is shown qualitatively in Figure 7–5, with a typical back corona limit indicated. Above this limit the precipitator will not work properly.

The poor performance of some electrostatic precipitators when used with western coals, coupled with the very stringent county or

Figure 7–6.   Diagrammatic representation of water, air, and gas circuits in a steam generator.

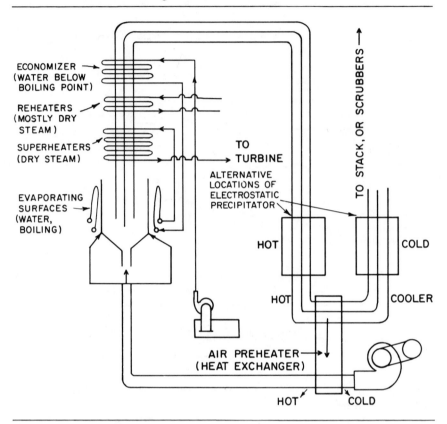

state limits on SO_2 emissions in western states such as New Mexico, Nevada, and Wyoming, has led to the use of scrubbers principally for particulate removal. The high alkalinity of the fly ash from western coal has acted in place of a reagent in these scrubbers so that the SO_2 is simultaneously reduced.

The back corona limit may be raised by sizing the equipment for low current densities, which means that its size must be increased. Alternatively, the precipitator can be designed to operate on the upstream, hot side of the air preheater, instead of the normal downstream, cold side (Figure 7–6). In this location the flue gases have a lower density, so that either the passages must have a larger cross-section, or the flow velocities must be larger. The construction materials and design provisions of a hot precipitator will naturally be more expensive than those that are satisfactory for a cold unit.

A hot precipitator (with gas inlet at about 700°F or 370°C) has been designed for a 650 MW unit burning Wyoming coal [Friedrich and Pai 1977]. Whether this approach, or others involving high-intensity electrostatic fields in special configurations [EPRI 1978], proves to give a successful long-term solution is still an open question. At present, the likelihood is stronger that, to meet any future standards limiting the emission of respirable particulates, fabric filters will be required.

7.3.2 Fabric Filters (Baghouses)

Baghouses are multiple scaled-up versions of the bag filters on domestic upright vacuum cleaners (Figure 7–7). The bags are typically just under one foot in diameter (290 mm) and 30–40 feet long (9–12 m). They hang vertically under a tension load, open end down, with spacers at intervals to keep the bags open. Several thousand may be installed, at about 15-inch or 390 mm spacing, the gas flow being upward through the inside of the bag. The gas must be dry, and steam drying coils or a gas bypass are usually installed for light-load startup conditions. The maximum temperature capability of the bag materials, which are usually some form of coated glass or polyester fabric or, more recently, of a woven or felted aramid fiber, is up to 550°F (285°C) but normal operation is at about 320°F (160°C). This precludes locating the baghouse on the hot side of the air preheater.

The design pressure loss is typically 4 inches of water, about eight times that of a typical electrostatic precipitator. When a baghouse is first put into operation, the pressure drop will be less than this, and filtration efficiency will also be under specification, although still high (well above 99 percent). As a filter cake develops on the inside of the bag, the pressure drop increases and the filtration efficiency also increases. The gas flow is switched from sets of bags according to a fixed schedule or when the pressure drop reaches some preset value, and the bags are shaken, or exposed to reverse-flow pressure pulses, or some combination, to dislodge the dust to collection hoppers. In steady-state conditions, with some bags always out of circuit and those in circuit having a range of filter-cake thicknesses, the fil-

Figure 7-7.   Baghouse compartment (diagrammatic representation).

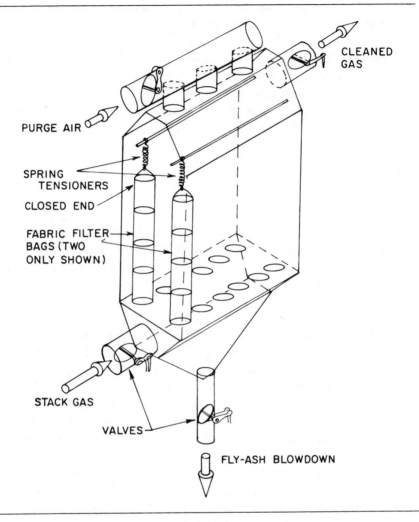

tration efficiency can be over 99.8 percent on total particulate mass, and 99.9 percent for particles larger than 0.36 μm.

The large baghouse installations are still regarded as experimental, and in some cases bag breakage is frequent, costly, and troublesome. But in some units bags have lasted for over three years [Frederick 1979] and there is no obvious reason why all units should not do as well once experience on optimum procedures is built up.

7.3.3 Alkali Spray Plus Fabric Filtration

A new development is the combination of particle filtration with sulfur removal through the use of an upstream alkali spray into the hot flue gas. The gas is cooled but the moisture added is low enough in mass flow for the gas to be dry at the end of the sprayer. The cooled treated gases then pass into a baghouse, where the alkali particles are filtered out and further contact the gas while on the filter fabric.

This process is still experimental, although it is being given a full-scale trial at Coyote, North Dakota.

7.4 REDUCTION OF NITROGEN OXIDES

Oxides of nitrogen (NO_x) are formed during the combustion process through the reaction of oxygen in the air with nitrogen in the air and with organically bound nitrogen in the fuel. The equilibrium concentration of nitrogen oxides in combustion products is a strong function of temperature. Little NO_x is formed outside of the flame zone and of the high-temperature combustion products. The reactions are quite slow. Whereas most high-temperature reactions take place almost instantaneously, nitrogen and oxygen react slowly and break down equally slowly.

High temperatures in combustion processes cannot be avoided. However, the formation of NO_x can be greatly diminished by transferring heat from the combustion products as rapidly as possible. Thus several small burners with associated nearby water-cooled surfaces to which the gases can transfer heat by radiation and convection will produce lower NO_x levels than a single large geometrically similar burner.

The fluidized-bed combustor will therefore produce low NO_x levels because of the almost instantaneous heat transfer first to the bed medium, for example sand or limestone, and then to the steam-raising tubes, which are normally immersed in the bed.

Excess air has a double effect on NO_x production. A small degree of excess air increases NO_x production because it makes more atmospheric nitrogen and unused oxygen available for reaction. At higher levels of excess air this effect is superseded by the cooling effect of the excess air, which greatly reduces NO_x formation. To use cold excess air for this purpose obviously is wasteful of energy, and it is more usual to bring about a reduction in flame temperatures through the addition of exhaust gases or steam to the combustion zone.

There are several other methods of reducing NO_x emissions, all generally in the design or experimental stages. (1) Two-stage combustion, similar to stratified-charge arrangements in automobile engines, can be used to burn the fuel first in a very rich mixture (in fuel low in oxygen), and therefore cool, and then in a lean mixture (in fuel high in oxygen), which again will be cool if some heat has been transferred out of the flame in between these stages. (2) The stack gas may be brought into contact with ammonia, which converts some of the NO_x to N_2, activated charcoal, copper oxide sorbent, and possibly a catalyst (used experimentally with oil but not with coal). (3) Various scrubbing processes may be used. It is possible that one scrubber can handle sulfur and NO_x simultaneously.

Unofficial reports state that the capital costs of the dry or wet stack-gas-cleaning equipment for NO_x are between \$60 and \$100 per kilowatt in Japan. The removal efficiency of the dry processes is between 40 and 70 percent. The performance of the scrubbers on NO_x is not known, but would be expected to be somewhat better. The costs are summarized in Table 7-2.

7.5   FUEL CLEANING

There are many possibilites for producing clean fuels, principally liquid and gaseous, from coal, although none is yet competitive with petroleum fuels, which were in mid-1979 still relatively inexpensive. Coal fuels were used, however, in Germany during World War II, and they are used in South Africa today.

The methods may be grouped as follows.

> Liquefaction
> > Solvent refining
> > Petroleum-type fuel
>
> Gasification
> > Low-Btu gas (moving bed)
> > Medium-Btu gas (slagging moving bed)
> > Low-Btu gas (entrained)
> > Medium-Btu gas (entrained).

All involve the removal of almost all the sulfur. The lowest projected costs are for the medium-Btu entrained gas, for which the busbar power costs have been estimated in a study by the Electric Power Research Institute [EPRI 1978] to be only 10 to 20 percent greater than those for low-sulfur coal. This process is, however, not yet proved in practice.

7.6 FUTURE TECHNOLOGIES

The potential availability of clean gaseous and liquid fuels derived from coal opens up the prospect of employing technologies that require clean fuels and that may bring other benefits as well.

7.6.1 Fuel Cells

Fuel cells are a form of direct energy conversion devices because they can produce electrical or mechanical energy without going through a heat-engine cycle. This means that their conversion efficiencies are not limited by the second law of thermodynamics, and that high efficiencies can be theoretically obtained without the need for high-temperature processes (although some fuel cells do operate at molten-salt temperatures). Therefore, the production of nitrogen oxides is almost always negligible.

The fundamental processes are those of diffusion and ion exchange, and they are therefore virtually silent and produce simple by-products. There is no obvious advantage to large size. For this reason the efficiency is maintained at large turn-down ratios (low loads). United Technologies Corporation (Pratt and Whitney) is using these advantages of fuel cells to market them for residential locations, taking advantage of the far lower distribution costs and the unobtrusiveness of the installations. The efficiencies quoted for installations using refined distillate fuel are high, over 40 percent, which is comparable to the best of present steam-electric generation, but much lower than the potential efficiency of fuel cells—70 to 80 percent—because the losses in fuel preparation are included. With the production of clean fuel from coal, these losses could be charged to the conversion process, making the fuel cell even more attractive.

Table 7-2.   Capital costs for ten clean ways to burn coal.

	Heat Rate (Btu/kWh)	Base Cost ($/kW)		Contingency (%)	Uncertainty (%)	Total Cost[a] ($/kW)
Conventional Steam Plants						
Low-sulfur coal	9,000		290	+10	±10	375–460
High-sulfur coal with alkali scrubbing	9,500	PP: SR:	290 50 —— 340	+10 +20	±10 ±20	485–625
High-sulfur coal with regenerative scrubbing	10,000	PP: SR:	290 150 —— 440	+10 +20	±10 ±20	575–740
Atmospheric fluidized-bed combustion	9,500	PP:	340	+20	+25, −15	450–665
Solvent-refined coal	9,000 BC: 10,000	PP:	290	+15	±15	375–500
Petroleum-type fuel	9,000 BC: 13,400	PP:	190	+10	±10	250–300
Low-Btu gas, moving-bed, dry ash Lurgi process	BC: 13,600	PP: SR:	190 390 —— 580	+10 +20	±10 ±15	760–1,000
Medium-Btu gas, slagging moving-bed process	BC: 11,300	PP: SR:	190 255 —— 445	+10 +20	±10 +25, −15	585–800
Low-Btu gas, atmospheric, two-stage	BC: 10,600	PP: SR:	190 210	+10 +20	±10 +25, −15	525–710

Process	Heat rate	Cost			Range
Medium–Btu gas, pressurized, two-stage entrained process	BC: 9,800	PP: 190 SR: 155 345	+10 +20	±10 +25, −15	490–600
Combined-Cycle Plants					
Petroleum-type fuel	7,500 BC: 11,200	PP: 160	+15	±15	185–250
Low–Btu gas, moving-bed dry ash Lurgi process	7,500 BC: 9,500	PP: 160 SR: 335 495	+15 +20	±15 ±15	650–875
Medium–Btu gas, slagging moving-bed process	7,500 BC: 9,100	PP: 160 SR: 215 375	+15 +20	±15 +25, −15	490–695
Low–Btu gas, atmospheric, two-stage entrained process	7,500 BC: 8,400	PP: 175 SR: 180 355	+15 +20	±15 +25, −15	460–650
Medium–Btu gas, pressurized, two-stage entrained process	7,500 BC: 8,150	PP: 160 SR: 130 290	+15 +20	±15 +25, −15	375–530

a. Includes IDC and startup at 30% (except 22% for combined-cycle petroleum-type fuel plant).

PP = power plant.

SR = sulfur removal system.

BC = basis coal (coal conversion and power generation).

Source: *Combustion* 48 (19), (April 1977): 23–25, "Ten ways to clean coal and their busbar costs," a review of an article in the *EPRI Journal*, November 1976.

7.6.2   Cogeneration with Fuel Cells

Cogeneration is being advanced as an important method of conserving fuels. Unfortunately, it also leads to a reduction in flexibility because there are few places where electricity and heat are needed at the same place and in the right amounts. However, the places where the two are needed at the same place and time are in cities that have pollution problems.

The fact that there is no advantage to making fuel cells of large size suggests that fuel cells can be used with advantage in office complexes. The reduction in efficiency to below 80 percent comes from heat generation in the fuel preparation, and this heat can be used with advantage for heating the building.

In the proposed United Technologies fuel-cell installations some fuel is burned at the preparation stage. This leads to some nitrogen oxide emissions, but there are only a small fraction of those that would come from other sources producing the same amount of electricity and heat.

7.6.3   Gas Turbines

The term "gas turbine" covers a large range of different engines, all heat engines, all using gaseous working fluids and turbomachinery for compression and expansion, but otherwise each serving a different purpose in a different way. The most familiar gas turbine is the jet engine, the only gas-turbine variation on which considerable development funds have been spent. Research principally for aircraft engines has raised the high-temperature capability of the expansion turbines to at least $1,400°C$ ($2,500°F$, approximately) with experimental turbines running at $300-800°C$ ($550-1,440°F$) higher than this temperature through the use of turbine blade-cooling systems.

The development of uncooled ceramic turbines is proceeding for application to automotive as well as electricity-generation turbines, to withstand $1,400-1,500°C$.

The ideal application of the high-temperature gas turbine to power generation would be in a cycle of low pressure ratio with heat regeneration from the turbine exhaust to the compressed air. Such a cycle can yield very high thermal efficiencies, well over 60 percent at these temperatures. Gas turbines have some advantages with large size, but not to the extent of steam plants; smaller, more dispersed generating stations are likely, leading to greater possibility of using the exhaust heat for district heating, for instance. They are the only type of heat engine to burn fuel at a less-than-theoretical mixture strength, and should thereby produce fewer nitrogen oxides. With clean fuels the other gaseous pollutants will be negligible.

A combination of gas turbines and fuel cells seems likely to produce the bulk of fossil fuel-generated electricity in the future, with gas or oil produced from coal, and with environmental and health costs transferred to the coal gasification plant; these costs will be lower than in 1980.

7.7   INTERMITTENT CONTROL OF EMISSIONS

In this section we examine several alternative strategies for reducing emissions of a power plant (for example by switching to a high-cost low-sulfur fuel) during certain critical meteorologic conditions. The health damages are accounted for by six alternative models. It is concluded that intermittent control is a low-cost way of meeting short-term SO_2 standards. However, its use would bring about a worsening of long-term exposures in comparison with the employment of flue gas desulfurization.

7.7.1   Survey of Intermittent-Control Systems (ICS) [1]

The regulatory standards set by the U.S. Environmental Protection Agency assume that there is a threshold for each pollutant below which no significant public health or environmental hazard exists. There are slightly higher standards for SO_2 concentrations that are present only for a short time than for concentrations present for a long time. There is a strong incentive to reduce SO_2 concentrations to the standard level, but of course there is now no incentive to reduce SO_2 concentrations to below the standard. This is recognized in the EPA regulations [EPA 1973b]. Moreover, there are no EPA standards for sulfate concentrations at the present, and it can be

1. These are often called closed-loop systems, dynamic-emission control systems, SO_2 emission limitation systems, and — more popularly — fuel-switching schemes.

argued that there is no reason why industry should take sulfates into account. This is further discussed in Chapter 9.

Although tall stacks are moderately inexpensive and effective at dispersing pollution, there are power stations, such as those near airports, that cannot use tall stacks. For these power stations it is possible to use an intermittent-control system whereby power is reduced, or low-sulfur fuel is burned, during nondispersive weather conditions. One of the first intermittent-control systems was installed by Commonwealth Edison Corporation of Chicago in the late 1950s on a small electricity-generating station near Midway Airport [Fancher 1977]. The airport approaches prevent any tall chimney stacks being used; instead the power stations used six short ones. Under some meteorologic conditions a downdraft occurs and the SO_2 concentrations near the ground are then appreciable. Arrangements were made with a meteorologic service to predict, on a day-to-day basis, whether or not the weather conditions were conducive to downdraft, and if so, the power station switched to burning natural gas. This experience is not immediately applicable to 1980 conditions because almost all power stations now use tall stacks and downwash is not a problem (except perhaps for stations with a wet-lime stack-gas scrubber such as at Battersea, London, where the wet gas is liable to downdraft) and also because natural gas is in short supply and it is likely that electricity-generating stations will soon be prohibited from burning it for base-load operation.

Although many others are being planned in 1980, ICS measures are currently being implemented at only a small number of installations on the American continent, (1) at Tennessee Valley Authority electricity-generating power plants [Montgomery et al. 1973]; (2) at other electricity-generating power plants including Ontario (Canada) Hydro, Long Island Lighting Company [NAS 1975, p.497], and Commonwealth Edison Corporation of Chicago (Kincaid Plant, Springfield, Illinois) [Fancher 1977]; (3) at a number of western smelters operated by American Smelting and Refining Company (ASARCO), Kennecott Copper Co., and others [NAS 1975]; and (4) at Dow Chemical Co., Midland, Michigan.

There are two basic classes of ICS systems. In a Class I system, the meteorologic model estimates the impact of the sulfur—as measured subsequently by the SO_2 monitors. This model is checked against observation. By using the forecasting model, a shift is made. At TVA, a shift can be made from high-sulfur to low-sulfur fuel (with a 0.5

hour delay to empty the bunker) or load may be switched to another power station. At smelters or chemical plants where a continuous output is not essential, the plant output can be reduced.

Class II systems incorporate readings on SO_2 monitors to provide a feedback. The TVA's Paradise Steam Plant uses this feedback in addition to meteorologic models to control emissions. This feedback has delays built into it—time to empty the bunker at TVA (this is reduced if a conveyer belt system is used to bring the coal into the power station instead of the bunker used at TVA), and time for the SO_2 to blow to the measuring station.

We here note the performance of two systems. At the TVA Paradise plant, the number of violations of the air quality standards has been reduced from 7 to 2 for the 3-hour SO_2 standard and from 8 to 0 for the 24-hour SO_2 standard. Table 7−3 presents these figures.

The performance of the ICS at the ASARCO smelter in Tacoma, Washington can be judged from the number of violations of local ambient standards as a function of time. This is shown in Table 7−4 [Welch 1974]. It should be noted that in 1976 improved continuous (permanent) emission controls were installed that roughly halve the peak concentrations.

It is clear that these systems can have considerable success in reducing SO_2 concentrations, and many future systems are planned. Some plans do not improve air quality but allow fuel of higher sulfur content to be used without violating the air quality standards. These may worsen the effects of sulfates.

The costs of fuel switching to low-sulfur fuels are many. First, there is a difference in cost of fuels. For instance, in the summer of 1975 there was a large differential in price between high-sulfur and low-sulfur fuels—about $3 per barrel of oil or $20 per ton of coal. This corresponds to up to 50 cents per pound of sulfur. This arises because low-sulfur fuel is in short supply. If we had fuel switching we would use less low-sulfur fuel and this price increment would tend to diminish, saving more money for those who still use high-sulfur fuels. Therefore, it is not possible with a simple analysis to assess this cost.

Various states are considering endorsing fuel-switching policies. These are states that have tough air-control regulations that presently ban use of fuels with high sulfur content. As an example, we refer to an independent study prepared for the New England Energy Policy Council [NEEPC 1975]. This study endorses fuel switching,

Table 7–3. Summary of SO_2 emission-limitation conditions September 1969 through November 1972, Paradise Steam Plant.

3-hr Average SO₂ Conc. Range (ppm)	Before 1/1/68– 9/19/69	After 9/19/69– 6/25/71	24-hr Average SO₂ Conc. Range (ppm)	Before 1/1/68– 9/19/69	After 9/19/69– 6/25/71
0	51,097	55,961	0	3,064	3,752
.01–.10	10,574	9,363	.001–.02	4,248	4,043
.11–.20	443	275	.021–.04	400	305
.21–.30	102	59	.041–.06	92	73
.31–.40	37	28	.061–.08	24	19
.41–.50	16	10	.081–.10	14	9
.51–.60	7	2	.101–.12	6	3
.61–.70	2	0	.121–.14	4	2
.71–.80	1	0	.141–.16	7	0
			.161–.18	1	0

Source: Adopted from Montgomery et al. 1973.

Table 7–4. Number of violations of *local* ambient standards at ASARCO smelter (5 monitoring stations), Tacoma, Washington.

Time Period	1.0 ppm 5 Minutes[a]	0.4 ppm 1–hour[b]	0.25 ppm 1–hour[c]	0.1 ppm 24–hours[b]	Total Violations
1969	364	273	370	28	1,035
1970	257	175	173	8	613
1971	31	29	29	1	90
1972	25	16	5	0	46
1973	10	10	4	0	24
1974 (Jan–Sept)	5	6	2	0	13

a. Allowed one exceedance in 8–hour period.

b. Not to be exceeded.

c. Allowed two exceedances in 7 days.

Source: Data from R.E. Welch, published in NAS 1975, p. 504.

but it does not have specific recommendations for guidelines. The reason for the lack of specificity becomes clear when we read the full study—the question of whether SO_2 or sulfates are the relevant parameters is mentioned, although it is hidden in later pages and in the appendixes.

It seems obvious that an application of SO_2 dispersal models to power stations on the coast would show that there are many days when high-sulfur fuel may be burned with almost no health hazard because the sulfur blows out to sea.

If we use the idea that SO_2 concentrations are an indicator of public health impact, but assume a linear–no threshold model then fuel-switching schemes that do not lower long-term or short-term averages may be ineffective. As we can see from simple SO_2 dispersion models, or from experimental data, that peak concentrations can be reduced without lowering mean concentrations. At the Kincaid plant of Commonwealth Edison, output reduction is anticipated from five hours in each of fifteen days in the year, or 1 or 2 percent of total sulfur emissions. This will have, at most, a 10 percent effect on the average SO_2 concentrations—barely important enough to justify the effort.

Fuel switching is also not so useful if sulfate concentrations are believed to be the indicator of public health, whether or not a threshold theory is used. The rate of conversion of SO_2 to sulfates is slow enough, and the absorption of sulfates by the ground slow enough, that sulfate concentrations do not vary as rapidly with time and distance as do SO_2 concentrations. Fuel switching based on meteorologic conditions local in time and space will therefore be less effective.

But plants and trees are more likely to be effected by high short-term SO_2 concentrations. For example, Linzon [1971] showed that levels of SO_2 below NAAQS might damage the eastern white pine tree. At any given location the ambient concentrations could be well below the annual and daily standards. However damage may result from short-term fumigation. The fact that SO_2 absorbs on the ground and on vegetation may be an advantage to public health (because it takes SO_2 out of the air) but may be a disadvantage to individual plants. Therefore, intermittent SO_2 control with a short feedback time may be necessary to preserve plant life, especially during the sensitive early part of the growing season.

7.7.2 Cost Advantage of ICS

ICS systems with either tall stacks or flue gas desulfurization (FGD) have specific advantages. The greatest advantage that an ICS with tall stacks has over one with FGD is that it costs about one-tenth as much. An EPRI report [Yeager 1975] offers the following figures.

	ICS	FGD
capital costs	\$4−\$10/kw	\$100/kw
operating costs	0.15−0.4 mills/kwh	3−6 mills/kwh

Unlike flue gas desulfurization, however, ICS with tall stacks can reduce only the short-term SO_2 concentrations. In the final analysis, reliance on tall-stack technology will permit annual ambient SO_2 concentrations to rise to equal the Clean Air Act standards over a large region; this seems undesirable.

If power companies and others use ICS as an interim technique until research can improve other methods of cleaning gases, we might enjoy some immediate reduction of any deaths, injuries, or property damage related to peak levels of SO_2 concentrations, provided that power companies do not use the improvement to increase burning of high-sulfur fuel. Whether or not there can be a large reduction of damage through ICS depends to a large extent upon whether a threshold exists below which concentrations of SO_2 are innocuous. Figure 1−2 indicates that this threshold might exist. We then calculate that although ICS with tall stacks will always reduce short-term damage of SO_2 a threshold at the primary standard for SO_2 will offer great reduction of damage.

7.7.3 Reduction of Hazards Through the ICS

The effectiveness of ICS in reducing health damage depends on our model of the relationship between exposure and effect. Our current health standards assume a threshold below which no damage occurs. Since we do not know the true model, in this exercise we test three ICS schemes on six alternative health models. For these models the effects range from an exacerbation of health damage to a six-fold improvement. Therefore, any imputed benefits of an ICS program are highly dependent on the true relationship between exposure and health effect.

Our models of health damage are:

(A) Damage is linear with the mean daily concentration of SO_2 with a straight line going through the origin of coordinates. [No threshold: $D = kc$]

(B) Damage is linear with mean daily concentration of SO_2 with a threshold of the *mean* at a concentration at 0.14 ppm. Note that 0.14 ppm is the mean daily NAAQS. [$D = k\,|\bar{c} - 0.14\,|$]

(C) Damage is linear with mean daily concentration of SO_2 with a threshold of the *instantaneous* concentration at 0.07 ppm. [$D = k \cdot \overline{|c - 0.07|}$]

(D) Damage is proportional to the daily average of the square of the concentration. [$D = k\overline{c^2}$]

(E) Damage is proportional to the square root of the daily average of the square of the concentration. [$D = k\sqrt{\overline{c^2}}$]

(F) Damage is proportional to the sulfate (SO_4) level and not to the SO_2 level. ·

Relations (A)–(F) are plotted in Figure 7–8. In Chapter 6 we showed that the first three of these are good fits to the data; model D is more appealing and matches our general instincts of pollutant damage to people, but it fits the data less well. At concentrations above 2 ppm we expect all the curves to flatten off, because everyone is affected, leading to the usual sigmoid curve.

A quick examination of the data around a typical power plant shows that if the sulfur content of the fuel is kept down enough to keep SO_2 concentrations below the annual average SO_2 standard, there will be little problem in meeting the daily average. However, without an ICS there may be a violation of the 3-hour standard on several occasions. If indeed the health hazard depends only upon a long-term average concentration and not upon the short-term peaks, there would be no point to the ICS system for human health.

We take SO_2 concentration variations similar to those that occurred at the TVA's Paradise Steam Plant in 1968–1969 on ten occasions and idealize them. These should be representative of what an ICS system can do in practice.

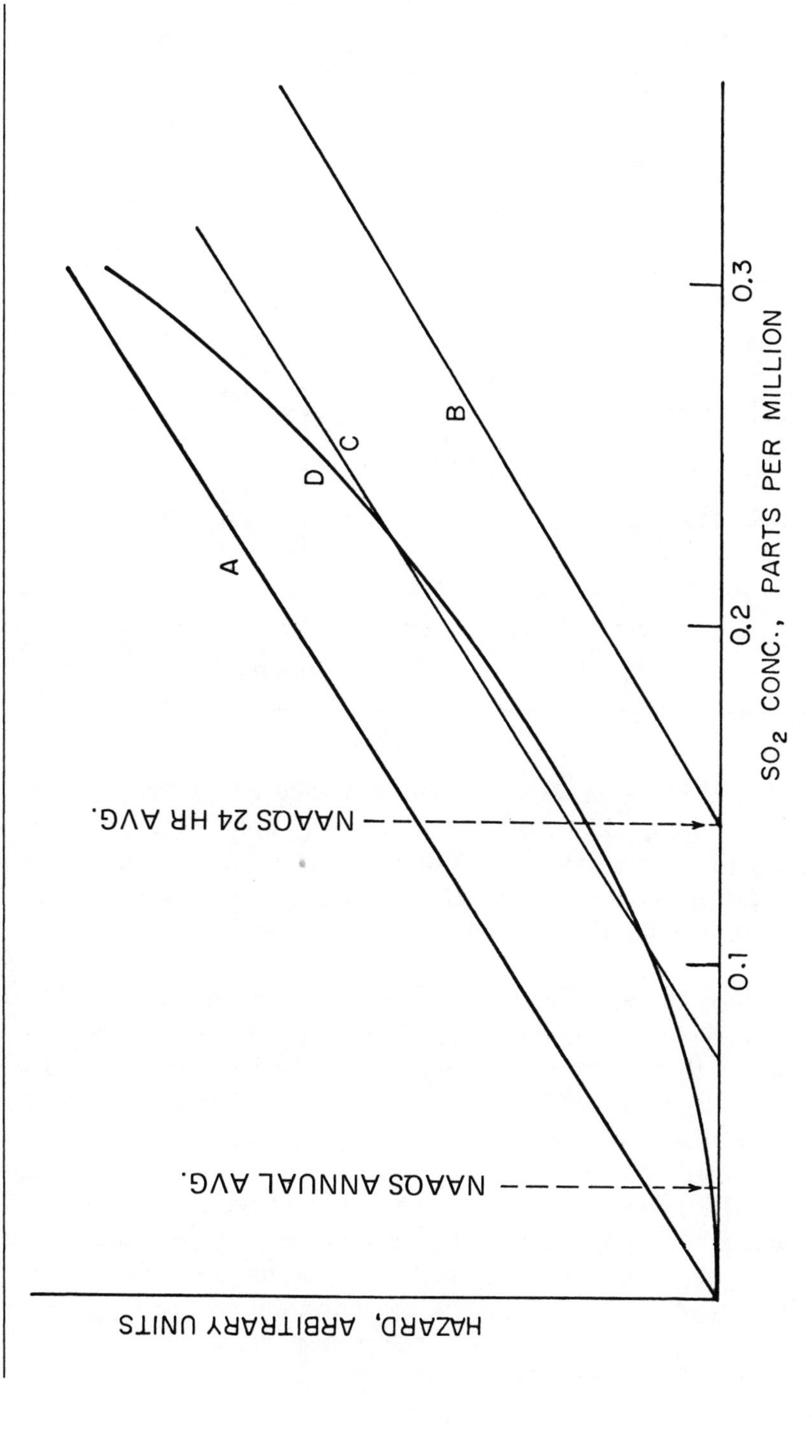

Figure 7–8.   Alternative hazard models.

Figure 7−9.   Idealized distributions (in time) of SO_2 concentrations.

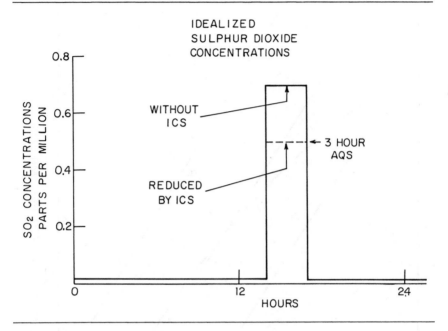

The idealization we take is shown in Figure 7−9. From Table 7−3 we see that peaks of this magnitude (0.7 ppm for 3 hrs) occurred three times in a 1.5-year period. We assume that during the rest of the day the concentration is 0.03 ppm—the *annual* average NAAQS, and we consider this day only.

7.7.3.1   For each of these models A to F of hazard versus concentration we will consider three simple schemes that a power station can use to meet the NAAQS for SO_2. For simplicity, we will consider only violations of the mean 3-hour secondary air quality standard for SO_2. The three schemes are as follows.

Scheme I. If meteorologists forecast a peak in SO_2 concentration, then the power station will reduce power production so that the mean 3-hour SO_2 concentration will not violate the 3-hour standard (0.5 ppm). In practice, this method would mitigate SO_2 hazards for all plots (A to F) of hazard vs. concentration because the power station would emit less sulfur into the atmosphere.

Scheme II.   If meteorologists forecast a peak in SO_2 concentrations, then instead of reducing power production, the power station will burn precious low-sulfur fuel. To compensate for using the more expensive fuel, it will burn less expensive high-sulfur fuel (during times when SO_2 concentration is expected to be low) in such amounts that the total sulfur emissions are constant.

Scheme III.   If meteorologists forecast a peak in SO_2 concentration, then the power station will burn the expensive low-sulfur fuel to meet the mean 3-hour NAAQS for SO_2, as in schemes I and II. During nonpeak times, however, it will burn fuel with a very high-sulfur content. This scheme can result in an increase in damage depending on which plot of hazard with concentration is correct.

7.7.3.2 Calculation for hazard model A.   We will assume that health hazard is proportional to the SO_2 concentration averaged over one day, and assume the idealized SO_2 concentration of Figure 7—9. If the hypothetical power station which causes these peaks reduced each peak to the mean 3-hour NAAQS for SO_2, then for 3 hours the reduction is 5/7. The SO_2 concentration is reduced over this 3 hour period by 7/5 = 1.40. But the average reduction over the whole 24 hours is a little less (= 1.28). This is shown in Table 7—5 column a, row 1. If the 0.7 ppm peak lasts for 6 hours, the reduction becomes a factor of 1.34.

The loss of power output, averaged over the whole 24 hours is a factor

$$24 / (21 + 5/7 \times 3) = 1.036 \quad \text{or} \quad 3.6\%.$$

For scheme II, the power station will mix fuel to meet these constraints. We assume that it has available the normal coal with 2% sulfur content (cost about 80 cents per 10^6 BTU in 1975), some solvent refined low sulfur coal with ½% sulfur content ($2.40 per 10^6 BTU), and some high sulfur coal with 4% sulfur content (60 cents per 10^6 BTU). These costs are estimated for solvent refined coal on p. 373 of NAS 1975.

During the 3 hour peak, 38% of the fuel burned must have a low sulfur content allowing 62% to be normal ($0.38 \times$ ½% $+ 0.62 \times 2\%$ = 5/7 $\times 2\%$). To compensate for the amount of low sulfur fuel burned, the power company is allowed to burn (during the remaining 21 hours) a mixture with 4% of the high sulfur fuel $[0.04 \times (4\% - 2\%)]$ 21 = $[0.38 \times (2\% - ½\%)] \times 3$.

Table 7-5. Scheme I.

Reduction of hazard to public health if ICS is used to reduce SO_2 levels that violate the primary air quality standards to levels that just meet the SO_2 primary air quality standards[a] (divide health hazard by the number shown to determine NAAQS effect).

Hazard Model (fig. 7-8) (Relation of Hazard to Concentration)	(a) Decrease in Damage if 3-hr Peak is Reduced to Meet the Mean 3-hr NAAQS for SO_2	(b) Decrease in Damage if 6-hr Peak is Reduced to Meet Mean 3-hr NAAQS
A. Linear with mean daily SO_2 level	1.28	1.34
B. Linear with threshold at the 24-hr SO_2 NAAQS[+] (.14 ppm)	According to this threshold model, there is no hazard either before or after the reduction	6.0
C. Linear with threshold at .07 ppm	2.3	1.7
D. Daily mean of square of SO_2 concentration	1.93	1.95
E. Proportional to root mean square of mean 3-hr level SO_2	1.39	1.40
F. Linear with respect to SO_4 level	1.04	1.08

a. The mean daily SO_2 concentration indicates health effects for the mean 3-hr standard only. The text illustrates how these numbers were calculated. NAAQS refers to the Primary Air Quality Standards of the Clean Air Act Amendments of 1970.

The average SO_2 concentration in this 21 hour period is then increased by only 3.6%, so that the decrease in damage is almost the same as in scheme I. The entries, therefore, in the first rows of Tables 7-5 and 7-6 are similar.

For scheme III (Table 7-7), the power station will burn very high sulfur fuel for 21 hours in the day. This still leaves a decrease in health damage under model A due to the ICS.

Table 7-6. Scheme II.

Decrease in health hazard of Scheme II is used to meet the mean 3-hour PAQS at the hypothetical power station (divide by factor shown to get new hazard figure) (reduction of 3-hr peak only).

Hazard Model (fig. 7-8) (Relation of Hazard to Concentration)	Decrease in Damage
A. Linear with SO_2 level	1.27
B. Linear with threshold at 3-hr PAQS for SO_2 (.14 ppm)	Both zero (No change)
C. Linear with threshold at .07 ppm	1.47
D. Daily mean of square	1.93
E. Root mean square of 3-hr concentration	1.39
F. Linear with respect to SO_4 level	No change

Table 7-7. Scheme III.

Decrease of hazard to public health under various relations of SO_2 concentrations to damage if ICS is used to permit increased emission of sulfur (divide original health hazard by this factor to get ICS-caused hazard). Decrease in damage if Scheme III is used to meet NAAQS.

Hazard Model (fig. 7-8) (Relation of Hazard to Concentration	Decrease in Damage if 6-hr Peak is Reduced to Meet Mean 3-hr SO_2 NAAQS
A. Linear with SO_2 level	1.16
B. Linear with threshold at 24-hr PAQS for SO_2 (0.14 ppm)	1.92
C. Linear with threshold at .07 ppm	1.47
D. Daily mean of square of SO_2 concentration	1.89
E. Proportional to root mean square of mean 3-hr level for SO_2	1.37
F. Linear with respect to SO_4 level	0.63

7.7.3.3 Calculation for hazard model B. (The reduction of the peak reduces the daily average to below the threshold level of 0.14 ppm except for the rare case of a 6 hour long peak.) Also in scheme III the burning of high sulfur fuel can bring the average up significantly.

7.7.3.4 The damage calculated under model F will be proportional to the total sulfur emission since the local weather conditions that give use to a high peak in SO_2 concentrations do not apply to $SO_2 \rightarrow SO_4$ conversion;

Under scheme I the reduction in health damage is the same as the reduction in electricity output, 1.036. For scheme II the health damage remains unchanged since scheme II is designed to keep the total sulfur emissions unchanged. In scheme III, there is a marked increase in health damage. In practice, scheme III is impossible, because the long term average SO_2 concentration rises to over 0.06 ppm and this violates the annual average NAAQS.

As we look at these tables we see that scheme II is reasonable for all the hazard models A–F. We can ask what it might cost the power station to use this scheme to reduce three hour peaks. The *fuel* cost can be calculated as follows. The *increase* for the 3 hour peak is ($2.40 \times 0.38 + 0.62 \times $9.80)/$0.80 = 1.72. This is compensated by a small *decrease* for the rest of the 21 hours, to get an overall increase of 19%. The question then arises that we do not answer; is this cost worthwhile?

7.7.4 Meteorologic Models

All the calculations of weather or pollutant dispersion in this report are deliberately elementary calculations using naive models, to emphasize the general features of the arguments we are making. The arguments do not depend upon the last degree of sophistication of a model. However, in the final application of any of these procedures, it is desirable to use the most accurate model that is possible.

As an example of the sophistication now possible we note the procedure used by the TVA in its ICS system (Table 7–8). As an example of the accuracy of such systems, we show the errors in the use of the AIRMAP system of Environmental Research and Technology to the Greater Boston area (Figure 7–10). These models are adequate for prediction and control of SO_2 concentrations up to 20 miles from the source. Beyond this distance the concentration is small, so that the models would suffice if we merely want to reduce concentrations below a threshold.

However, for sulfate pollution we must trace pollutants hundreds and even thousands of miles. The basic models for these studies are different and are incapable of representing detailed local trends. For example, a typical model takes measured winds at 1,000 feet, assumes uniform mixing vertically, and a constant attenuation coefficient to simulate ground absorption.

Much detailed work is in progress to correlate these models with observed long-range (interstate and transfrontier) pollution levels. We note in particular the studies by OECD discussed in Section 2.3.3.

Figure 7–10. Errors in forecasting of SO_2 concentrations in the greater Boston AIRMAP network (lines represent percentage of observations with errors of less than the specified concentrations (ppm). Almost 100 percent of observations were within 0.02 ppm). From Gaut, [1975].

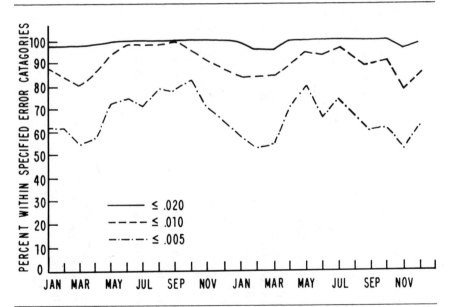

Table 7–8. Meteorologic criteria for ICS measures at TVA's Paradise Steam Plant.

No.	Criteria	Parameter Values	Remarks
1.	Potential temperature gradient between stack top, 183 m, and 900 m	$\frac{\Delta\Theta}{\Delta Z} \geq 0.64°\text{K}/100\text{ m}$	Identifies the mean atmospheric stability throughout the layer involved in the plume heigh calculation
2.	Potential temperature gradient between stack top, 183 m, and 1,500 m	$\frac{\Delta\Theta}{\Delta Z} \geq 0.51°\text{K}/100\text{ m}$	Identifies the mean atmospheric stability throughout the layer involved with the thermally induced mixing of the plume effluent
3.	Difference between daily minimum and maximum surface temperature	$T_{max} - T_{min} \geq 6°\text{K}$	Identifies the potential magnitude of insolation as it affects the rate of development of thermally induced mixing
4.	Maximum daily surface temperature	$T_{max} \leq 298°\text{K}$	Identifies the temperature used in calculation of the maximum mixing height
5.	Maximum mixing height	$\text{MMH} \leq 2,000\text{ m}$	Identifies the maximum level of the mixing height when SO_2 concentrations might exceed the ambient standard levels
6.	Maximum mixing height and plume centerline heights	$\text{MMH} \geq H_e$	Identifies the related height at which thermally induced mixing will penetrate the plume and uniformly mix the effluent to the surface

7.	Time for mixing height to develop from plume centerline to critical mixing height	$T \geq 3{,}960$ sec	Identifies the potential for persistence of concentrations above the ambient standard levels. The critical mixing height (CMH) is the upper limit of the thermally induced mixing layer when concentrations could exceed the ambient standard levels
8.	Mean wind speed between stack top, 183 m, and 900 m	$2.5 \leq \bar{u} \leq 8$ m sec^{-1}	Identifies the wind speed used in plume rise and CMH calculations
9.	Cloud cover	$CC < 80\%$	Identifies the availability of insolation which affects the thermally induced mixing

Source: Tennessee Valley Authority, *Summary of TVA Atmospheric Dispersion Modelling: Conference on the TVA Experience at the International Institute for Applied Systems Analysis*, Schloss, Laxenberg, Austria: October, 1974.

7.8 SITING

We argue in this section that the selection of sites for electricity-generating and coal-gasification plants, for example such that the emissions affect very few people, can greatly reduce overall population exposure levels. Of course there are always competing uses for potential sites (e.g., preservation of pristine areas). There are also economic considerations such as transportation costs and proximity to markets.

7.8.1 Qualitative Advantages of Downwind Siting

As noted in Section 7.1, a tall stack can ensure that the SO_2 is dispersed enough that the peak concentrations are always below a supposed threshold. There is no particular preference for one site over another, except to ensure that the power stations must be far enough apart that the high concentrations do not overlap.

However, if it is assumed that sulfates, nitrates, and particulates are important, and that these drift far enough that power plant plumes overlap and their concentrations add, then siting becomes vital. This is particularly true if it is assumed that there is no threshold concentration below which health is not damaged.

The obvious location for a power plant then becomes in the middle of a low-population zone, downwind of major population centers, and with no populated areas further upwind.

We might, for example, consider a site on the northeastern seaboard of the United States, 100 miles north of the urban area of Boston—Washington, D.C. The prevailing wind from the west will generally blow the sulfates and particulates out to sea, where there is no population to be exposed. To be sure, the wind will sometimes blow from the east and return some of the pollution. Under average weather conditions there are two effects: first, the prevailing winds blow offshore and the pollutants go out to sea, where there are no people; second, there are indications that sulfates are absorbed more easily on seawater than on land.

Over land, sulfates adsorb with a deposition velocity of 0.2 cm/sec or less (see Chapter 2). Over water, sulfates are more quickly absorbed. For example, the particulate $NH_4 SO_4$ has the same size of condensation nuclei onto which the ^{137}Cs adsorbs. Slade [1968] shows that the deposition velocity of ^{137}Cs onto water, soil and grass are, respectively, 0.9 cm/sec, 0.04 cm/sec, and 0.2 cm/sec. Clearly, water is a much more effective sink of sulfates than is land.

7.8.2 A Model for Coastal Siting

In this section we develop a crude and didactic model of east coast power-plant siting to illustrate the potential reduction in pollutant exposure when we consider power-plant location. A dramatic reduction in exposure occurs because of 1) pollutant absorption by the ocean; 2) an extremely low population over the ocean; and 3) prevailing winds that carry pollutants out to sea. Of course, there are many competing uses for scarce coastal property and other complexities that we do not consider in this crude model (e.g., coastal recirculation patterns that can keep pollutants trapped in the densely populated areas of the east coast). A serious siting program designed to mitigate exposure would need to consider these other factors.

We observe that the plume from a power plant diffuses vertically about 1,000 m for every 30 km that it travels horizontally in the usually steady wind. This 1,000 m is about 1/20 of the total height of the tropopause. Thus, in one hour, 5 percent of the plume will have diffused down into the sea, where we can assume that the sulfates are immediately absorbed. In calculations we take three alternative rates for water adsorption: 0%/hr, 5%/hr, and 10%/hr. We assume that SO_2 converts to sulfates slowly enough that we can smoothe out the geographical spread of sources on land; and by the time of conversion the sulfate will have been blown nearly a hundred miles by an average wind.[2] Therefore, there should be only a small local effect of a power plant. We also assume a uniform population over land and none over sea.

We assume that sulfates are not absorbed by land, but stay in the air over the land until a rainfall of at least 2 mm washes them out (once every ten days). We assume that the damage done by sulfates is proportional to the time which the sulfate-laden air spends over land. We also assume that the winds, though varying with time, vary little with the distance from the pollutant source.

2. Note that earlier we correctly took the average wind velocity, which is lower than the average speed because of the changes in direction.

Table 7–9. Factor of improvement by which average sulfate concentration is reduced as compared with continuous and uniform emission over nonabsorbing land. The power plant is assumed to be on the seacoast, and the sulfates are assumed to be absorbed by the sea at varying rates.

Wind Conditions and Assumed Control Conditions	Assumed Absorption on Sea (% per hr)	Factor of Improvement
January 1974, Logan Airport Winds		
Continuous emission of pollutants	0	19.8
	5	27.8
	10	31.7
Emission only when wind is blowing offshore	0	58.2
	5	77.5
April 1974, Logan Airport Winds		
Continuous emission of pollutants	0	8.3
	5	16.8
	10	21.5
Emission only when wind is blowing offshore	0	28.0
	5	43.6
September 1974, Logan Airport Winds		
Continuous emission of pollutants	0	2.9
	5	4.8
	10	5.6
Emission only when wind is blowing offshore	0	6.0
	5	9.4

Two Check Calculations to Show the Effect of the Prevailing Wind

September, 1974—Logan Airport winds assumed exactly opposite; continuous emission of pollutants	0 5 10	1.5 2.2 2.4
September, 1974—Logan Airport winds assumed exactly opposite; emission only when these winds blow offshore	0 5	6.8 9.2

We have taken winds measured at ground level at Logan Airport, Boston [Commerce 1974] on the edge of the ocean in three different months. We then compare the damage from a single imagined coastal power plant, to a single imagined power plant located inland. Using these assumptions, we present in Table 7−9 factors of improvement by locating a power plant on the coast compared with one inland for various parameters.

If we assume that the ocean does not adsorb pollutants (0%/hr) then Table 7−9 shows the effect of having winds blow offshore away from populated areas. From Table 7−9 one can see that if the power plant continuously emitted sulfates near Logan Airport, then the offshore winds would have reduced concentrations by a factor of 2.9 during September 1974. In April 1974, the offshore winds behaved such that sulfate damages caused by the hypothetical power plant located near Logan Airport would be one-eighth of what they would be had the same winds occurred at an inland power plant. This is a huge reduction. The reduction occurs because offshore winds blow the sulfates over the uninhabited sea.

If the sea reduces airborne particulate concentration at the rate of 5 percent per hour, this process will produce a further significant reduction in sulfate-related damages caused by a hypothetical continuously emitting power plant near Logan Airport. In January 1974, this removal process would have produced an additional 1.4-fold improvement; in April 1974, the additional improvement would have been 2.0, and in September 1974, the additional improvement factor would have been 1.8.

If we assume that the sea adsorbs sulfates that pass over at a rate of 10 percent per hour instead of 5 percent there is a further reduction, but not proportional. In September 1974, the factor of sulfate reduction increased by 1.6 when the ocean removed particulates at 5 percent per hour instead of 0 percent per hour. If the ocean cleanses not at 5 percent per hour, but at 10 percent per hour, then the additional 5 percent is three-fourths as effective as the first 5 percent. A similar trend holds for January and April 1974.

7.8.3   Effects of Intermittent Control

It is conceptually possible to limit the power station operation to times when only offshore wind conditions exist. Allowing these emissions only during offshore winds presents the maximum possible reduction of sulfate-related damages. One would achieve these levels of improvement only if one could stop emissions every time the wind blows inland. Assuming that there is no reduction by the sea the maximum reductions range from a factor of six in September 1974 to 58 in January 1974.

If we consider the situation where the ocean reduces the number of airborne particulates at the rate of 5 percent per hour (the most likely figure), then the lowest calculated factor by which an ICS can reduce sulfate-related damages is 9.4 (emission only during offshore wind, September 1974). We conclude that in a coastal situation, with constant power plant operation, one can expect the action of the winds and the ocean to reduce sulfate-related damages by at least a factor of five. With ICS withholding emissions during onshore winds, using just the cases we examined, one can expect that in any month (we assume September 1974 as the worst case) that the sulfate-related damages will fall by a factor of 9.4. If the ICS does not withhold emission of sulfates every time the wind blows onshore, then the improvement factor will be between 4 and 9.4.

In order to verify the effect of the prevailing offshore winds, and to separate this effect from that of the lack of population over the sea, in the last two cases of Table 7–9, we assume that the wind directions are exactly opposite to the real wind directions. The improvement factors for these two cases are half as large as the corresponding improvement factors for the two preceding cases in which we used the true direction of the winds. This situation holds because the prevailing wind direction at Logan Airport is from the west.

This suggests, therefore, that power stations on the coast of New England can burn high-sulfur fuel with more safety than power stations inland, especially if burning of high-sulfur fuel is restricted to times of offshore winds.

A power station a few miles inland might be allowed to burn high-sulfur fuel when the wind is blowing offshore and the plume dispersion is such that the plume does not reach ground level until it has crossed the coast.

Of course, these calculations are only quite crude, and should be improved before being used as a basis for decision. The model of Meyers et al. [1978], discussed in Section 2.3.4, could, for example, be used for this calculation.

7.8.4   Siting and Political Compromise

Many factors prevent application of this siting principle, which is otherwise desirable from a public health point of view. There are questions of regional equity: for example, why should the residents of Utah suffer to provide electricity for the power-hungry gluttons of the West? Second, there are environmental as well as public health questions involved. There is more resistance to spoiling an unspoiled landscape than a city; for this reason the Sierra Club has urged, and as discussed in Chapter 9, the courts have ruled, that air may not be degraded where it is presently pure.

This is a conflict between the desire for public health and the desire for a clean environment—one of the many conflicts that our society and its politicians must resolve. One approach is the pollution pricing-rebate proposal discussed in Chapter 8.

Finally, it is important to realize that mitigation of air pollution by careful siting of power plants should work well whether the agent of pollution is SO_2, sulfates, nitrates, or particulates. Siting therefore has certain advantages over methods that are specific to a particular pollutant, because we are in the unfortunate situation of not knowing precisely what the active agent or agents is or are.

This siting approach may work for electricity-generating plants, or for coal-gasification plants, but it is not a general recipe for industry that has to use coal in or near a population center. One suggestion is that low-sulfur fuels be reserved for these cases.

7.9   FOREST FILTERING

We end this section with a suggestion based on the physical concepts of the last chapters. The aim is to prevent respirable particulates from reaching population centers, and to prevent SO_2 from converting to sulfates and then reaching population centers.

This suggests that a large power plant might be built in a forest with a stack only 10 m high. The SO_2 and particulate concentrations near the plant would be very high, but any surviving vegetation would absorb the SO_2 (for which the deposition velocity is high—0.8 to 3 cm/sec) before it converts to sulfate and disperses. Particulates would also be filtered. The forest would, of course, have to be owned by the power company.

If SO_4 has far more serious health effects than SO_2, we may be exacerbating the problem by allowing SO_2 to convert to SO_4. We could reduce much conversion by a return to shorter chimney stacks. In order to avoid high population exposures, the plants would have to be in unpopulated areas. This leads to a suggestion that the power plant effluent be emitted sideways, cooled to allow no plume rise, in a forest, to filter the sulfates and particulates. An alternative to a forest might be a power plant offshore. We note that the measurements suggest a higher deposition velocity for SO_2 than for sulfates both on water and on land.

This is admittedly an extreme and possibly absurd suggestion but it would reduce sulfate concentrations and health impacts. However, current regulatory structures (Chapter 9) prevent higher emissions so that some siting approaches cannot be considered.

8 RISKS, REGULATIONS, AND REBATES

In this chapter some of the risks of our everyday life are surveyed qualitatively and, very approximately, quantitatively. The apparent risks to individuals from the burning of fossil fuels are assessed. The health damage in the United States is estimated to be on the order of 50 billion dollars per year from air pollution. It is argued that the preferable approach for reducing this expense is to tax emissions, or surrogates for the emissions (e.g. the sulfur content of fuels). The monies collected could be returned to individuals, when the sums involved justify the expense of making the transfers, and to cities and towns when the amounts are small (e.g. less than $20 per capita per year). A suggested schedule for the introduction of emission charges and rebates is proposed.

8.1 RISK/COST/BENEFIT ANALYSIS

That life is a risky business is obvious. Exactly what the risks of an individual action are may not be clear to those who contemplate the action. Steps taken to implement this action thus might be less logical than perceived by the planners. A variety of articles [Pochin 1975; Wilson 1975; Lowrance 1976] compare different risks in life in order to make decision-making by individuals and by society clearer and more accurately addressed to society's needs.

This book is addressed to the risks of fossil fuel burning. There are uncertainties and some will therefore tend to abandon risk analysis for this reason. Kramer [1976] says "All of the uncertainties make conclusive 'scientific' benefit/cost calculations impossible."

We sharply disagree with the words as stated. When uncertainties are properly included, analysis can be scientific and sometimes conclusive. The purpose of risk analysis should not be to obtain a precise number for entering into an equation. The most important purpose is to guarantee that the analyst has thought the problem through logically. Numbers are useful and important if and only if their uncertainty is clearly stated. We noted in Chapter 6 that the mean estimate of the death rate in the continental United States from fossil fuel burning is 53,000 per year and the average mortality risk from fossil fuel burning is $53,000/215,000,000$ or 2.5×10^{-4} per year. This number could be close to zero; it might be double. But it is not ten times the risk or 2.5×10^{-3} per year. These risks are worth reducing; they are not worth as much investment as reducing the risk caused by tobacco cigarette smoking, for instance, even though the latter is voluntary.

No one has adequately quantified, or even discovered a workable procedure for quantifying, environmental risks as distinct from public health risks. To this extent Kramer's objection is valid. But most of the public concern about polluted air is related to public health, and for these Kramer's objection is not valid.

The benefits of fossil fuel burning have not been well quantified — the benefits of cheap and convenient transportation and cheap energy that have been widely assumed to be great and overriding, but this assumption is being brought into question [Schumacher 1973]. Accordingly, there is now some uncertainty in assigning benefits. However, there is less uncertainty in calculating the cost for reducing a risk, keeping the benefit constant. This sort of calculation is the daily work of industry. Therefore, even though a risk-benefit calculation may not be possible, a risk–cost calculation may be useful.

The risk and the cost are expressed in different units: the risk in mortalities per year; the cost in dollars per year. These are incommensurate until further assumptions are made. There is almost universal agreement that comparisons between these incommensurate quantities must be made by individuals and by society as a whole, not by a scientific elite. For that reason, the calculations that follow must be considered illustrative only.

The U.S. electorate, through the congressional enactment of its Clean Air Acts (see Chapter 9) has decided that polluted air is a risk to health. Furthermore, it has been decided that presently clean areas will be kept at low levels of pollution—levels below the air quality standards.

This decision is roughly equivalent to choosing, for prudent public policy, the upper bound of our risk estimate and not the lower bound of zero. We will therefore take the prudent course in the next chapters, and follow the arithmetic to logical conclusions. To the extent that the conclusion is hard to accept, we must work hard to understand this number better.

The important question arises: To reduce a risk, how much should society pay and how much is society willing to pay? Much depends upon how the risk is perceived; is it voluntarily accepted or involuntarily accepted? In a classic paper Starr [1969] suggests that the public accepts risks voluntarily 100 times as dangerous as the involuntary risks. One of the cheapest ways of reducing a risk is to buy and use a good seat belt for one's car; buying and using a seat belt corresponds to an expenditure of $5,000 per life saved [Wilson 1975]. Some people find seat belts inconvenient, so only 20 percent of Americans use them; the mandatory installation of airbags is suggested. An airbag costs about $100 when installed initially. The total expenditure in the United States is then about $1 billion per year assuming 10 million new cars per year. They would probably save 10,000 lives a year, leading to a cost of $100,000 per life saved. This still seems cheap, but some people still object to their use. For involuntary risks, costs of $1,000,000 and more are suggested.

We would like to apportion our money to save as many lives as possible. We cannot spend all the gross national product ($2 trillion) to save one life; more productive occasions than this arise. For this reason we suggest a number close to $1 million for every hypothetical (because our calculated risk is uncertain) life saved.

The Nuclear Regulatory Commission was probably the first regulatory commission in the United States to have to face up to the problem of the value of life. In a decision after a three-year hearing [NRC 1975], the NRC suggested that if exposure to radiation can be reduced at a cost of $1,000 per rem, it should be so reduced. The risk of radiation, according to the BEIR report [BEIR/NAS 1972] corresponds to 10^{-4} per man rem. This is calculated on a linear, nonthreshold basis. One thousand dollars per rem corresponds to $10,000,000 per life saved. The NRC considered this to be a temporary figure and suggested a large, long public hearing, probably with other agencies involved, to decide on this number. Meanwhile it chose $1,000 per rem as being a round number larger than any other presented in testimony at the hearing.

The cost of reducing a risk has also been addressed recently by the International Commission on Radiological Protection [ICRP 1977], which discusses the cost for saving a life in terms of its own new unit, the Sv or sievert. Translated into older units, the authors quote numbers from $10 to $250 per rem, or with the risk factors we (and they) assume ($10^{-2}/Sv = 10^{-4}/man$ rem) between $100,000 and $2,500,000 per life saved (conservatively calculated). Our figure of $1,000,000 falls near the middle of this range.

Another way of looking at the same problem is to realize that if money must be spent on control equipment, lives will be lost in the process. These are secondary effects of the decision process. It is a well-known feature of a decision process that if the primary effects are small, the secondary effects must be carefully examined, however difficult that may be.

Thus, about half of any expenditure on reducing occupational exposure might be expected to be on capital equipment—often construction equipment. In construction work, people die in all sorts of accidents from bulldozer accidents to falling off roofs. The oft-quoted example is that three people died in building the Brooklyn Bridge. The total number of workers killed in construction work in the United States was 2,200 in 1975 [NSC 1976].

The total receipts of the construction industry were $164 billion in 1972 [Stat Abst. 1978]. But this contains a great deal of duplication, because of such items as subcontracts. If we assume that this represents $80 billion of primary construction contracts, we can deduce, with some uncertainty, that for every $36 million spent in construction one life will be lost. Thus, for this secondary effect alone, no expenditure more than $72 million total ($36 million capital) should be made merely to save one hypothetical life, or $72 million a year to save more than one life per year, because it will result in a net loss of life in society as a whole, and even in the subset of working men.

Our figure of $1,000,000 to save a life may be low; but other distinguished men think it high. Thus, Nobel Laureate Joshua Lederberg says, "We might be willing to double our health expenditures for 20% improvement in health; this would imply a willingness to invest $400,000 to prevent a death, which is on the high side of present-day political judgments" [NAS 1974b, see also Mishan 1976].

McCarroll [1979], quoted in Chapter 1, eloquently pleads for not spending too much on air pollution control, and cites some advantages of cheap electricity to public health. Indeed, there are many cases in medicine where lives can be saved for $100,000 or less. In 1977 an artificial kidney cost $30,000 and an intensive care unit often cost only $20,000 per life saved. An average cost of cancer treatment was about $50,000 and saved perhaps 30 percent of all cases, corresponding to $150,000 per life saved. But, we believe that if the number is correctly set, overexpenditure on pollution control will be automatically avoided.

It is also useful to try to imagine how we would best spend money to save lives. One might well spend $1,000,000 on 20 full-time police to reduce automobile accidents or a more strict regulation of automobile speed limits, imprisonment for those with a high concentration of alcohol in the blood, and so on. If each pair of police saved one life a year an expenditure of $1 million could save ten lives a year. As Kletz [1977] pointed out: "There is nothing humanitarian in spending lavishly to reduce a hazard because it hit the headlines last week and ignoring the other [hazards]."

Once society has agreed on numerical factors—whatever they are and however tentative—to compare risks, costs, and benefits, we can proceed to calculate more readily how best to regulate these risks and how to use our resources in the most effective manner to reduce them.

We first show how this can proceed in a way averaged over the whole United States. About 22 million tons of SO_2 are emitted every year in the United States by fossil fuel burning, and its distribution was discussed in Sections 2.3.4 and 3.1. The data compiled by the Brookhaven National Laboratory are given in Table 2−9, accounting for 22 of the 27 million metric tons emitted in the United States per year.

We estimated using Table 2−9 that when 22 million metric tons of SO_2 are emitted from 1,088 typical sources, the integrated sulfate exposure is $1,524 \times 10^{-6}$ persons $\times$ ($\mu g/m^3$). The damage function we estimated was 3.5×10^{-5} deaths/$\mu g/m^3$ or 53,000 deaths from this distribution of sources. Allowing $1,000,000 per death we find

$$\frac{53,000 \times \$10^6}{22 \times 10^6 \text{ ton} \times 1,000 \text{ kg/ton} \times 2.205 \text{ lb/kg}}$$

$$= \$1.09/\text{lb SO}_2 \quad ,$$

which is $2.18/lb or $4.82/kg elemental sulfur. This is an average number that will change if the distribution of SO_2 emissions changes. The health-damage cost per person per year per $\mu g/m^3$ sulfate exposure is

$$\frac{\$53 \times 10^9}{1,524 \times 10^6} = \$35 \quad .$$

Some may charge that the figure of $1 million per mortality is arbitrary, and that it is large. We have discussed previously the various valuations of human life and have shown that this figure is considered by some to be, in fact, low. What is important to consider here is that, because we have made no account of the costs of morbidity, each $1 million per death must also be reckoned to include the valuation of the effects of the lives of a number of people who became seriously ill but who do not succumb; a larger number of people whose illnesses are not major but are perhaps chronic; and a very large number of people whose health is impaired, but not seriously. In many fields, such as the valuation of traffic accidents, the estimated costs of morbidity greatly exceed those of mortalities. This may or may not be true here. But as the figure of $1 million per mortality was taken as a justifiable value from previous studies, it can be regarded as conservative when it is taken to include the valuation of the average morbidity per mortality. Cohen [1980] has summarized what society actually pays in a variety of situations. Numbers vary from $3,600 to over $200,000,000 per life saved in the United States.

Others have used smaller figures, down to five cents per pound of sulfur, corresponding to a reduced willingness to pay to save human life. Of course, this $2.18 per pound of sulfur is an average figure, and is not disaggregated into sulfur emitted into low- and high-population areas; but it is nonetheless useful so that we may gauge the dimensions of the problem. Although the long-range transport of sulfur oxides averages the risk over large areas, we should expect that in the eastern United States the risk value of sulfur would be much higher, say $5.00 per pound, whereas in the sparsely populated areas of the U.S. West the risk value would fall to an almost negligible level. Later we shall use 60¢ per pound and $3.00 per pound to make representative calculations for low- and high-density areas.

If we follow this logic, then industry would be forced to spend money to reduce sulfur emissions generally if it can reduce them for a marginal cost of about $2.18 per pound. But if the cost is $3.00 per pound for a particular industry, is that industry freed of all obligation to reduce its emissions? Under the regulatory approach there is a difficult problem of equity. An alternative approach, that industry should pay a pollution charge as discussed below, is fairer and more easily administered in specific circumstances (see Section 9.1.1). As we shall see, this leads to a natural way of ensuring that the disaggregation of costs and risks occurs in a rational manner.

We shall endeavor to show that a pollution charge is a logical concept. However, in any one case it may be unacceptable for a number of reasons, generally categorized as political or "ethical." Two objections are that human life is beyond price and must not be valued in the same accounting table as goods and services and other costs; and that specifying a pollution charge is, in effect, allowing rich and powerful bodies to buy the right to impose their discharges on poor people. We introduce the concept of pollution charge transfer to counter this second objection. Whether or not pollution charges are used, the pollution damage arithmetic is still valid, and whatever the decisions, whatever the method of regulation, the effects should be compared with this simple yardstick.

8.2 STANDARDS FOR CONTROL AND REGULATION

There is a range of alternative actions that can be taken by government when its citizens complain about pollution. To start with, it can, and usually does, do nothing at all. There may be very few people involved—perhaps a few families living downwind of a polluting plant. Existing laws may be uncertain in their application, and to write new legislation to cover just one case might seem to be bad law. Then again there is the widespread belief that "where there's muck there's brass," in the words of a perhaps mythical Yorkshire colliery owner, looking with pride at his belching smokestack and referring to the money and jobs that were accompaniments to the smoke. There is some justification for this belief, but all too often we do not have the information necessary to determine the health and ecological benefits of pollution reduction. Furthermore, we do not anticipate that this condition of uncertainty will disappear.

At some point, politicians listening to complaints about the level of pollution on the one hand, and to grumbles and threats from industry about the serious effects of government regulation on its ability to generate wealth and employment on the other, may decide that some action to limit pollution is justified. The most likely first step is the establishment of standards that may be applied to the environment, to discharges, to supplies such as fuels, and to equipment such as automobiles. In some cases, standards applied to the environment may be merely goals—for instance, the goal to have zero discharge into the nation's waterways by the mid–1980s.

8.2.1   Concentration and Emission Standards

As stated previously, it is a tautology that health effects are related directly to pollutant concentrations and only indirectly to pollutant emissions. Therefore, the goal of regulation should be reduction of pollutant concentrations. But this raises several questions: pollutant concentrations are harder to monitor than pollutant emissions on large sources; and if there are two or more polluters, then it is impossible, merely by measuring a concentration, to decide which polluter is responsible for the health hazard.

The first of these problems is solvable. The air dispersion models discussed in Chapter 2 are sufficiently accurate that the emissions may be measured and regulated in such a way as to keep the concentrations below a specified level. The most sophisticated controls are those used to regulate the radioactivity emissions from nuclear power stations. The basic regulatory standard is the dose level applicable to an unclothed individual residing in the open air continuously at the site boundary. This is related to ambient concentrations and emissions of radioactive gases by computer programs. It is the emissions of radioactive gases that are monitored to keep the calculated levels below the imposed standard. This procedure, the setting of a general air quality-concentration standard and the derivation of a particular emission standard, has been suggested for the air pollutants of concern in this book.

The second problem rarely arises for nuclear radiation because sources are few and separated. However, the discussions of long-range pollution in Chapter 2 showed that the pollution from many sources is mixed. Which source should be blamed for a violation of the air quality standard? At the present time it is usually the largest source that is blamed. We suggest procedures whereby polluters can be blamed roughly in proportion to the increase in (population-weighted) concentration, and therefore to the hazard that they cause.

The Clean Air standards specify pollution levels that may vary from one locality to another, and that may affect government regulation of discharges according to whether the current air quality is above or below the specified standard.

Liquid or gaseous discharges may have quality standards—for instance, the maximum concentration of particulates may be specified—or the total mass discharge per day may be limited. In motor vehicles, the maximum allowable discharge of either carbon monoxide (CO) or of nitrogen oxides has been set in terms of grams per mile.

Fuels such as coal and oil may have limits to the amount of sulfur in them for them to be burned in certain areas or for certain uses.

8.2.2 The Concept of Threshold

For many of these standards there is an implicit concept of a threshold below which, it may be supposed, the damage to health and perhaps to property will be negligible. There is a widely held belief that health damage from pollutant concentrations is analogous to burn damage from hot surfaces. If the flesh comes into contact with a warm surface, no damage whatsoever will normally occur. If the temperature of the surface is raised, a range of temperatures is reached in which the skin will suffer burns of increasing severity, dependent to some extent on the intimacy and length of contact with the surface, but the burns will heal completely if they are subsequently treated with care. Burns produced from contact with surfaces at temperatures above this range may result in permanent damage or even death.

The body reacts in a similar way to other injuries, from cuts and bruises, for instance, so that it is a natural extrapolation to assume that exposure to carcinogens, say, can be treated similarly. However, it is pointed out in Chapters 4, 5, and 6 that there is increasing evidence that for carcinogens in particular, and very likely for many other categories of pollutants, there is no threshold below which irreversible damage may not occur.

It is therefore desirable that such emissions be reduced to very low levels. But it does not seem desirable that the concept of threshold disappear for legislative purposes. To achieve zero discharge at all times and at all places may be virtually impossible. Increasingly sensitive instruments will become available with which inspectors will be able to demonstrate that some discharges, even if they are infinitesimal, will have occurred. If the legislation does not prescribe a threshold, the inspectors or the courts will do so, or otherwise there would have to be a wasteful expenditure of effort in prosecuting, defending, and trying innumerable rather trivial cases. There is also little point in devoting society's resources to reducing to zero the exposure to one type of pollutant when there may be a natural background level that is relatively high and quite unaffected by such measures.

How then can we ask for pollution to be reduced when we do not have regulation to reduce it? It is here that the pollution charge is important. We have shown in Chapters 4, 5, and 6 that we cannot scientifically demonstrate a threshold concentration below which pollutants produce no health effect; but we can legislate a threshold, or a no-effect level, below which no government action would be taken. Such a level might be a level below which no health effects have been proved, with an appropriate safety factor.

To some extent, this procedure is used for setting the air quality standards in the United States. However, we note here several problems. First, there is no uniformity in the setting of such standards.

Second, a standard has been set only for those pollutants for which a reasonable body of observational data exists (SO_2, NO_2, O_3, particulates, CO) and no standard has been set for sulfates, which may be a worse problem. Third, as emphasized in Chapter 7, reliance on rigid standards will tend to lead to a situation where all air just meets the standards. If, as we suggest may be the case, the health hazard is proportional to concentration and no threshold exists, the enactment and enforcement of health-based standards may then worsen the effects of air pollution.

Roberts [1978], discusses how bureaucratic inefficiency, lack of incentive, and delays have prevented improvement of air quality. He suggests many improvements. But his suggested improvements would improve air quality only insofar as the standards would be violated less frequently. Our suggestions in many ways go further.

8.3 POLLUTION PRICING

In the preceding discussion of legislative standards, below which no violation of pollution regulations is deemed to occur, we presuppose that the method of government regulation in which standards are stipulated and penalties are prescribed for violations has been chosen.

There are, however, alternatives to the setting of standards. One alternative is the pricing of, or charging for, pollution [Wilson and Chen 1975]. In this approach there would be no clean air standards; there would instead be rather high limits (or even no limits) to the amount of pollution that may be legally discharged. But the amount of pollutants emitted would be carefully measured in some acceptable way, and the polluter would be billed at a set rate per unit mass emitted. The amount that has been suggested for sulfur emission is in the range of 10¢ to $6 per pound, based on risk considerations such as those discussed in Section 8.1.

Such a policy has gained increasing acceptance overseas, especially in Germany, and at least in academic circles in the United States as one having certain advantages in equity, in allowing industry freedom to plan and to choose among alternatives, and in achieving a fast reduction of pollution levels [Kneese and Bower 1968; Congressional Research Service 1977].

The rationale behind pollution emissions charges is that clean air and water are scarce resources because so many wastes are being generated in our modern industrial economy that they exceed the limited carrying capacity of air and water. It is a well-established fact of basic economics (the theory of public goods) that when scarce resources are made available at no cost or under no control, they will be overused. Thus pollution is basically a problem of scarcity of clean air and water. Because the market economy has in the past proved to be a useful and efficient method of dealing with scarcity problems in our society, it could be applied to the pollution issue. W.L. Hoskins [1971] asserts that to solve the scarcity problem of air and water requires "the definition of the rights to use environmental resources and the provision of a method for exchanging such rights." The provision can be in the form of emissions, or residuals, charges, or taxes.

Freeman and Haveman also wrote that charges work by "reproducing the effect of private markets by charging a price or fee to those who would use the common property resource" [1972].

8.3.1 Lower Costs

Economically, it is clear that enforcement at a certain standard is not always the most efficient way of reducing pollutants in terms of social benefit and social cost of its implementation. Solow [1971] gives the following illustration:

> If two factories producing different commodities both contaminate the same stream to the same extent, it might seem natural to require each of them to reduce its contamination by, say, 50 percent.

But the two factories have different incremental abatement costs (they use different production methods). A cheaper alternative that gives the same total reduction in contamination would be to require the factory with the smaller incremental abatement cost to reduce by more than 50 percent:

> Since it is the total amount that matters, the cheaper possibilities of reduction should be exploited first. . . . By charging the two factories an amount proportional to their emission of pollutants, they themselves would see to it that the reduction in pollution occurred in the cheapest possible way.

8.3.2 Simple Administration

In practice, commented Freeman and Haveman [1972], "the number of violations (whether or not detected) depends on the cost of compliance with the regulations, the penalties associated with being caught in a violation, and the probability of being detected in violation and having the penalties imposed." To set the limits is to encourage efforts to find loopholes in regulations and to require a large number of enforcers if the limits are to be adequately enforced.

Once the charge system is set up, "there is much less room for litigation and other evasive tactics, such as lobbying, pressure-group persuasion and legal maneuvering" [Lambelet 1972]. Also, because payment of a tax or charge does not of itself cause unemployment by shutting down an industry, this system needs little interpretation by administering officials. "The bureaucratic component is lessened" [Solow 1971]. There will be an incentive for the polluters to abate pollution as quickly as they can because taxation, once in effect, is an immediate economic cost that the polluters will try to reduce. The incentive to use new technology in production or new waste treatment facilities to cut down on pollution is thus spontaneous. The control agency needs no information on device installation and operation of the plants or factories except for the amount of pollutants emitted [Freeman and Haveman 1972].

As summed up by a report for the Senate Committee on Public Works for the case of sulfur emission:

An emissions charge appears to be a well-suited policy instrument for inducing efficient sulfur-emissions control. The application of an emission charge on SO_2, perhaps at the level of the estimated incremental cost of the pollution consequences from the average power plant, would provide a strong, immediate, and across-the-board incentive to undertake emissions-controls activities. At the same time . . . it would permit flexibility of response. This flexible response would be achieved in a decentralized manner without the necessity of administrative agencies and the courts trying to decide every individual case in an adversary atmosphere. The latter approach invites delays and, frequently, arbitrary decisions, and establishes the incentive to hire lawyers rather than to proceed with emission control. Emissions charges exert a persistent incentive to act whereas variances and delays in imposing requirements allow the emitter free use of environmental resources, with no incentive to act as long as these (free resources) can be obtained. Moreover, a charges policy would have desirable efficiency characteristics. It would tend toward an application of controls first at those locations where costs per unit of SO_2 reduction are lowest. In the longer run it would provide a powerful spur for the development of more efficient technologies [NAS 1975].

Reliability of electric-generating facilities and industry generally will increase with the use of pollution charges. If pollution control equipment is out of action, the industry could still operate even though the pollution would be greater. At the moment, industry must close or seek a variance to continue to operate. The use of a pollution charge constitutes an automatic variance but with an automatic payment. There would be no waiting for the bureaucratic or political process to operate.

At the same time pollution would be less. The economic pressure on bureaucrats and politicians to grant variances is now great; when a variance is granted there is no pollution charge, and only a political incentive exists to repair or to install the pollution control equipment. Unfortunately, the economic incentive is to leave the pollution control equipment inoperative for as long as possible. An emission tax would provide a financial incentive to expedite the repairs.

8.4 PROBLEMS OF ADMINISTRATION AND FORECASTING

The problem of enforcing pollution emission controls was realized by Senator Edward Muskie in a Senate discussion of the Energy Supply and Coordination Act.

An important classification in the conference report relates to enforcement of interim procedures to ensure compliance. Senate conferences insisted that the Environmental Protection Agency's determination that emissions from coal converters would not cause primary standards to be exceeded must be articulated in emission limitations or other, precise enforceable measures for regulating what comes out of the stack. The conference report on this bill underscores the fact that it is not ambient standards which are enforced but emission limitations or other stack-related emission-control measures. Ambient standards are only a guide to the level of emission controls which must be achieved by specific sources. In 1970, we recognized that a control strategy based on a determination of ambient air-pollution levels in relation to each individual source would be unenforceable. Existing clean-air implementation relies specifically on the application of enforceable controls against specific sources. We have continued that procedure in this law [Congressional Record 1974].

8.4.1 Small and Large Polluters

When there is only one polluter in a geographical area, it is obvious how to relate the ground concentration to emission. This particularly applies to large electricity-generating stations in rural areas. For these instances, the EPA, in its conference procedures, assumes on good faith that the large polluter will take the responsibility to meet air quality standards. This is very difficult for sulfates in contrast to SO_2. The sulfates, as we have noted, are widely dispersed, and there are almost no places in the United States where we can claim that only one polluter is the cause of the sulfate concentration. Any discussion of sulfate control must face this problem squarely.

8.4.2 Forecasting

The intermittent control schemes are aimed at reducing short-term concentrations of SO_2. Because it may take a few hours for the emission from a given chimney stack to reach a given sampling point, it is not sufficient to rely upon feedback from the measurements themselves. Weather forecasting must also be used. Who is to produce this weather forecast: the polluter or government? What incentive is there for this to be done properly? When ICS were discussed, there was much concern that if the forecasting were left up to the polluter, there would be too much reluctance to initiate costly power reduction as often as desirable. Indeed, Downing and Watson [1974] argue that it is cost effective for industries to violate the law (pollution standard) on frequent occasions; often violators will not be found out, and often the penalties for transgression will be small. Nor might the situation be appreciably different if a public official made the forecast: if the cost of power reduction is great, the potential for political pressure or bribery is also great. However, we are asking for cost effectiveness for society as a whole, which can be appreciably different.

Few intermittent control systems have been tried and only for the largest industries. None of the problems discussed in this section have materialized. But as more systems come into operation, and as the threat of more expensive and stringent control technologies recedes, the likelihood of evasion will increase.

We therefore note one advantage of a pollution charge: insofar as possible it will be related to actual emissions or actual concentrations and not to forecasts of them, and there should be no question of abatement because of erroneous forecasting "in good faith." The forecasting can then be left to whomever wishes to do it, and believes it can be cost effective.

8.4.3   Treatment of Sources

Because about half of the sulfate and particulate emissions comes from small sources that are predominantly in urban areas, the basic pollutant charge would be set at a rate appropriate to a dense urban area—about $3 per pound of sulfur in the fuel. This would apply to both stationary and moving sources. Most sources would not be monitored individually, and the charge would be assessed at the wholesale level. This charge could be reduced to $1 per pound of sulfur, with appropriate reductions in rebates, for fuel sold in states of low population density upon request by the state.

The larger emitters would find it advantageous to request a modification of the flat charge per pound of fuel; they could request that meteorologic calculations be performed for their sources. They could then request a modification of the charge per pound of sulfur in the fuel based on the results of the meteorologic calculations, or, if they use stack gas suppression and if they supply emission monitors, they could ask for charges based on the actual emissions. A yearly administrative charge would be established for this modification.

The schedule of charges and the method of computation, including upgrading to allow for inflation, could be reviewed automatically every five or ten years.

8.5   PRACTICAL IMPLEMENTATION SCHEDULE AND COSTS FOR THE INTRODUCTION OF EMISSION CHARGES AND REBATES

We will assume that the U.S. Congress would pass into law an act that would give monitoring responsibility to the Environmental Protection Agency, with a committee from the National Academy of Sciences advising on health and other damage costs. A similar arrangement was set up for the Clean Air Act. The EPA would assess the amounts to be paid by the emitters, and the billing and the distribution of the compensatory rebates* would be carried out by the

---

*Compensatory rebates are discussed in Section 9.1.3.

Internal Revenue Service according to agreed formulas. The schedule and associated costs could be as follows.

1. The computer program relating emissions to topography and meteorology, developed at the Brookhaven National Laboratory [Meyers et al. 1978], or some refinement of this program, would be used to establish initial charge rates, in cents damage per kilogram of emitted sulfur, for each major point source. The time and cost for this step should be very small—at the most perhaps six months and $50,000. Rebate contours by zip code (or other code compatible with IRS data) would also be determined.

2. Each major point source would be equipped with stack gas-monitoring and recording equipment that would relay readings to local regional offices or recording stations of the EPA. For an estimated 1,200 major sources and a maximum cost of $50,000 per source, the cost would be less than $60 million. This step could be accomplished in eighteen months.

3. Monthly, the local EPA would report the total emission of each source in the region to the federal EPA, which would multiply the quantity by the charge rate for each source and transmit the data to the IRS for billing.

4. Annually, the total charges collected minus the administration and monitoring cost (which perhaps should be limited by legislation) should be returned to individuals by the federal Internal Revenue Service at the time of the tax return. The level of rebate per person should be determined by the corresponding person's residential address or zip code, which should in turn correspond to certain known mean levels of sulfate, SO_2, and particulates.

5. Ideally, similar charges and rebate schemes could be applied to all pollutants (especially particulates and nitrogen and carbon oxides). All sources, including mobile sources, should be charged for emissions and the total revenues collected minus the administration cost should be returned to individuals. Economic optima and equity could then be achieved. In fact, smaller sources will probably pay a sulfur tax on the sulfur in the fuel they purchase.

6. The undesirably severe impact on industry because of rapid transients should be avoided by introducing the emissions charges by increments (for example, the charge could be zero for the first year, 25 percent of the total for the second year, and 50 percent during the third year). The schedule should be decided by Congress.

7. It is our present suggestion that sulfur be used as a surrogate for all air pollution and that therefore charges be based on the mass

of sulfur discharged. This charging arrangement would have to be reviewed—perhaps every five or ten years—as abatement procedures improve and the pollution mix changes, as scientific knowledge improves, and as inflation erodes the value of the charge.

8.6 POLICY QUESTIONS

In the previous section we have advocated and in the next chapter we discuss in detail a pollution charge as a way of providing incentives for pollution control. We are not alone in this advocacy. Why is this approach not adopted? We here outline some possible reasons. (1) Industry is afraid that a pollution charge will be added as well as a regulation and not instead of a regulation. (2) People believe that a pollution charge is inequitable. (3) Environmentalists object to a "license to pollute." (4) The claim is made that a pollution charge is complex. (5) Pollution charges are incorrectly believed to be inflationary.

It is instructive to consider a crude argument from Lave about why we have our present policy. We can construct a table (Table 8–1) showing in the left column the possible types of action, and grading them in terms of simplicity, efficiency, and perceived equity. Then we can assign grades 1–6 (low grades being desirable) to the possible actions, and add. Of course, assigning the grades is subjective and each reader will make his or her own assignment.

The point, however, is that with the grades suggested by Lave, the national emission standards, the present approach to pollution control, have the lowest number (highest grade). Pollution charges are generally perceived as complex and inequitable. Little consideration has been given to the disposition of the collected funds. If properly set, with feedback of the collected charges, we believe that neither statement is correct.

By 1980, the major short-term effects have been reduced by standards which have led to the construction of, for example, tall chimney stacks. This makes application of a pollution charge for the remaining long-term effects simple to apply and equitable in its application, contrary to previous perceptions. If pollution charges are set according to the scheme outlined in Section 9.2, we believe that the scores in the last row are more reasonable.

Table 8–1. Scoring methods of pollution control.

	Simplicity	Efficiency	Equity	Sum of Scores	
No action	1	6	6	13	Grades suggested by Lave (low grades being more desirable)
Pollution charges	6	1	5	12	
Concentration Standards – National	4	3	3	10	
– Local	5	4	2	11	
Emissions Standards – National	2	5	1	8	
– Local	3	4	2	9	
Pollution charges outlined in section 9.2	3	3	1	7	Grades suggested by the authors of this book

9 POLLUTION PRICING

The pollution pricing approach appears to have many advantages in regulating emissions resulting from coal burning. We will therefore review in detail past implementations of pollution pricing applied to other aspects of environmental damage. Also we will describe the results of studies of alternative methods of pricing sulfur emissions and of rebating the taxes collected back into the economy.

Many economists, scientists, and legislators have suggested that the use of a residual charge (or pollutant tax, or emissions charge) on each unit of pollutant discharged would be the best strategy for controlling pollution.

The prospects of acceptance of pollution pricing (emissions charges) have been greatly improved by the publication of the quoted and highly authoritative report for the Senate Committee on Public Works, by the National Academy of Sciences and other bodies which stated [NAS 1975]:

> When first suggested the idea of emissions charges was greeted with some skepticism by many policy makers and environmentally concerned persons. For various reasons industry was also opposed. In recent years this policy option has gained increasing acceptance among conservationists, environmentalists, policy makers here and abroad, and even industry, as indicated by a recent committee on economic development report.

A follow-up report for the same committee specifically on pollution taxes and effluent charges [Congressional Research Services 1977] reiterates these views.

Johnson and Brown [1975] state that the adoption of effluent charges as a policy is a simple procedure.

> Typically the charges begin at low levels with explicit steps for increases in the future. Charges are keyed to a few pollutants and may expand to cover more pollutants as the charge system becomes accepted and techniques for measurement are developed.

9.1 POLLUTION PAYMENTS

Chapman [1973], employing mathematical modeling on the consequences of taxing sulfur emissions, concluded that emissions and damage would be significantly reduced with this method as compared with alternative methods of regulation. A system that charges for industrial wastes entering the sanitary and storm sewer system of an unidentified eastern U.S. city was reported by Ritter and Podolick [1973]:

> Industries discharging high-strength waste, for which treatment costs are proportionally higher, are required to pay a greater portion of total operating costs than those companies discharging wastes of domestic or near strength.

The result:

> ... Revenues obtained from application of the surcharge are proving significant in helping to offset costs of treating the high-strength industrial wastes. Equally important is the fact that the strength of wastes received at the treatment plant has been reduced.

This case provides an actual example of the effectiveness of pollution charges. In addition, EPA regulations require cities to collect "user charges" from the users of federally funded municipal sewage systems. The charges depend on the costs of the services rendered.

In France, the problem of water pollution is controlled by six River Basin Agencies, which charge fees to industries and municipalities. The charges depend on the abstraction of water or the amount of pollutants. For example, the charge is higher for a polluter at the upstream end of the river basin. In 1974, the water pollution charge ranged from $0.50−2.00 per population equivalent (a measure of biological pollution) in the Seine−Normandie river basin and

$1.00−1.50 per population equivalent for the city of Paris (EST 1974]. Such charges obviously act both to reduce pollution and to encourage downstream location, where both the perceived and the social costs of pollution are least.

The Organization for Economic Cooperation and Development made studies of air pollution charges for Norway, Sweden, and Denmark. In our view the suggested charges (less than 10 cents per kilogram of sulfur) were much too low and bore little relation to the estimated health costs.

In Germany, a widespread system of effluent charges was introduced in a large number of water-use associations under the provincial governments, starting in 1904 in the Emscher River Basin. Complex formulas were used in many cases to define the damage resulting from unit flow of effluent from various types of plants— slaughterhouses, paper plants, sauerkraut plants, and so forth. It is obviously a far more difficult political step to establish relative charges for all possible pollutants than it is for just one—sulfur—as is being proposed here. The use of effluent charges spread and were apparently successful as a policy instrument, at least up to the time of the Nazi regime.

The German government is currently proposing to make water quality management a federal, rather than a provincial, responsibility. (There has been divided responsibility at times and there have been some previous attempts to centralize control under different governments.) The effluent charge system will probably be confirmed and extracted, to judge from a policy statement published in 1971 and quoted by Johnson and Brown [1975]:

> In a free-market system, one of the generally recognized principles is that all costs are to be attributed to the products or services which cause them. In other words, cost allocation is made on the basis of the originator principle. According to the principle, the cost of damage done to the environment must be borne by the individual or organization which is responsible for the damage. . . . However, under the present system of sharing the cost of damage to the environment, the originator principle is largely ignored, and this cost is imposed on the community without any reference to the product or service which caused it. The community is compelled to accept damage to its environment and must raise money to eliminate it. This means that products and services which do not carry the burden of the cost of protecting the environment can frequently be offered at lower prices than would otherwise be the case. This, in turn, may even increase the demand for such goods or services, causing capital to be misdirected and a diminution of economic efficiency.

This excellent statement of the beneficial incentives that follow from effluent charges, and the negative incentives that follow from sharing the cost of pollution among the general population, is illustrated by the well-known economic allegories of the tragedy of the commons and of the shared lunch [Baxter 1974]. If a dozen people eat lunch together and agree in advance to share the bill equally, each person has an incentive to eat more expensively than usual. For his *extra* expense will be shared among twelve people. If each person has a separate check, he has a financial incentive to keep the cost low.

We repeat that the pollution charge must be adequate to compensate for pollution and paid to the people who suffered from it; in this case the citizens of Holland downstream from Germany.

9.1.1 Limitations of Pollutant Charging Proposals

The literature reviewed has shown strong arguments for pollutant charging, to the extent that it is implied to be a universal solution. However, we believe that this is not the case. Emissions charges are suitable only for those cases where the administrative costs of collecting and enforcing the charges are small compared with the funds collected. As an arbitrary rule, we advocate that if the total costs of measuring and administering emission charges are estimated to be more than 20 percent of the charges to be collected, it is better for government to set and enforce a standard. The level at which the standard is set can be estimated by approximating abatement-cost and population-disbenefit curves (see Section 9.2.1).

This argument leads to the possibility that a pollution-charge method of control might eventually be changed to a pollution standard if the economic optimum level of pollution falls, because of improvements in technology or reductions in costs of pollution abatement. The regulatory body might then require that emissions be below limits that will vary with locations; that below those limits there be no charges; and that a more normal system of fines and court procedures be used for the presumably rare cases of polluters exceeding the limits.

The question of when to switch from charging polluters to enforcement has been discussed by Downing and Watson [1974]. They assert that the net benefit of control (benefit of control minus the total administration and monitoring cost) for each alternative (charging and enforcement) "be compared and the scheme with the largest net benefit chosen." This analysis must then be carried out from time to time to determine if the enforcement scheme has a greater net benefit than the charging scheme. At present, Downing and Watson concluded that the indication is that "an effluent-fee scheme would be optimal in controlling fly-ash emissions from coal-fired power plants."

Whether or not the pollutant charge approach is appropriate to pollutants that can cause major irreversible changes in the environment or in health is debatable. Obviously it cannot be used alone: a maximum limit that, if exceeded, would bring about legal action would also be required. However, it is likely that the costs of monitoring and charging for such necessarily low-volume pollutants as mercury or arsenic would probably make the pollutant charge method unattractive.

9.1.2 Determination of Charge Level (Charge-Concentration Relation)

The question arises of how much to charge per unit of pollutant. One might think that the charge should be set to equal the marginal cost of social damage, which will equal the marginal cost of abatement at optimum pollution level [Freeman and Haveman 1972]. There may also be an argument for the charge to be higher, on the grounds that because there will be a net increase in efficiency, it is reasonable for the population to share the gains with the polluters.

Although the abatement cost for a pollutant-emitting source can easily be estimated, the damage function is usually difficult to establish in monetary terms. An alternative is to set the charge "at the marginal cost of waste reduction at the level of control required to attain the desirable standard"[1] that is first established by the government. Even if the marginal cost of waste reduction is not known with accuracy, the control agency could adjust the charge until the desired level of pollution is achieved [Freeman and Haveman 1972].

---

1. The fact that the "desirable standard" is most probably not the optimal standard is a drawback of this approach.

In practice the agency would need to predict the long-term response in addition to measuring the short-term changes. Thus even if the damage function cannot be determined and a more or less arbitrary standard would have to be set, it would still be preferable to achieve this standard by means of emission (or effluent) charges, as "it will ensure efficiency in the sense that the arbitrary standard will be achieved at a minimum cost" [Lambelet 1972].

If no technology were available for abatement, even arbitrary pollution charges in such a situation will be an incentive for the polluter to invest funds in research and development of new waste treatment technology. On the other hand, without this incentive, the polluter may be reluctant to risk investment in pollution control research and may use the unavailability of abatement technology as an excuse to continue polluting.

We can see from the foregoing that (1) the damage function (cost of externalities) cannot be exact; and thus (2) the determination of the charge-concentration relationship is essentially political.

A condition of pareto-optimality will exist if the polluter behaves rationally. That is to say, in a system of control involving emission charges together with transfer of the money collected to the people most affected by the pollution, the polluter will perceive choices over a wide range extending from doing nothing and paying high charges, which will be transferred to local affected individuals or communities, to completely eliminating pollution at high cost and paying no charges. The polluter will take the step perceived to yield maximum net profit or minimum net cost. The "external" public will be fully compensated whichever choice the polluter makes (Figure 9–1) and whether or not the polluter behaves rationally. It should matter little to the average pollution-affected individuals exactly where the new equilibrium point is compared with the status quo ante, because they will have a lower level of pollution than would otherwise be the case and at the same time will receive compensation rebates for the remaining level of pollution.

The apparent optimum pollution level reached should be of no concern to government so long as all substantial external costs are covered. For instance, it is very likely that the cost relation for pollution abatement will drop considerably as new technology is developed in response to market demand. The economic optimum level of pollution would then fall, and the amount of compensation received by the affected community would also fall. If the tax-concentration

Figure 9–1.   Comparison of the costs of pollution abatement, taxes, and administrative costs, showing how an "optimum" pollution level can be derived.

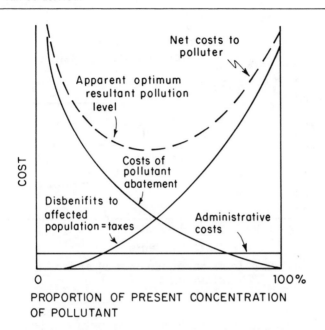

relation has been well chosen, the population affected should be, on average, neither distressed nor overly delighted that their "pollutant bonuses" were decreasing simultaneously with their pollutant levels.

This point serves to emphasize that the choice of the tax-concentration relation is essentially political. Politicians will need the guidance of scientists, particularly of biomedical scientists, who can estimate health and other effects as a function of pollutant concentrations. But the final judgment is purely political. There is no exact science of the pricing of externalities.

There are secondary external costs associated with disruption when changes in regulations are made very rapidly. Such transients should be avoided by introducing pollutant taxes by increments extending, usually, over several years on a predetermined schedule (Figure 9–2).

In fact, the gradual imposition of emission charges enables political decisions on the appropriate level of charges to be made rather

Figure 9–2.   Qualitative prediction of emission of pollutant concentrations as the pollution charge is introduced.

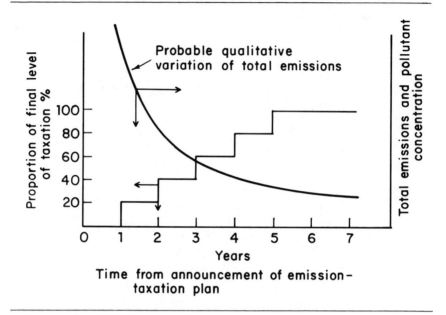

effectively. As the charges are increased toward a level previously judged to be appropriate, politicians will receive feedback from the polluters, and from the population experiencing both the pollution and the compensatory bonuses. One side will want the charges to go no higher, and the other will presumably demand that the rate of increase of the charges, and therefore of the bonuses, be increased and to go to an eventually higher-than-planned level. This is the stuff politics is made of, and politicians are best able to judge the level of charges at which the best interests of society are met. Because of the large curvature of the typical optimization-concentration curve (see Figure 9–1), precision is not required and the inequities consequent upon being a little removed from the optimum—an optimum that cannot be arrived at except politically—are small.

9.1.3   Use of the Revenues

On the subject of the use of the revenues received from pollution charges, an interesting question arises. One opinion is to use it "to further improve the environment or to assist genuine hardship cases or to accomplish socially desirable ends of any kind" [Solow 1971]. Another similar suggestion, in the case of water pollution, is to apply the revenue toward the cost of public investments in reservoirs or downstream treatment facilities [Hoskins 1971].

We would favor the distribution of the revenue to the individuals or communities affected by the pollution in some approximate proportion to the integrated exposure, at least in those cases where the funds distributed are sufficiently large to be a welcome addition to a household or community income.

If we are agreed that revenue collected should approximate the damage done, how do we compensate those that suffered the damage? One way is to give them, as in the first two cases, some investments that are probably well intended for their good. But will the investments really be in the best interests of those who suffered the damage? Must the government decide how to compensate those that suffered?

One view of the function of government is that it exists to promote equity among its citizens [Rawls 1957]. If so, citizens should be awarded exactly what they earn, be compensated for exactly how much they have been damaged, and punished for exactly how much damage they have done. Because the polluters were made to pay for what damage they have done, should not those who suffered the damage be compensated? One use to which individuals could put pollutant-compensation funds is to move to a less polluted area. If the cost of relocation is less than the cost of damage they suffered, moving will mean a net increase in total social benefits [Coase 1960] and is thus preferred. We believe that those who suffer the pollution damage should have the right to make such a choice.

In his study of the consequence of taxing sulfur emission, Chapman [1973] concluded that the electricity rate increases induced by a pollution tax would be regressive, a situation that seemed unfair to the lower-income group. But Chapman quotes A.M. Freeman III as having shown that "lower-income groups are disproportionately exposed to sulfate and particulate pollution. . . . It is probable that reduction of sulfur emissions brought about by a tax would benefit lower-income groups." In other words, even in the absence of compensatory payments, there could be a net gain for this group, although not in cash. Feedback of the pollution charges in proportion to the local pollutant concentrations would restore complete equity to this group, because in general the only people who at present cannot afford to move out of high-pollution areas are those with low incomes. Low-income people then experience a proportionately larger share of pollution costs than does the higher-income segment of the population. Alternative mechanisms of rebates to pollution-affected people are discussed in Section 9.5.

9.1.4   Estimated Maximum Costs of Implementing a Pollutant-Pricing Policy

We will show why sulfur and particulate pollution from stationary sources is suitable for the application of emission charges, using the results from the risk-benefit analysis of Sections 8.1.1 and 8.1.2.

We have estimated the total health-related damages to be $53 billion per year. The national cost of material damage by sulfur oxides has been estimated as $8.3 billion in 1968 [Barrett and Waddell 1973b]. We can thus estimate the material damage by sulfur oxides from stationary sources as $6.5 billion in 1968, using the proportion of emissions from stationary sources given by the Nationwide Inventory of Air Pollutant Emissions. At an average inflation rate of 7 percent per year (which unfortunately is lower than actual current inflation), the material damage in 1979 would be $17.5 billion. The total annual damages from sulfur oxide emissions in the United States could then be estimated to be about $70 billion in 1979 dollars.

In what follows, we will assume that damages from fine-particulate emissions are included in these figures. Particulates and sulfates appear to have a synergistic relationship, and attempts to account separately for the effects of particulate could lead to double-counting.

Neglecting the small stationary sulfur emission sources, we will estimate the number of major stationary sources. Assuming that the sulfur oxides from stationary sources are mostly SO_2, the total sulfur emitted was $22 \times 10^6 \times (1/2) = 11 \times 10^6$ metric tons. Using the average fuel consumption of stationary sources as 3×10^5 tons per year (equivalent to a 150 Mw power plant) and using the figure of 3 percent as the average sulfur content, the estimated number of sources = $11 \times 10^6/3 \times 10^5 \times 0.03$) or about 1,200. This number compares satisfactorily with 1,088 reporting sources used earlier in the chapter.

We can now estimate the monitoring and administration costs, and can find if these would be a major percentage of the total damage caused by sulfur emissions from stationary sources.

Let us take what seems to be extremely high estimates for the monitoring cost: an additional staff of 1,000, at an average salary of $20,000 with an overhead of 100 percent. We assume a first cost of the monitoring equipment of $50 million, and maintenance and depreciation costs of $13 million per year. Then the total cost is $40 million + $13 million per year = $53 million per year = 0.075 percent of the estimate of the pre-existing total damage ($70 billion).

We can see that the estimated total cost of administering the charge scheme would be much less than the total estimated damage (and therefore eventually the estimated revenue), even if the administration costs are as high as these estimates, which is unlikely. We have estimated that the incentives introduced will be likely to reduce the total damage by a factor of about five, to 20 percent of its unregulated value, but even if the same large staff is required at this point the cost will be only about 0.4 percent of the damage. The cost of the emission charge scheme is small and thus, by our criteria, preferable to the enforcement of standards for low pollution levels.

9.2 PROPOSED CHARGE SCHEMES FOR SULFUR AND PARTICULATE EMISSIONS

It will be appreciated from the foregoing discussion that an application of the emission charge principle involves decisions in three areas: (1) the measures and formulae used to calculate the charges; (2) the level or rate of the charges; and (3) the disposition of the money collected.

Some alternatives in each category have been mentioned. In this section, some very approximate calculations will be made or quoted to illustrate the scale of the transfer payments and thereby to enable the incentives on polluters to reduce their emissions to be estimated for some representative charge levels.

9.2.1 A Flat Charge on the Sulfur Content of Fuel

This scheme has been in use in Norway [OECD 1977]. The level of charges there, at about 3.5 cents per pound of sulfur, is much too

small to correspond to the level of property damage and health hazard. Moreover, they are too small to induce fuel substitution even by
a fuel of lower sulfur content. The main function of the charge is the
collection of taxes for subsidizing the government's expenditures.

This approach would be effective in reducing pollutant concentrations if a maximum charge of, say, 30 cents per pound of sulfur in
the fuel were made. At this level, it would be economical for large
stationary sources to switch to expensive low-sulfur fuel. Any reduction from this basic charge would have to be on application and justification by the polluter, presumably after a public hearing. For
instance, it is possible that a seashore power plant could have its
charge reduced by perhaps 20 percent because the sulfur causes presumably zero damage in the sea. Any power plant operator who felt
that the charge of 30 cents per pound of sulfur in the fuel was unjus-
tified could ask for a public hearing.

A flat charge on sulfur content, set at a level that would compensate for national damage, would be a major improvement in equity
and effectiveness over the present pollution-control strategy because
polluters would on the whole pay for damages caused. North and
Merkhofer [NAS 1975], using very conservative and now outdated
figures, calculated that the costs of health and other damage per
pound of sulfur were 55 cents in urban New York and 21 cents in
nonurban areas. We showed earlier that the average national charge
could be conservatively estimated to be $2.18 per pound of sulfur. In
the work to follow we have used $3.00 and 60 cents per pound of
sulfur as representative possible final charges for urban and nonurban
sources. However, there would inevitably be cases where power plant
operators would argue vigorously that the sulfur they emit is much
less than the sulfur contained in the fuel; that the population density
around the power plant is low; and that the meteorologic conditions
are such that only the population in a certain direction and distance
could be affected by the sulfur. There could be lengthy hearings to
work out the local damages to determine the charge level. A charging formula that varies with locality and climate conditions would
reduce the scope for argument.

9.2.2 Transfer Payments (Rebate) for a Charge on Sulfur in Fuel

The average annual per-person transfer payments will be simply the total estimated health damage valuation that is recovered by taxing elemental sulfur content at an average of $2.18 per pound, divided by the national population:

$$\frac{\$53 \times 10^9}{217 \times 10^6 \text{ people}} = \$244 \text{ per person} .$$

In fact, it would be very unlikely that a rebate of this magnitude would be reached. The sulfur tax should, as affirmed earlier, be introduced by increments over several years: as relief from paying the tax would be granted to those sources that installed sulfur-removal equipment, there would be strong incentives to introduce technology with marginal sulfur-removal costs up to the future probable tax level. There would be large improvements in technology, and an attempt to predict the final tax and rebate level could be only a guess. One guess is that at least 80 percent of the sulfur would be removed by one means or other (reduced to 1/5 of original level), so that, assuming a linear relation of health damage to sulfur emissions, the rebate would reach:

$$\frac{\$244}{5} \sim \$49 \text{ per person} .$$

It takes approximately 0.750 lb (340 g) coal for each kilowatt-hour of electricity for an average plant [ASME 1973]. We will calculate the charge for 3 percent-sulfur coal with no sulfur removal equipment used, and for charge rates of 60 cents and $3.00 per pound of sulfur. These charges will then be identical whether based on sulfur in the fuel or in the stack.

Sulfur emitted $= 0.03 \times 0.75 = 2.25 \times 10^{-2}$ lb/kwh $= 10$ g/kwh .

Charge per kwh $= 2.25 \times 10^{-2} \times 60 ¢ = 1.35 ¢$/kwh for $60¢$/lb tax

or $\phantom{= 2.25 \times 10^{-2} \times 60 ¢} = 6.75 ¢$/kwh for $3/lb tax .

If the mean retail cost of electricity is 4.5 cents per kilowatt-hour, and if the charge is "passed through" without change, the percentage increase in cost of electricity = (1.35/4.5) = 30 percent for $60¢$/lb tax and 150 percent for a $3/lb tax. Those percentages will be reduced as increased fuel prices increase the mean retail cost of electricity.

9.2.3 A Flat Charge on Each Pound of Sulfur Emitted

A charge on sulfur emitted is an improvement over a charge on sulfur in the fuel because it would encourage fuel substitution or the use of sulfur removal equipment, depending on relative price and performance. Economic efficiency would thus be better served.

An equal rate for all power plants has the virtue of simplicity. Measurements of sulfur emission would have to be made reliably and inexpensively. Sulfur monitoring equipment is available commercially [Wolf 1974].

The costs of this simplicity are a certain degree of inequity and a reduction of incentives. This charge scheme does not differentiate according to local differences in damages — differences in ground-level concentration of sulfur on varying population distributions. Therefore there is no incentive for new power plants to locate in areas where fewer people would be affected. Power plants would be more likely to continue to be located near the consumers (more densely populated areas). The densest sulfur concentration is likely to continue to be in low-income as well as in industrial areas.

The scale of transfer payments is the same as for the fuel sulfur flat-charge proposal.

9.2.4 Charges Based on Integrated Pollutant Exposure

By making the emission charges proportional to local estimated integrated health effects, a function of population distribution, and of topographic and meteorologic conditions, equity and incentives are maximized at the cost of some complication. Thus charges would vary with region. Meteorologic data, population distribution, and dispersion models would all be used in the calculation of a damage function. Each stationary source could thereby be given a relative damage rating that when multiplied by a cost factor will result in dollars or damage per pound of sulfur emitted. The cost factor would be uniform over the country and would be chosen politically. The relative damage rating for any one stationary source would remain relatively unchanged, being influenced by meteorologic data, which presumably would not vary, and by population distribution around the source.

Trials of this approach must await the development of more complete computer programs and population and meteorologic data.

9.2.5 Charge per Pound of Sulfur and Fine Particulates Emitted, Weighted by Population Density

The local effects of air pollution are already mitigated by the use of tall chimney stacks and dispersion. Then the damage function of Sections 8.5.3 and 8.7 is probably related to sulfates and fine particulates. The population within 100 miles is probably the population at risk. The radius of 100 miles might be modified to be some convenient administrative boundary such as a state line.

The calculation of an estimated ground-level mean concentration of a pollutant at a distance from a point source has been summarized in Chapter 6. In a full application of the emission charge approach using the integrated pollution exposure as a measure of damage, the effects of all point sources on the pollutant concentration at each location would have to be estimated and added. Computer models are being developed to perform this complex series of calculations using the locations of all major point sources, and mean meteorologic data, as inputs. Then the variable inputs required are the actual rate of pollutant emitted from each point source, and the population distribution to a rather fine mesh—perhaps in square miles or finer.

The integrated pollution exposures so calculated can be converted immediately to charge rate per pound or kilogram of pollutant emitted from each source, and, if required, to a rebate per person in each locality, once the health effect or damage function is specified as a multiplying factor. We have maintained elsewhere that this multiplying factor should be arrived at through the political process, should be uniform across the country, and should be imposed by small steps over a period of several years.

9.3 CHARGE RATES

In any application of emission charges it is desirable, as stated previously, that the full charge rates should be approached by steps over a period (see Figure 9-2), in this case by periodic increases of the cost factor. Polluting industries would be given notice one year before an initial charge is applied and, perhaps five years before the estimated total charge is reached. Industry would thus be given adequate time for installing any treatment facilities, tall stacks, or other abatement procedures to cut down their eventual charges.

The impact on electricity prices would be greatly reduced by this gradual imposition of charges. By the end of five years the sulfur emissions would have probably been reduced enough so that the impact of the charge on the emitter would remain small.

Meteorologic conditions and population distributions will vary little from year to year, and it would probably be sufficiently accurate if both sets of data were updated at each census (every ten years). Only the total emissions need to be monitored on a year-to-year basis, and the total charge in that year would be determined by this information.

Any of the charge schemes described in Sections 9.2.1 to 9.2.5 can easily be used to charge large stationary sources such as electricity generating plants. Each pollutes enough that the individual charge would be much greater than the administrative cost of measurement, collection, and distribution. But it is important for fairness' sake to charge small polluters also. The charging scheme should be simple, even if at the cost of some equity. A suitable candidate is the system discussed in Section 9.2.1 whereby a charge is levied on the sulfur in the fuel, plus a levy on an average particulate load. This charge could be collected at the wholesale level (gasoline distributors, fuel-oil distributors, and coal distributors). This charge might vary a little from state to state in accordance with population density, as suggested in Section 9.2.4.

For the intermediate-sized polluting bodies the more equitable system of Section 9.2.5, in which actual sulfur and fine-particulate emissions are measured, would be adopted. In general this would lead to a smaller charge than application of the first formula; in order to prevent small polluters seeking the formula of 9.2.5 a fixed administrative charge would be added.

Small industrial sources could claim exemption from paying a tax on sulfur in the fuel only if they could prove that substantially all the sulfur is removed from the discharges.

It is important, however, that justifiable charges never be waived; otherwise there will always be an incentive to industry to delay any procedure, whether it be to move to a new site, to burn low-sulfur fuel, or to install stack gas scrubbers, that leads to pollution reduction. We mentioned earlier, and Roberts [1978] has emphasized, that the absence of any penalty other than a shutdown has led to a slower clean-up of urban air than was technically possible.

Likewise it is important, both from the viewpoint of incentives and for simplicity of administration, that a simple, conservative procedure be established for the charge rates, and that the onus be on the polluter to prove that the emission control employed justifies a modification in charge rates.

9.4 INCENTIVES AND SUBSIDIES FOR POLLUTION CONTROL FACILITIES

Another form of government action that may be taken to reduce pollution levels is to provide incentives in such forms as direct subsidy or tax reduction for the installation or use of pollution control devices or processes.

Tax exemption has been increasingly used as an incentive for industry to build pollution control facilities. In 1966 a federal incentive in the form of a 7 percent general investment credit for pollution control facilities was passed into law. This was later replaced in its entirety by the 1969 Tax Reform Act, which granted an accelerated amortization of pollution control capital expenses.

In many states there also exist one or more ways of economic relief for pollution control facilities such as grants-in-aid, property tax exemptions, income tax credit, sales and use tax exemptions, and accelerated amortization.

In general, such payments are made from general tax funds.

Johnson and Brown [1975], in reviewing European policies for reducing water pollution, point out that compensation rebates are a logical part of the effluent-charge approach.

> Those damaged by the residual pollution must be compensated in some way by the polluters since the polluter-pay principle means that the polluted ought not to be hurt. In no country using effluent standards do polluters directly pay the polluted. No country using effluent standards, therefore, believes in the polluter-pay principle absolutely. . . . When improved water quality is desired and yet rigid adherence to the PPP is rejected, there must be some form of subsidy or transfer of resources to existing polluters. . . . The essential part of this discussion is that the absolutist notions of the polluter pay principle and of subsidies to polluters are antithetical to each other.

9.4.1   Rebate Formulas

Annual mean sulfate levels vary greatly from urban to nonurban areas, and from west to east across the United States, as shown in Figures 6–1 and 6–2. In the northeast, urban concentrations are greater than 13 μg/m^3, and nonurban concentrations are generally from 5 to 9 μg/m^3. Most areas in the Midwest, by contrast, have urban concentrations normally below 7 μg/m^3, and nonurban levels range from 1.0 to 2.9 μg/m^3. Rebate formulas would help to equalize any discrepancy caused by a flat rate system.

Four alternative rebate formulas have been examined. The rebate funds may be divided equally per capita across the country, or the amount may be set to be proportional to sulfate mean concentration at the location of principal residence. For either of these alternatives, the funds may be returned directly to each individual, or they may be given to the local political subdivision, most likely the city or town.

We believe that a combination of equity and efficiency would be best served by dividing the country into areas of high concentration— mean annual sulfate concentrations of 5 μg per cubic meter and above—and of low concentration. In the high-concentration areas the rebate would be approximately proportional to concentration at each individual's principal place of residence, and it would be returned to individuals through the Internal Revenue Service once per year as a reduction, or a quasi-refund, of income tax.[2] In the low-concentration areas the rebate would be uniform per capita and the funds would be returned to the governmental bodies of the cities and towns. Whether or not the cities or towns should be required to devote the funds to uses that compensate the public in some direct way for air pollution damage—as subsidies to pulmonary medical-care facilities, for instance—would be a political decision.

---

2. There would consequently be additional administrative costs in sending checks to people who normally do not file income tax returns. On the other hand, if all transfer payments (social security, welfare, etc.) were made by one agency such as the IRS, as seems to some people desirable, there would be virtually no additional administrative costs in handling the pollution rebates.

9.4.2 Equal Per Capita Rebate

As stated in Section 9.3.2, earlier, under the health damage costs assumed, the average national per capita rebate would be $244 per year if there were no reduction in the present level of emissions.

However, with the strong incentives that industry would experience to reduce emissions, the final average per capita rebate for the nation as a whole might be of the order of $50 per year. Large areas of the country would have low levels of sulfur and particulate emissions, and in these areas an equal per capita rebate of, perhaps, $20 could be established. In the remaining comparatively smaller but more highly populated areas of greater pollution, a rebate approximately proportioned to local pollution concentration could be established.

If the estimated rebate is above about $20 per capita, it would seem worthwhile to make the distribution to each individual (including family dependents) through the Internal Revenue Service income-tax system because no more than a few percent of the funds would be diverted to administrative costs. If the equal per capita rebate is reckoned to be less than $20 per year, it would probably be preferable to distribute the funds to cities and towns in proportion to the resident population. The rebate mechanism actually chosen would, of course, be arrived at by the political process.

9.4.3 Variable Rebate Contours

When the estimated final rebate is substantial, say over $50 per person per year, a more equitable scheme than one giving equal rebate to all residents of a region would be to set the rebate level as a function of the ambient sulfate levels. A considerable amount of earlier work can be used to establish possible equal-rebate contours. In 1968 the Department of Health, Education and Welfare published a series of reports on the air quality control regions for various urban areas. The report for the Boston area is used here for the purpose of illustration [NAPCA, 1968]. The reports made use of a diffusion model to establish the equal-concentration contour lines and presented them in graphic or map form. These graphs, plus all considerations of a nonengineering nature, served as a guide to the necessary size of the air quality control regions.

Figure 9–3. Wintertime SO₂ concentrations in ppm in the Boston area; from NAPCA 1968.

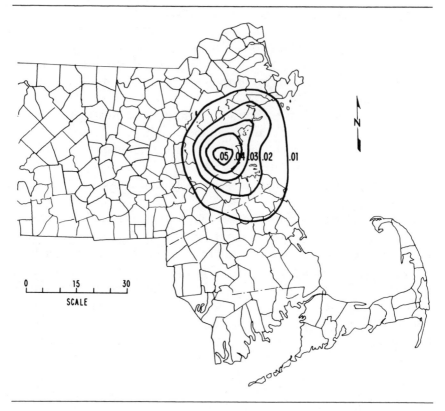

The contour map, Figure 9–3, shows a tendency to a leveling out of the concentration level of SO_2 at the 0.02 ppm contour. We can thus assume that outside the 0.01 ppm contour, the concentration level stays fairly constant.

Because no sulfate contour map is available, we approximate sulfate level by the following procedure. Finklea et al. [1975] found that, for urban areas in the northeastern United States, $C(SO_2)$ and $C(SO_4)$ (annual average) can be roughly modeled by the formula

$$C(SO_4) = 9 + 0.04\ C(SO_2)\ (\text{in } \mu g/m^3)$$

(Parts per million should be multiplied by 2,800 to convert to $\mu g/m^3$.) From these, the isopleth contours (in ppm SO_2) can be converted to isopleth contours in $\mu g/m^3$ SO_4 (Table 9–1, top).

Table 9—1. Relation among SO_2, SO_4, and rebates. Conversion of SO_2 to equivalent SO_4 levels.

$C(SO_2)$ (ppm)	$C(SO_2)$ $(\mu g/m^3)$	$C(SO_4)$ $(\mu g/m^3) = 9 + 0.04\,C(SO_2)$
0.01	28	10.1
0.02	56	11.2
0.03	84	12.4
0.04	112	13.5
0.05	140	14.6

Relation Between Sulfate Level and Per Person Rebate	
$\mu g/m^3$ SO_4	Rebate per Person per Year ($)
10.0	350
11.0	385
12.0	420
13.0	456
14.0	490
15.0	525

The total health damage cost per person was calculated earlier to be given by $35 times the SO_2 level in ($\mu g/m^3$). Then the isopleth contours can then be converted to constant rebate contours (Table 9—1, bottom).

If the administrative cost per person does not exceed $1—2, which seems likely, it would be only a small percentage of the total rebate, so that returning the rebate to every individual would be efficient as well as desirable.

For nonurban areas of the U.S. Midwest, where the annual average SO_4 concentration is presently between 1 and 3 $\mu g/m^3$ (corresponding to annual rebates of $35 and $105 per person, respectively), the administrative cost would be a larger proportion of the total rebate. This would particularly be the case if the introduction of incentives to emitters in the western states greatly reduced the already low SO_4 (and particulate) concentrations in the Midwest, reducing the annual rebates perhaps to the $5—$15 range. In this case, it would seem preferable to transfer rebates as a lump sum to the state or local government. It could be used, for instance, for subsidizing respiratory disease hospitals.

Material damage has been purposely left out of the personal damage-cost calculation. Perhaps damage to public buildings and facilities at least should be taken into account. One alternative, then, would be to return the material damage proportion of the rebate to local governments for maintenance of public facilities and recreational areas. A second alternative would be to return the material damage rebates as equal per-capita payments to individuals, on the grounds that the costs of all goods and services are increased. Because the amounts involved are smaller than the health damage costs, the small inequities involved (because the increase in costs of goods and services will be higher in some areas than in others) should be politically acceptable.

9.4.4   Disadvantages of Subsidizing Abatement Facilities

By subsidizing polluters to build waste treatment facilities, there may be little incentive for the polluters to find an efficient way of abatement. Polluters may even have an incentive to increase pollution on purpose to get a larger subsidy. In any event, there is little logic in a situation where the taxpayers pay for those who cause damages upon them [Solow 1971].

A report for the Senate Committee on Public Works cited by Solow [1971, pp. 189–199] strongly emphasizes that

> an essential device in limiting wasteful consumption of energy is to see to it that its price is equated to its marginal social cost of production. In fact, our pricing of electricity falls far short of this requirement in many ways.

In other words, the report suggests that electric power is presently underpriced, and that to meet marginal social costs it should be made more expensive. Therefore, the inclusion of external costs in the price of electricity would be acceptable.

However, two paragraphs later the report fails to appreciate the logic of its own argument that energy is underpriced, by suggesting that government subsidies are needed in addition.

> While efficient pricing of energy is one very important means for reducing wasteful use of electricity and, therefore, for reducing sulfur-dioxide emissions, it is by no means the only device, nor is it necessarily sufficient. Improved effectiveness of fuel utilization typically requires capital outlays

either by individuals or businesses. Capital shortages and high interest rates both inhibit investment in energy-conserving fixed assets. These considerations argue for government assistance to individuals and businesses in such forms as tax rebates or low-interest loans for energy-conserving investments. At the minimum, non-economic institutional barriers which make it difficult for many to obtain financing, even at market rates of interest, should be overcome.

We believe that this contrary argument represents neither reality nor the majority opinion of the authors of the Senate document.

9.5 COMBINATION OF REGULATION AND CHARGE

From the foregoing review it may well be concluded that society may best be served by a combination of regulation and charge. Regulation is necessary to reduce emissions to below dangerous levels. In certain cases, emission charges would provide the incentive for polluters to reduce their emissions to below this amount.

Regulation alone is appropriate when the cost of conforming is small, or, conversely, when the cost of measuring emissions and collecting the charges is relatively high. We suggest that regulations be retained (1) to ensure suppression of 99 percent of particulates; (2) to ensure that chimney stacks are high enough that short-term air quality standards are not violated; and (3) to preserve greenbelts and rural areas and other quality-of-life amenities. This will avoid clashes between environmental and health goals [Baxter 1974].

We propose that pollution charges be used to bring the average levels of pollution to below the primary air quality standards. Then the pollution charge becomes simple and equitable to administer.

10 LEGAL PROBLEMS OF POLLUTANT CONTROL

In this chapter we will outline historically the way in which the United States has developed laws to cope with the problems of air pollution as they have been perceived at specific times. We show how the central dilemma we pose in this book—should we, and how should we, control low levels of pollution—has been discussed by Congress and the courts, and how there still remains a large area for interpretation. Because long-range transportation of pollutants is an important issue, we conclude with a summary of interstate pollution source cases.

10.1 THRESHOLD AND STANDARD VERSUS ZERO POLLUTION

It is becoming fashionable to assume that no threshold exists and that at low exposure levels the adverse effect—mortality or morbidity—is proportional to the exposure. The laws have not been consistently changed to meet this new perception.

This confusing situation is made worse by the scientific realization that sulfur dioxide (SO_2) concentrations are not good indicators of the change in public health hazard when technology changes. If indeed, as strongly indicated by the data, sulfate concentrations are better indicators, a struggle to meet SO_2 standards may be irrelevant to public health, if it is done in such a way as to keep the average sulfate levels constant—by tall chimney stacks, for example. Yet, so far, no standards have been set for sulfates.

A main purpose of this chapter is to outline how the present laws fail to cope with this dilemma. For this purpose we have made use of a review of the law and court cases up to the end of 1976 [Kramer 1976] and the Clean Air Act of 1977 and a report thereon [CIFC 1977].

The historic method of regulation has first been to establish an air quality standard at a sufficiently low level that no adverse health effect exists at or below this level, and to do nothing if no such standard can be met. We have no quarrel with the levels of these standards, but we do call attention to the confusing nature of their description. It is probable that adverse health effects do occur to people exposed only to levels below the air quality standards, and we have in earlier chapters presented evidence on this point. But the wording surrounding the establishment of these standards does not seem to admit this. Normally, in cases of this sort, courts have considerable leeway to exercise common sense. But in air pollution cases much of this leeway has been specifically removed by direction of the U.S. Congress.

10.1.1 History

The Board of Health of the Commonwealth of Massachusetts stated in October 1869,

> We believe that all citizens have an inherent right to the enjoyment of pure and uncontaminated air, and water, and soil; that this right should be regarded as belonging to the whole community; and that no one should be allowed to trespass upon it by his carelessness or his avarice, or even by his ignorance. This right is in a great measure recognized by the State, as appears by the general statutes.

The statutes in question were not quoted, and indeed neither they nor this fine statement were heeded for nearly 100 years. To some extent this state of affairs can be summarized in the phrase, "The polluter is innocent till he can be proved guilty."

As recently as 1935 these rights were regarded as secondary to maintenance of industrial prosperity. In that year a nuisance case was decided in Pittsburgh because some nearby valley dwellers had resorted to the law to put a stop to the pall of smoke being generated by smoldering mountains of waste coal. The coal-mining company won the case. Four decades of experience render poignant the ratio decidendi then articulated as rough-and-ready common sense:

> Much of our economic distress is due to the fact that there is not enough smoke in Pittsburgh and the Pittsburgh district. The metropolis that earned the sobriquet of the "smokey city" has not been living up to those vaporous laurels. The economic activity of the city that was known as "the workshop of the world," has decreased in proportion as its skies cleared of smoke. While smoke per se is objectionable and adds nothing to the outer aesthetics of any community, it is not without its connotational beauty as it rises in clouds from smoke stacks of furnaces and ovens (and even gob fires) telling the world that the fires of prosperity are burning—the fires that assure economic security to the workingman, as well as establish profitable returns of capital legitimately invested. [Versailles Borough 1935]

Of course we can regard the legal decision as a simple application of the risk-benefit analysis we recommended in Section 8.1. Air pollution control devices were not available in 1935, the expectation of life was less than it is now, and the preservation of a job was all important. But these background facts have changed, and the balance of risks and benefits has changed, both in fact and in our perception of them.

10.1.2 The Basic Statutes

There are two basic federal statutes that govern the control of air pollution: the Clean Air Act of 1963 with its amendments (in particular, the Clean Air Acts of 1970 and 1977) [42 U.S.C. 1857 et. seq.] and the National Environmental Policy Act of 1969 [42 U.S.C.A. 4332(2) (C)]. These statutes and the regulations implementing them emphasize two aspects, already noted, of air pollution control that are superficially distinct. Under the Clear Air Act, air quality standards (thresholds) are set, and there is a legal presumption that below the concentration levels of these standards there is no health hazard.

On the other hand, Section 102 of the National Environmental Policy Act enjoins each federal agency, before taking any action, to compare the environmental consequences of all available alternatives.

This can, and has been, taken to imply that there can be environmental consequences even at levels below the threshold, as suggested for example by a linear plot of health hazard versus dose, where air quality standards themselves are not violated. The consequence of this decision is that there will be no additional growth in areas of pure air while the dirty urban air, by receiving the pollution caused by growth, becomes dirtier.

There are several indications that the U.S. Congress intended that the air be made clean independent of the cost of doing so. We quote here one of these:

> The committee determined that . . . the health of the people is more important than the questions of whether the early achievement of ambient air-quality standards protective of health is technically feasible. . . .
>
> Therefore the committee determined that existing sources of pollutants should meet the standard of the law or be closed down. [U.S. Senate 1970]

In three separate court cases, this view of the law has been upheld.

> The material portions of the Clean Air Act itself do not mention economic or social impact, and it seems plain that Congress intended the Administrator [of the EPA] to enforce compliance with air quality standards even if the costs were great. . . . Congress has already made a judgment. . . . EPA and the courts are bound [South Terminal Corporation 1974].
>
> We have concluded that claims of economic or technological infeasibility may not be considered by the Administrator in evaluating a state requirement that primary ambient air quality standards be met in the mandatory three years [Union Electric Co. 1976].

We note that the regulations that according to these decisions must be enforced, regardless of cost, do not take account of the possible adverse health effects and environmental effects of concentrations of pollutant below the primary air quality standard.

There has been one challenge [Kennecott Copper 1972] to EPA's secondary standard for sulfur oxides, but not to the primary standard. Kennecott accepted that EPA must exclude economic, technological, and social factors from consideration in enforcing a standard, but attacked the standard itself, claiming that it was based on inadequate medical evidence. EPA withdrew the standard.

The primary air quality standard was set at a level below which EPA could not prove adverse health effects, as has been claimed by many industries [Consolidated Electric Utility 1974]. If this claim indeed corresponded to scientific fact, or even to unanimous scientific opinion, meeting this standard would be all that is necessary. However, early chapters of this book make it clear that this standard is not proved. Nor is it logical to take the position that sulfate particulates are not important for public health merely because EPA has not set a standard for them.

We have had, for a century, an analogous, highly successful demand for clean water independent of cost. As a result, every American has available clean water that can be drunk without boiling or filtering and that does not cause disease or death. The exact detail of what pollutants in water were causing troubles, and to what level they must be reduced, was not considered, and the best available control technology has been demanded and used. Yet society did not completely disregard cost. We could have distilled the water but we did not do so, because the increase in expense was much too great.

Although the courts have not had to decide explicitly how much to spend on air pollution control equipment, it is to be hoped that the law will be interpreted in this common-sense way.

10.1.3   Best Available Control Technology

It is obvious that of two methods of meeting the ambient air quality standards, reduction of emission or dispersion of pollutants, the former is preferable if no other criteria intervene, because it may also reduce pollutants for which there may be a health effect but as yet no standards.

In an important case [NRDC 1974] it was ruled that of the alternative methods of control, total reduction is the method of meeting the ambient air quality standards to be preferred over intermittent control systems, supporting our view that total sulfur emission is important.

The use of tall stacks to disperse air is not considered an adequate pollution control for new sources. The Fifth Circuit Court of Appeals [NRDC 1974] ruled that an industry may use tall stacks to meet air quality standards only if (1) "the constant emission-control requirements of the control strategy can themselves attain the standards," or (2) "if the best available (subject to economic and technological constraints) emission limiters cannot meet the air quality standards without the help of tall stacks." In the Sixth and Ninth Circuits, similar rulings were made [Kennecott Copper 1972; Big

Rivers 1975]. It ruled that the pollution source must install reasonably available control technology (RACT), instead of the best available control technology (BACT). Should the reasonably available control technology be inadequate, the polluter would be required to install tall stacks to meet the standards. In each of these cases the U.S. Supreme Court denied petitioners application for writ of certiorari (refused to consider the case).

One notices immediately the foggy definition of "reasonably available control technology." The gray area of definition offers the polluter the chance to implement less expensive control systems if the opportunity should arise. In the final analysis, then, the court must apply its best efforts and the true spirit of the air pollution standards to prevent unnecessary degradation of our air.

The EPA has issued regulations to implement these decisions [Federal Register 1976]:

> Reasonably Available Control Technology becomes a factor only after it has been determined on a case by case basis that retrofit of BACT on the existing sources would be unreasonable, considering the processes, fuels and raw materials to be employed; the engineering aspects of various types of control techniques, process and fuel changes; the cost of employing the available techniques; and the cost of certain control techniques due to plant location, configuration, basic process design and expected remaining life. It is expected that only some smelters and a limited number of coal-fired power plants will be able to justify application of RACT in lieu of BACT.

A review of costs of installing flue gas desulfurization (FGD) at existing coal-fired power plants revealed that the remaining life and capacity are major factors determining the reasonableness of an FGD installation. For existing power plants, the following criteria should be considered in determining reasonableness.

1. If a plant has ten to fifteen years or less of remaining life, and is projected to operate at an annual average of no more than 25 percent capacity, FGD installation might be considered unreasonable. In such instances, the source owner should provide the anticipated shutdown date, and that date must be included in the compliance schedule.

2. FGD may be unreasonable because of space limitations. An important feature of the 1977 Clean Air Act was its unequivocal endorsement of the Best Available Control Technology provision — indeed, this provision was extended. Some utility companies that had available a supply of low-sulfur coal preferred to satisfy the require-

ment for clean air by burning low-sulfur coal. This is no longer allowed, and any new source must use best available technology for reducing sulfur from stack gases [CIFC 1977, Section IIIb].

For the nation as a whole this procedure is clearly desirable, for the total amount of sulfur emitted in the United States will be reduced; but it does not address some problems of equity that more sophisticated policy can address. The House report on the 1977 Clean Air Act Amendments makes it clear that this support is in anticipation of a major switch of energy sources in the United States from oil to coal, with a likely increase in pollutant emission [CIFC 1977]. There follow strong recommendations of a committee of the National Academy of Sciences [NAS 1973b, p. 25] who state, "New mines, processing plants, and factories, should be obligated to adopt the best available technology."

Although we have no objections to these present applications of the concept of best available control technology to reducing air pollution, we have fundamental logical problems with the concept and believe that these problems will become increasingly evident in society. This is because as time goes on, we expect the best available control technology to reduce pollutants more, and at the same time to become more expensive.

As noted in Chapter 6, we believe that a strong case can be made for the claim that 2.5 percent of all deaths in the United States are related to air pollution. This is a mortality risk of about 2.5×10^{-4} per year, a large risk of involuntary cause of death for an individual, and we believe that it must be reduced. Almost any pressure to find a technological solution to bring the risk down is good—and the present demand for BACT is therefore tolerable. By the year 2020 the health effects from pollution may have been reduced by, say, a factor of two. At that time it may be desirable to reconsider whether Best Available Control Technology (which will surely be better than now) should still be employed to reduce it further. We suggest that a more sophisticated balancing of risks and benefits will be the way to make that decision.

10.1.4 Geographic Location and Nondegradation

We showed in Section 6.4.5 that public health can be improved by locating and relocating power plants away from, and downwind of, population centers. Possible locations in the United States could be the coast of Maine, on offshore islands, or in the sparsely populated western states. However, there is an instinctive national revulsion against polluting the air of our wilderness areas any more than it is now polluted, in addition to destruction of the wilderness itself by locating such a facility there.

Originally the EPA held that a new power plant need meet only the air quality standards. However, the U.S. Court of Appeals in 1972 ruled that air quality standards may not be made significantly worse by any federal action even though the air quality standards themselves are not violated. The consequence of this decision is that there will be no additional growth in areas of pure air while the dirty urban air, by receiving the pollution caused by growth, becomes dirtier. The Sierra Club raised the point, however, that if the national air standards are sufficient, harmful levels of pollutants will never arise. Presumably there is room for discussion of the precise meaning of the "significantly worse" and of the Sierra Club's use of the word "sufficient."

In response to this case, the EPA has issued new regulations that allow the air to be worse by 20 percent in urban areas [EPA 1978c], and not degraded at all in wilderness areas.

This particular question of nondegradation does not fall into the concept in this book of a pollution charge and must be included as an addition. In view of the strong national support for reducing the health effects of air pollution (even though, as we stressed in Chapter 6, the effects are hypothetical and not strictly proved) we believe that there should be some incentive for industry and utility companies to locate in more geographic areas where pollution causes lower health effects. There seems to be no such incentive in present law. This leaves some urgency to find a procedure for comparing and resolving the requirements (largely aesthetic) of nondegradation and the requirements of improvement of public health.

10.1.5   Legal Validity of a Pollution Charge or Tax

It seems clear that the states and the federal government have the right to tax, and so long as the tax is reasonably equitable, not punitive, there would seem to be no problem with a pollution charge or tax. For the range of charges we consider here, there is clearly no problem in establishing that it is equitable; the burden of proof is on someone who believes that a proposed tax is not equitable.

10.1.6 Transfrontier Pollution

A problem might arise if a state sets a stricter standard than the federal government. All our estimates concur that sulfate particulates cross state boundaries, and it is probable that as much as 3/4 of the pollution problems in East Coast cities come from out of state. What ability do these states have to prevent the health of their citizens from being adversely affected by other states?

This problem is worse in Europe, where pollution crosses national boundaries [OECD 1974]. Because it affects our particular problem, we might consider how and whom a state might sue to claim damages on behalf of its citizens for health effects, and for property damage from sulfate pollution.

First, the burden of proof is clearly different in a suit of this sort than in establishing that a tax is equitable. We would have to prove (1) that there is a reasonable case that sulfates cause health effects and property damage; we would then have to quantify these effects. The case we have might be enough to justify a tax, or persuade a legislature to enact restrictive legislation, but it might not be enough to win a tort case; and (2) that the defendant's sulfur was causing the problem or at least a large part of it. Even if the use of tort law were not possible, it might be possible to prevent sulfur emissions under the law about public nuisance.

There would be a strong case for the defense if the polluter was in compliance with state and federal (EPA) air concentration standards and federal (EPA) emissions rules, if any.

The U.S. Supreme Court has been very cautious about resolving problems arising among states, and it has demanded a high standard of proof. However, various cases have slowly brought the federal government into interstate pollution cases. We present here several of these cases.

10.2 A SUMMARY OF EIGHT INTERSTATE POLLUTION CASES

In this section we will present some of the legal aspects of interstate pollution cases by describing eight suits that have taken place in this century. Three of the suits involve interstate air pollution. The other five suits involve water pollution, but the courts' reasoning in each of these cases is applicable to subsequent air pollution suits. Two of the suits are the foundation of interstate pollution policy. The first is an air pollution case, *Georgia* v. *Tennessee Copper* [Georgia 1906] ; the second is *Texas* v. *Pankey* [FCLE 1972a].

The holding in *Texas* v. *Pankey* is based upon the outcome of *Georgia* v. *Tennessee Copper*, but the second case is unique because, unlike *Georgia* v. *Tennessee Copper*, which was based upon nuisance, *Texas* v. *Pankey* is based entirely upon "ecological rights." The courts dismissed four more cases for insufficient proof. In another case, who has jurisdiction? was the principal issue, and the last case deals with multiple defendants.

Georgia v. *Tennessee Copper* was a case in which proof that Tennessee Copper's SO_2 emissions threatened Georgia's vegetation "requires but few words." There was a "preponderance of evidence that the sulfurous fumes [that developed when the SO_2 contacted moist air] caused and threatened damage on . . . a considerable scale to forest and vegetable life [in Georgia] " [Georgia 1906].

The Supreme Court noted that Georgia "used every friendly office and . . . every means open to her" to persuade Tennessee Copper to reduce its emissions. The Court further noted that because states relinquished some of their rights when they joined the Union, "Georgia [would not only be] unable to punish constantly recurring criminal and injurious acts within her territory and upon her citizens, [but also] unable to enforce her laws. . . . " [Georgia 1906] without the aid of the Supreme Court, which found no alternative to injunction.

In *Texas* v. *Pankey* [FCLE 1972a] "Texas sought an injunction in the New Mexico federal district court against eight New Mexico ranchers who were co-representatives of a class" that proposed to control insect population with a pesticide. The plaintiffs alleged that the pesticide would pollute the water of eleven Texas municipalities when the runoffs of rains carried the pesticide into Texas waters. "The district court dismissed the case for lack of jurisdiction"

[FCLE 1972a] but the Tenth Circuit remanded the case back to district court, finding "Texas' quali-sovereign right to be free of out-of-state pollution [is] based on federal common law." This wording is similar to the wording in *Georgia* v. *Tennessee Copper*, where it was decided that federal common law, in addition to statutory law, applies to the nuisance of air pollution. The pesticide spraying had already occurred by the time the Tenth Court ruled. Because no further harm threatened, Texas abandoned further proceedings.

As in cases not related to pollution, the question of what court has jurisdiction was the main issue in the suit *Illinois* v. *Milwaukee* [FCLE 1972b]. Illinois sought an injunction to prevent four cities and two sewerage agencies in Wisconsin from violating Michigan law by discharging 200 million gallons per day of inadequately treated sewage into Lake Michigan. Illinois stated that because the suit involved agencies in Wisconsin, the State of Wisconsin should be made a defendant. In a suit between two states, the U.S. Supreme Court has original jurisdiction. Presumably, this suit would take place there. The Court's philosophy is, however, that "we incline to a sparing use of our original jurisdiction so that our increasing duties with the appellate docket will not suffer." Citing several cases and the federal district courts' "considerable interest in the purity of interstate waters," the Supreme Court ruled that although Wisconsin may be made a defendant, such action was not necessary. Upon the recommendation of the Supreme Court, Illinois then filed a complaint of public nuisance. We do not know of its outcome.

For these three cases, the authors upon whose writing we base this section showed that the courts deliberated (1) before they decided to create a "new cause for action," or federal common law applied to the nuisance that interstate pollution causes; and (2) before they decided who had jurisdiction. The first consideration in each case were the facts relevant to it. The courts believed that it is especially important to decide each case upon its own facts—not upon the decision of other pollution cases—because existing guidelines are scanty. The courts dismissed the following four cases for insufficient proof. In these cases too, the courts adhered to the pattern of considering relevant facts only before they decided.

In *New York* v. *New Jersey* in 1921 and in *New Jersey* v. *New York City* in 1931, water pollution by refuse was the issue. In the former case, New York claimed that a proposed New Jersey project would release sewage that river currents and the Atlantic tides would carry to the New York shoreline. After much consideration, the U.S. Supreme Court ruled that the sewage agency was a corporate agency of New Jersey, so that the state of New Jersey was a defendant. After establishing jurisdiction, the Supreme Court "dismissed the case without . . . prejudice, finding that a sufficient causal connection did not exist. . . . " The Court cited its earlier decision in *Missouri* v. *Illinois* (1906) in which it claimed "the threatened invasion of rights must be of a serious magnitude and it must be established by clear and convincing evidence" [FCLE 1972b].

Proof of causal connection may be difficult to establish, but establishing the connection is not impossible. In *New Jersey* v. *New York City* [1931], the U.S. Supreme Court employed a Special Master, who showed that New York City did indeed create a nuisance by dumping garbage into the ocean, garbage that eventually washed onto New Jersey's beaches. New York City had delayed incinerating its refuse, and this delay influenced the Court to alleviate New Jersey's problem.

In *Missouri* v. *Illinois*, Missouri claimed that Illinois' newly dumped sewage increased the number of typhoid fever cases and increased the bacteria content of the Mississippi River by 75 percent. Missouri also estimated the dollar value of the consequent damage. Both Missouri and Illinois used expert testimony to back their cases. The U.S. Supreme Court dismissed the case without prejudice, however. Its main reason was "Missouri had not shown that Chicago's sewage was the [direct] cause of the increase" of the number of typhoid cases [FCLE 1972b].

In *Chicago* v. *Commonwealth Edison*, the City of Chicago filed its first complaint in September 1970, alleging, "the Indiana plant violated city, state and federal regulations and therefore [its operation] constituted a statutory nuisance" [Chicago 1974]. The Illinois Court of Appeals, First District, dismissed the case on the grounds that the Indiana plant does not have to obey Chicago ordinances.

A year later, Chicago filed an amended complaint that alleged that the Indiana plant burned coal with 3 percent excess sulfur by weight. Chicago claimed that, furthermore, burning the coal resulted in SO_2 and particulate emissions that caused substantial, unreasonable, and irreparable injury to the residents of Chicago, injury that constituted a nuisance actionable under common law.

The trial court decided that Edison's actions were not a common-law nuisance because Chicago did not establish that Commonwealth

Edison's Indiana plant posed a nuisance to Chicago's residents. In its defense, Edison stated not only that it used low-sulfur coal and that it made a significant effort to reduce its SO_2 and particulate emmissions, but also that "the City of Chicago is not authorized to bring suits to abate alleged nuisances originating in other states." The trial court avoided the question of original jurisdiction because, in the first place, Chicago's evidence was insufficient.

Chicago appealed the trial court's decision in 1974, basing its case upon the testimony of three men. Of the three, two testified only that air pollution in general is harmful. The case rested upon the testimony of the third man, who, upon cross-examination, "revealed several assumptions and methodological errors which had the effect of exaggerating several [of his] estimations." In its defense, Edison employed an independent consulting meteorologist. The consultant showed "the Indiana facility's contribution to ground-level concentrations of sulfur dioxide and particulate matter were well below applicable federal health standards." The City of Chicago's case was not helped by the testimony of "smoke inspectors" who used the Ringlemann charts to estimate particulate density of Edison's plumes. They testified on behalf of Chicago, but upon cross-examination, it was revealed that the charts are unreliable because moisture affects the readings, and no effort to determine the plumes' moisture had been made.

The court made several considerations before it decided that "the Indiana station's operation [was not] an unreasonable interference with the [Chicago residents'] right to clean air." First, people testified that "concentration levels [of pollutants] below federal standards were not substantially harmful to the public health." Next, the court noted that the Indiana electric station is located in a highly industrialized area, so that "the City of Chicago could not establish that the Edison plant was the direct cause of harmful pollution in Chicago, distinguishing the facility from other emission sources located in the area."

How does the court decide in a case where several defendants contribute to create a single, indivisible harm whose liability the plaintiffs cannot attribute? This is the principal question in the case of *Michie* v. *Great Lakes Steel Division* (1974). In this case, thirty-seven Canadian residents with American citizenship claim that three corporations operate seven steel mills that emit noxious air pollutants that (1) violate the defendants' municipal air standards and (2) cause health and property damage estimated to be between $11,000 and $35,000 per defendant. The steel corporations based their case upon old Michigan law [Robinson 1875]: each is responsible for only his contribution to the total harm. They sought dismissal and hoped, furthermore, that the burden of distributing responsibility would be the plaintiffs.'

The federal district handling the case thought otherwise. For lack of any other suitable precedent, the court relied upon the decision of a multiple-automobile collision case, *Maddux* v. *Donaldson*. In that case, the court concluded, "if competent testimony showed that the injuries were separable and that liability for the injuries could be allocated, then the triers of the case must do so, regardless of the difficulty." Their majority opinion held that if on the other hand the injuries were not separable, then although there was no concert of action, the defendants would be jointly liable. The court denied the defendants' motion for dismissal. More recent information about this case has not appeared in the environmental law reviews, but the *Environmental Law Reporter* noted that two of the defendants will claim that other sources—corporations, homes, and individuals—contributed to the plaintiffs' damages, so that it will be even more difficult to assess responsibility for the damage.

The federal courts have a flexible tool, federal common law, which can complement and modify existing statutes to control interstate pollution, be it air pollution or water pollution. Not afraid of the complicated problems that interstate pollution cases present—there is an old saying, "hard cases make bad law," that would favor ignoring the issue—the courts and plaintiffs have worked to abate pollution. This is evident in the eight cases presented in this paper. But the courts are also mindful of setting bad precedents and of the need for equitable balancing, so progress has been and will be slow.

So far the sulfate problem and the related acid rain problem, as transfrontier pollution issues, are only now reaching courts. The eastern coastal states, New York, Connecticut, and Massachusetts, have tighter air pollution emission standards. The standards of some midwestern states are more lax, yet their pollution almost certainly drifts east with the prevailing wind. How long will the people of these eastern states be willing to pay the money for enforcement of their own tough laws, only to be foiled in their efforts by their upwind neighbors?

10.3 THE NATIONAL DILEMMA

In the preceding pages we have discussed the evidence that sulfates, nitrates, and particulates, severally or in combination from stationary or mobile fossil fuel-burning facilities, cause a human health hazard at 1980 ambient air pollution levels.

Although the evidence is circumstantial, and may remain so, it makes a consistent picture. It is possible therefore for a public health authority to insist that 50,000 persons a year die of air pollution-related problems in the United States. It is also possible for a utility company executive to argue that no one dies.

This dilemma has been met before in regard to radiation. Since 1928 the International Commission on Radiological Protection (ICRP) has recommended that it be assumed that low levels of radiation may cause harm, and that the proportional relation between cancers and incident radiation dose be assumed. This assumption is made with no direct data; radiation doses, given suddenly, of 50 rem have been shown to cause cancer, yet we discuss natural background doses 500 times smaller (0.1 rem/yr) and increments caused by nuclear electric power 100 times smaller still.

In contrast, doses of SO_2 or particulates of 500 $\mu g/m^3$ have been shown to give adverse health effects (Section 4.1). The air quality standard is only eight times lower, and we are considering increments of dose close to the air quality standard. It is possible to maintain, therefore, that adverse health effects from air pollution at ambient levels are more likely than adverse health effects from radiation at ambient levels, and society already has a precedent of no threshold but proportional relation for handling the latter.

Also with radiation an obvious anchoring point is the natural background dose; it seems intuitively reasonable to allow any extra dose that is a small fraction of the natural background and within the fluctuation of the background. The natural background for sulfur oxides and for particulates is low (tobacco cigarette smoke we do not here consider to be natural), and therefore this cannot be used for anchoring.

With this uncertainty how should we proceed? If it were cheap to reduce sulfate, nitrate, and particulate levels, it would be best to do so. But the cost to reduce sulfur emissions by a factor of four is probably $20 billion in capital cost plus $5 billion in operating cost per year. To save 50,000 lives a year it would probably be worthwhile, but it would all be wasted if there is no health hazard at ambient 1980 levels, or if the hazard is due to some other cause such as nitrates.

For that reason, we suggest that continuous attention be paid to those solutions, which are simple and moderately cheap, and which work independent of the source of the hazard. The most important ones seem to be these:

1. Siting of power plants downwind of major urban areas.

2. Use of other sources of fuel than coal

3. Gasification or liquefaction of coal with cleaning of the gas and liquid

4. Simultaneous control of fine particulates and all gases that can convert into particulates; new baghouse methods look hopeful here

5. Allocation of clean fuels in the best possible way to minimize the integrated population exposure

There is a need for legislative and economic incentives for industry to take any of these steps.

In conclusion, we state our view that it is an international disgrace that although we have been burning coal for 800 years, we still do not know how to burn it cleanly and we are not sure, in detail, of the effects of burning it dirtily.

BIBLIOGRAPHY

Adamson, L.F., and B.M. Bruce. 1979. "Suspended particulate matter: A report to Congress." Washington, D.C.: Office of Research and Development, USEPA; EPA-600/9-79-006.

AEC. 1968. "Meteorology and Atomic Energy." Washington, D.C.: U.S. Atomic Energy Commission.

Alarie, Y. 1973. "Sensory irritation of the upper airways by airborn chemicals." *Toxical. Appl. Pharm.* 24: 279-97.

Alarie, Y.; I. Wakisaka; and S. Oka. 1973. "Sensory irritation by sulfite aerosols." *Envior. Physiol. Biochem.* 3: 182-84.

Alarie, Y.; W.M. Busey; A.A. Krumm; and C.E. Ulrich. 1973. "Long-term continuous exposure to sulfuric acid mist in cynomolgus monkeys and guinea pigs." *Arch. Environ. Health* 27: 16-24.

Alarie, Y.C.; A.A. Krumm; W.M. Busey; C.E. Ulrich; and R.J. Kant II. 1975. "Long-term exposures to sulfur dioxide, sulfuric acid mist, fly ash and their mixtures, Results of studies in monkeys and guinea pigs." *Arch. Environ. Health* 30: 254-62.

Alarie, Y.; C.E. Ulrich; W.M. Busey; H.E. Swann, Jr.; and H.N. MacFarland. 1970. "Long-term continuous exposures of guinea pigs to sulfur dioxide." *Arch. Environ. Health* 21: 769-77.

Alarie, Y.; C.E. Ulrich; W.M. Busey; A.A. Krumm; and H.N. MacFarland. 1972. "Long-term continuous exposure to sulfur dioxide in cynomolgus monkeys." *Arch. Environ. Health* 24: 112-258.

Altshuller, A.P. 1973. "Atmospheric sulfur dioxide and sulfate distribution of concentration at urban and nonurban cities in the United States." *Environmental Science and Technology* 7: 709-12.

331

_____. 1976. "Regional transport and transportation of sulfur dioxide to sulfates in the U.S." *J. Air Poll. Contr. Assoc.* 26: 318-24.

_____. 1980. Private communication. Environmental Science Research Laboratory, U.S. EPA.

Amdur, M.O. 1969. "Toxicological appraisal of particulate matter, oxides of sulfur, and sulfuric acid." *J. Air Poll. Contr. Assoc.* 9: 638-44.

_____. 1971. "Aerosols formed by oxidation of sulfur dioxide. Review of their toxicology." *Arch. Environ. Health* 23: 459.

_____. 1973. "Animal Studies." In *Proceedings of the Conference on Health Effects of Air Pollutants.* National Academy of Sciences, National Research Council, Assembly of Life Sciences, Washington, D.C., October 3-5, 1973. Report prepared for the Committee on Public Works, United States Senate, pp. 175-205. U.S. Government Printing Office.

Amdur, M.O., and M. Corn. 1963. "The irritant potency of zinc ammonium sulfate of different particle sizes." *Amer. Ind. Hyg. Assoc. J.* 24: 326-33.

Amdur, M.O., and D. Underhill. 1968. "The effect of various aerosols on the responses of guinea pigs to sulfur dioxide." *Arch. Environ. Health* 16: 460-68.

Amdur, M.O.; J. Bayles; V. Ugro; M. Dubriel; and D.W. Underhill. 1975. "Respiratory response of guinea pigs to sulfuric acid and sulfate salts." Paper delivered at the Symposium on Sulfur Pollution and Research Approaches, May 27-28, sponsored by the U.S. EPA and Duke University Medical Center.

American Thoracic Society. 1978. *The Health Effects of Air Pollution.* New York: Medical Section, American Lung Association. [See Shy et al. 1978.]

Andersen, I.; G.R. Lundquist; P.L. Jensen; and D.F. Proctor. 1974. "Human response to controlled levels of sulfur dioxide." *Arch. Environ. Health* 28: 31-39.

Angel, J.H.; C.M. Fletcher; I.D. Hill; and C.M. Tinker. 1965. "Respiratory illness in factory and office workers." *British Journal of Diseases of the Chest* 59: 66-80.

Armitage, P., and R. Doll. 1954. "The age distribution of cancer and a multistage theory of carcinogenesis." *Brit. Journ. of Cancers* 8: 1.

Armytage, W.H. 1961. *A Social History of Engineering.* Cambridge, Massachusetts: MIT Press.

ASME. 1973. *Prediction of the Dispersion of Airbourne Effluents.* New York: American Society of Mechanical Engineers.

Asmundsson, T.; K.H. Kilburn; and W.N. McKenzie. 1973. "Injury and metaplasia of airway cells due to SO_2." *Lab. Invest.* 29: 41.

Auerbach, O.; A.P. Stout; E.C. Hammond; and L. Garfinkel. 1961. "Changes in bronchial epithelium in relation to cigarette smoking and in relation to lung cancer." *New Engl. J. Med.* 265: 253-67.

_____. 1962. "Bronchial epithelium in former smokers." *New Engl. J. Med.* 267: 119-25.

Ayres, S.M.; S. Giannelli, Jr.; and H. Mueller. 1970. "Myocardial and systemic responses to carboxyhemoglobin." *Ann. N.Y. Acad. Sci.* 174: 268-93.

Ayres, S.M.; H.S. Mueller; J.J. Grefory; S. Giannelli, Jr.; and J.L. Penny. 1969. "Systematic and myocardial hemodynamic responses to relatively small concentrations of carboxyhemoglobin." *Arch. Environ. Health* 18: 699.

Baes, C.F.; H.E. Goeller; J.S. Olson; and R.M. Rotty. 1976a. "Carbon dioxide and climate: The uncontrolled experiment." *American Scientist* 65: 310.

____. 1976b. "The Global Carbon Dioxide Problem." Oak Ridge, Tennessee: ORNL-5194.

Barrett, L.B., and T.W. Waddell. 1973a. "Cost effectiveness of emission control." *J. Air Poll. Contr. Assoc.* 23: 173-79.

____. 1973b. "The cost of air pollution damage: Status report." Washington, D.C.: U.S. EPA, Pub. No. AP85, U.S. Government Printing Office.

Barter, C.E., and A.H. Campbell. 1976. "Relationship of constitutional factors and cigarette smoking to decreased 1-second forced expiratory volume." *Amer. Rev. Resp. Dis.* 113: 305-14.

Bartlett, D.; C.S. Faulkner; and K. Cook. 1974. "Effects of chronic ozone exposure on lung elasticity in young rats." *J. Appl. Physiol.* 37: 92-96.

Bates, D.V., and M. Hazucha. 1973. "The short-term effects of ozone in the human lung." In *Proceedings of the Conference on Health Effects of Air Pollutants.* National Academy of Sciences, National Research Council, Assembly of Life Sciences, Washington, D.C., October 3-5, 1973. Report prepared for the Committee on Public Works, United States Senate, pp. 175-205. U.S. Government Printing Office.

Bates, D.V.; C. Burnham; and L.D. Pengelly. 1970. "Problems in studies of human exposure to air pollutants." *Can. Med. Assoc. J.* 103: 833-37.

Bates, D.V.; G.M. Bell; C.D. Burnham; M. Hazucha; J. Mantha; L.D. Pengelly; and F. Silverman. 1972. "Short-term effects of ozone on the lung." *J. Appl. Physiol.* 32: 176-81.

Baxter, William F. 1974. *People or Penguins — The Case for Optimal Pollution.* New York: Columbia University Press.

Beard, R.R.; R.J.M. Horton; and R.O. McCaldin. 1974. "Observations on Tokyo-Yokohama asthma and air pollution in Japan." *Pub. Health Rep.* 79: 439-44.

Bedi, J.F.; L.J. Folinsbee; S.M. Horvath; and R.S. Ebenstein. 1978. "Human exposure to sulfur dioxide and ozone: Absence of a synergistic effect." *Arch. Environ. Health* 34: 233-39.

Bedrosian, P.A.; D.G. Easterly; and S.L. Cummings. 1970. "Radiological survey around power plants using fossil fuel." Montgomery, Alabama: EPA, EERL 71-3.

Beil, M., and W.T. Ulmer. 1976. "Wirjung von NO in MAK-Vereish auf atem mechanic un Acelytcholinemp-finflichkeit bei Normalpersonen." *Intern. Arch. Occup. Environ. Health* 38: 31-44.

BEIR/NAS. 1972. "The effects on populations of exposures to low levels of ionizing radiation." Report of the Advisory Committee on the Biological Effects of Ionizing Radiation, National Academy of Sciences–National Research Council, Washington, D.C.

Bell, A., and J.L. Sullivan. 1963. "Air Pollution by Metallurgical Industries." Department of Public Health, New South Wales, Sydney, Australia.

Bell, K.A.; W.S. Linn; M. Hazucha; J.D. Hackney; and D.V. Bates. 1977. "Respiratory effects of exposure to ozone plus sulfur dioxide in Southern Californians and Eastern Canadians." *Amer. Ind. Hyg. Assoc. J.* 38: 696–706.

Big Rivers. 1975. Big Rivers Electric Corp. v. EPA, 523 F. 2d 16 (6th Cir.).

Billings, C.E., and W.R. Matson. 1972. "Mercury emissions from coal allen combustion." *Science* 176: 1232.

Blair, W.H.; M.C. Henry; and R. Ehrlich. 1969. "Chronic toxicity of nitrogen dioxide: II Effect on histopathology of lung tissue." *Arch. Environ. Health* 19: 186–92.

Blot, W.J., and J.F. Fraumeni. 1975. "Arsenical air pollution and lung cancer." *Lancet* 2: 142–44.

Blumenthal, D.L., and W.H. White. 1975. "The Stability and Long Range Transport of Ozone or Ozone Precursors." Paper presented at the 68th Annual Meeting of the Air Pollution Control Association, Boston, Massachusetts, June 15–20.

Bolin, B. 1970. "The carbon cycle." *Scientific American,* September, p. 125.

Bolton, N.E.; J.A. Cantor; J.F. Emery; C. Feldman; W. Fulkerson; L.D. Holett; and W.S. Lyons. 1973. "Trace Element Measurements of the Coal-Fired Steam Plant, Progress Report." Oak Ridge, Tennessee: Oak Ridge National Laboratory, ORNL–NSF–EP–43, March.

Bondareva, E.N. 1963. "Hygenic evaluation of low concentrations of nitrogen oxides present in atmospheric air." In *USSR Literature on Air Pollution and Related Occupational Diseases. A survey,* vol. 8, B.S. Levine ed. Washington, D.C.: U.S. Public Health Service.

Boucher, R.C.; P.D. Pare; N.J. Gilmore; L.A. Moroz; and J.C. Hogg. 1977. "Airway mucosal permeability in ascaris suum-sensitive rhesus monkeys." *J. Allergy Clin. Immun.* 60: 134–40.

Boucher, R.C.; V. Ranga; P.D. Pare; S. Inoue; L.A. Moroz; and J.C. Hogg. 1978. "The effect of histamine and methocoline on guinea pig tracheal permeability to HRP." Submitted to *J. Appl. Physiol.* and discussed at the FASEB Symposium, April, 1978.

Boucher, W.S.; T. Takahashi; H.J. Simpson; and T.H. Peup. 1979. "Fate of fossil fuel carbon dioxide and the global carbon budget." *Science* 206: 409.

Bozzo, S.R.; K.M. Novak; F. Galdos; R. Hakoopian; and L.P. Hamilton. 1977. "Mortality migration, income and air pollution: A comparative study." Paper presented at the American Association of Public Health Annual Meeting, Washington, D.C.

Bozzo, S.R.; K.M. Novak; F. Galdos; and L.D. Hamilton. 1978. "Health effects of energy use: A temporal, spatial, and factor analysis approach." Upton,

New York: Biomedical and Environmental Assessment Division, National Center for Analysis of Energy Systems, Brookhaven National Laboratory.

Brant, J.W.A., and S.R.G. Hill. 1965. "Human respiratory diseases and atmospheric air pollution in Los Angeles, California." *Int. J. Air Water Poll.* 9: 219-31.

Brasser, L.J.; P.E. Joosting; and D. van Zuilen. 1967. "Sulfur dioxide: to what level is it acceptable?" Delft, The Netherlands: Research Institute of Public Health Engineering, Report G-300.

Breslow, J., and J. Goldsmith. 1958. "Health effects of air pollution." *Amer. J. Pub. Health* 48: 913-17.

Broecker, W.S. 1975. "Climatic change: are we on the brink of a pronounced global warning?" *Science* 189: 460-63.

Broecker, W.S., and Y.-H. Li. 1970. "Interchange of water between the major oceans." *Journal of Geophys. Res.* 75: 3545.

Brown, E.B., and J. Ipsen. 1968. "Changes in severity of symptoms of asthma and allergic rhinitis due to air pollutants." *J. of Allergy* 41: 254.

Bryan, W., and M.B. Shimkin. 1942. "Quantitative analysis of dose response data obtained with three carcinogenic hydrocarbons in strain C3H male mice." *J. Nat. Cancer Inst.* 3: 503.

Bryers, R.W. 1972. "Cleaning up the air we breathe." *Heat Engineering* (Foster-Wheeler Corp.), July-September.

Buck, S.F., and D.A. Brown. 1964. "Mortality from lung cancer and bronchitis in relation to smoke and sulfur dioxide concentration, population density and social index." London: Tobacco Research Council, Research Paper 7.

Buechley, R.W. 1975. "SO$_2$ levels, 1967-72 and perturbations in mortality: A further study in the New York-New Jersey metropolis." Report to the National Institute of Environmental Health Sciences, Contract No. 1-ES-52101. Research Triangle Park, North Carolina: National Institute of Environmental Health Sciences.

Buechley, R.W.; W.B. Riggan; V. Hasselblad; and J.B. VanBruggen. 1973. "SO$_2$ levels and perturbations in mortality: A study in the New York-New Jersey metropolis." *Arch. Environ. Health* 27: 134-37.

Buell, G.C. 1970. "Biochemical parameters in inhalation carcinogenesis." *Inhalation Carcinogenesis* (U.S. AEC Symposium Ser.) 18: 200-28.

Buell, P.; J.E. Dunn; and L. Breslow. 1967. "Cancer of the lung and Los Angeles-type air pollution—prospective study." *Cancer* 20: 2139-47.

Bulfanini, M. 1971. "Oxidation of sulfur dioxide in polluted atmospheres—a review." *Environ. Sci. and Tech.* 5, no. 8: 685.

Burrows, B.; A.L. Kellogg; and J. Buskey. 1968. "Relationship of symptoms of chronic bronchitis and emphysema to weather and air pollution." *Arch. Environ. Health* 16: 406.

Burton, G.G.; M. Corn; J.B.L. Gee; C. Vasallo; and A.P. Thomas. 1969. "Responses of healthy men to inhaled low concentrations of gas-aerosol mixtures." *Arch. Envir. Health* 18: 681-92.

CAG. 1978. Carcinogen Assessment Groups Preliminary Report on POM Exposures." Washington, D.C.: U.S. EPA, July 14.

Callendar, G.S. 1938. "The artificial production of carbon dioxide and its influence on temperature." *Quarterly J. of Roy. Met. Soc.* 64: 223.

Calvert, T.G. 1973. "Modes of formation of the salts of sulfur and nitrogen in an NO_x-SO_x-hydrocarbon polluted atmosphere." Proceedings of the Conference on Health Effects of Pollutants, Assembly of Life Sciences, National Academy of Sciences and National Research Council. U.S. Government Printing Office, No. 93-15.

Cancer Registry of Norway, The. 1964. "Cancer Registration in Norway." Oslo: Norwegian Cancer Society.

Carnow, B.W. 1978. "The 'urban factor' and lung cancer: cigarette smoking or air pollution?" *Envir. Hlth. Persp.* 22: 17-21.

Carnow, B.W., and P. Meier. 1973. "Air pollution and pulmonary cancer." *Arch. Environ. Health* 27: 207-18.

Carnow, B.W.; M.H. Lepper; R.N. Shekelle; and J. Stamler. 1969. "Chicago air pollution study—SO_2 levels and acute illness in patients with chronic bronchopulmonary disease." *Arch. Environ. Health* 18: 768.

Carnow, B.W.; R.M. Senior; R. Karsh; S. Wessler; L.V. Avioli; L. Green; and D. Rockoff. 1970. "Role of air pollution in chronic obstructive pulmonary disease." *JAMA* 214: 894-99.

Carpenter, S.B.; T.L. Montgomery; J.M. Leavitt; W.C. Colbaugh; and F.W. Thomas. 1971. "Principal plume dispersion models—TVA power plants." *J. Air Poll. Control Assoc.* 21: 491.

Cassell, E.J.; D.W. Wolter; J.D. Mountain; J.R. Diamond; I.M. Mountain; and J.R. McCarroll. 1968. "Environmental Epidemiology. Part II: Reconsiderations of mortality as a useful index of the relationship of environmental factors to health." *Am. J. of Pub. Health* 58: 1653.

Cavendar, F.L.; W.H. Steinhagen; C.E. Ulrich; W.M. Busey; B.Y. Cockrell; J.K. Haseman; M.D. Hogan; and R.T. Drew. 1977. "Effects in rats and guinea pigs of short-term exposures to sulfuric acid mist, ozone, and their combinations." *J. Toxicol. Envir. Health* 3: 521-33.

Cederlöf, R. 1966. "Urban factor and prevalence of respiratory symptoms and angina pectoris—A study of 9,168 twin pairs with the aid of mailed questionnaires." *Arch. Environ. Health* 13: 743-48.

Cederlöf, R., and J. Colley. 1974. "Epidemiological investigations on environmental tobacco smoke." *Scand. S. Resp. Dis.* [Suppl] 91: 47-49.

Cederlöf, R.; L. Friberg; and T. Lundman. 1977. "The interactions of smoking, environment, and heredity and their implications for disease etiology. A report of epidemiological studies in the Swedish Twin Registries." *Acta Medica Scandinavia* (Supplement) 612 (September): 7-128.

Cederlöf, R.; L. Friberg; Z. Hurbec; and U. Lorich. 1975. "The relationship of smoking and some social covariables to mortality and cancer morbidity. A ten year follow-up in a probability sample of 55,000 Swedish subjects age 18 to 69, Parts I and II." Stockholm, Sweden: Karolinska Institute, Department of Environmental Hygiene.

CEQ. 1975. "The Sixth Annual Report of the Council on Environmental Quality." U.S. Government Printing Office.

Chakrin, L.W., and L.Z. Saunders. 1974. "Experimental chronic bronchitis pathology in the dog." *Lab. Invest.* 30: 145–54.

Challen, P.J.R.; D.E. Hickish; and J. Bedford. 1958. "An investigation of some health hazards in an inert-gas tungsten-arc welding shop." *Brit. J. Ind. Med.* 15: 276–82.

Chambers, L.A. 1977. "Classification and extent of problems." In *Air Pollution,* A. Stern, ed. New York: Academic Press.

Chapman, D. 1973. "A sulfur-emission tax and the electric utility industry." Cornell Agricultural Economics Staff paper No. 73–17.

Chapman, R.S.; B. Carpenter; C.M. Shy; R.G. Ireson; L. Heiderscheit; and W.K. Poole. 1973. "Prevalence of chronic respiratory disease in Chattanooga: Effect of community exposure to nitrogen oxides." Research Triangle Park, North Carolina: National Environmental Research Center, EPA in-house technical report.

Chapman, R.S.; V. Hasselblad; C.G. Hayes; J.V.R. Williams; and D.I. Hammer. 1976. "Air pollution and childhood ventilatory function. I. Exposure to particulate matter in two southeastern cities, 1971–1972." In *Clinical Implications of Air Pollution Research,* A.J. Finkel and W.C. Duel, eds. Acton, Massachusetts: Publishing Sciences Group.

Chapman, T.T. 1965. "Air pollution and the forces expiratory volume in chronic bronchitis." *Irish Journal of Medical Science* 472: 189.

Charles, J.M., and D.B. Menzel. 1975. "Ammonium and sulfate ion release of histamine from lung fragments." *Arch. Environ. Health* 30: 314–16.

———. 1976. "Augmentation of sulfate ion absorption from the rat lung by heavy metals." *Pharmacolog.* 18: 125.

Chicago. 1974. Chicago v. Commonwealth Edison, 7 ELR 1480.

Chou, J.J., and J.L. Earl. 1972. "Lead isotopes in North American coal." *Science* 172: 150.

Chou, J.J., and R. Peto. 1977. "Mortality among doctors in different occupations." *Brit. Med. J.* 1: 1433–36.

Chow, C.K.; C.J. Dillard; and A.L. Tappel. 1974. "Glutathione peroxidase system and lysozyme in rats exposed to ozone or nitrogen dioxide." *Environ. Res.* 7: 311–19.

Chrisp, C.E.; G.L. Fisher; and J.E. Lammert. 1978. "Mutagenicity of filtrates from respirable coal fly ash." *Science* 199: 73–75.

Chu, E.W., and R.A. Malmgren. 1965. "An inhibitory effect of vitamin A on induction of tumors to the forestomach and cervix in Syrian hamsters by carcinogenic polycyclic hydrocarbons." *Cancer Res.* 25: 884.

CIFC (Committee on Interstate and Foreign Commerce). 1977. House Report 95-294. U.S. Government Printing Office.

Ciocco, A.J., and D.J. Thompson. 1961. "A follow-up of Donora ten years after: Methodology and findings." *Amer. J. Pub. Health* 511: 155.

Clarke, A.J.; D.H. Lucas; and F.F. Ross. 1970. "Tall stacks: how effective are they?" Second International Air Pollution Conference, Washington, D.C.

Coase, R.H. 1960. "The problem of social cost." *The Journal of Law and Economics* III (October).

Coffin, D.L.; E.J. Blommer; D.E. Gardner; and R.S. Holzman. 1968. "Effect of air pollution on alteration of susceptibility to pulmonary infections." In *Proceedings of the Third Annual Conference on Atmospheric Contaminants in Confined Spaces*, pp. 75-80. Dayton, Ohio: Wright–Patterson Air Force Base, Aerospace Medical Research Laboratories.

Cohen, A.A.; S. Bromberg; R.W. Buechley; L.I. Heiderscheit; and C.M. Shy. 1972. "Asthma and air pollution from a coal-fired power plant." *Am. J. Pub. Health* 61: 1181-88.

Cohen, B.L. 1979. "The role of radon in the comparison of effects of radioactivity releases from nuclear power, coal burning and phosphate mining." In press.

———. 1980. "Society's valuation of life saving in radiation protection and other contexts." *Health Physics* 38: 33.

Cohen, C.A.; A.R. Hudson; J.L. Clausen; and J.H. Knelson. 1972. "Respiratory symptoms, spirometry, and oxidant air pollution in non-smoking adults." *Amer. Rev. Resp. Dis.* 105: 251.

Cohen, D.; S.F. Arai; and J.D. Brain. 1979. "Smoking impairs long-term dust clearance from the lungs." *Science* 204: 514.

Colley, J.R.T., and D.D. Reid. 1970. "Urban and social origins of childhood bronchitis in England and Wales." *Brit. Med. J.* 2: 213.

Colley, J.R.T.; J.W.B. Douglas; and D.D. Reid. 1973. "Respiratory disease in young adults: Influence of early childhood lower respiratory tract illness, social class, and air pollution and smoking." *Brit. Med. J.* 3: 195-98.

Colucci, A.V. 1976. "Sulfur oxides: Current Status of Knowledge." Prepared by Greenfield, Attaway, and Tyler, Inc. for EPRI, EPRI EA-316.

Comar, C. 1978. "Air pollution: A pragmatic view." *EPRI Journal*, October.

Combustion. 1977. "Ten ways to clean coal and their busbar costs." 48: 23-25.

Commerce. 1974. Local climatological data, Logan Airport, January, April, September. Boston: National Oceanic and Atmospheric Administration. U.S. Department of Commerce.

Comstock, G.W.; R.W. Stone; Y. Sakai; T. Matsuya; and J.A. Tomascia. 1973. "Respiratory findings and urban living." *Arch. Environ. Health* 27: 143.

Congressional Research Service. 1977. "Pollution taxes, effluent charges and other alternatives for pollution control." A Report for the Committee on Environment and Public Works, U.S. Senate, Serial No. 95-5.

Congressional Record. 1974. Proceedings and Debates of the 93rd Congress, Second Session. June 12: s10409.

Consolidated Electric Utility. 1974. The Consolidated Electric Utility Case, Ohio: Environmental Protection Agency, Hearing Examiner's Report (Rosenweiss, Brown and Lapp), Case No. 73-AP-120 et al.

Cornwall, C.J., and P.A.B. Raffle. 1961. "Bronchitis-sickness absence in London transport." *Brit. J. Ind. Med.* 18: 24.

Crouch, E.A.C., and R. Wilson. 1979. "Interspecies comparison of carcinogenic potency." *J. Tox. and Environ. Health* 5: 1095-1118.

Crouch, E.A.C.; R.J. Eden; I.J. Bloodworth; E. Bossanyi; D.S. Bowers; C.W. Hope; W.S. Humphrey; J.V. Mitchell; D.J. Pullin; and J.A. Stanislav. 1977. "World Energy Demand to 2020." London: World Energy Conference.

Daman, E.L., and R.E. Sommerlad. 1972. "Cleaning up the air we breath." *Heat Engineering* 45: 145.

Davis, D.D.; G. Smith; and G. Klauber. 1974. "Trace gas analysis of power plant plumes via aircraft measurement: O_3, NO_x, and SO_2 chemistry. *Science* 186: 733-36.

Dawson, S.V., and E.A. Elliott. 1977. "Wave-speed limitation on expiratory flow—unifying concept." *J. Appl. Physiol.* 43: 498-515.

Dawson, S.V., and M.B. Schenker. 1979. "Health effects of inhalation of ambient nitrogen dioxide." *Amer. Rev. Resp. Dis.* 120: 281-92.

Deal, E.C., Jr.; E.R. McFadden, Jr.; R.H. Ingram, Jr; R.H. Strass; and J.J. Jaeger. 1978. "Heat loss and vaporization of water and exercise induced asthma." *Amer. Rev. Resp. Dis.* (Annual Meeting Supplement) 117: 328.

Dean, G. 1959. "Lung cancer among white South Africans." *Brit. Med. J.* 2: 852.

_____ . 1966. "Lung cancer and bronchitis in Northern Ireland." *Brit. Med. J.* 1: 1506-14.

Deane, M. 1965. "Epidemiology of chronic bronchitis and emphysema in the United States 2. Interpretation of Mortality Data." *Med. Thor.* 22: 24-37.

Deane, M.; J.R. Goldsmith; and D. Tuma. 1965. "Respiratory conditions in outside workers: Report on outside plant telephone workers for San Francisco and Los Angeles." *Arch. Environ. Health* 10: 323-31.

Derrick, E.H. 1970. "A comparison between the density of smoke in the Brisbane air and the prevalence of asthma." *Med. Jour. of Australia* 2: 260.

DOE (U.S. Department of Energy). 1976. "A National Plan for Energy Research, Development and Demonstration: Creating Energy Choices for the Future. Vol. I: The Plan." Energy Research and Development Administration, ERDA 76-1. U.S. Government Printing Office No. 052-010-00478-6.

_____ . 1978. "A comprehensive plan for carbon dioxide effects research and assessment. Part I: The global carbon cycle and climatic effects of increasing carbon dioxide." May.

Dohan, F.C. 1961. "Air pollutants and the incidence of respiratory disease." *Arch. Environ. Health* 3: 387.

Dohan, F.C., and E.W. Taylor. 1960. "Air pollution and respiratory disease: A preliminary report." *Amer. J. Med. Sci.* 240-337.

Doll, R. 1978. "Atmospheric pollution and lung cancer." *Envir. Health Persp.* 22: 23-31.

Doll, R., and R. Peto. 1976. "Mortality in relation to smoking: 20 years observations on male British doctors." *Brit. Med. J.* 2: 1525-36.

_____ . 1977. "Mortality among doctors in different occupations." *Brit. Med. J.* 1: 1433-36.

Doll, R.; R.E.W. Fisher; E.J. Gammen; W. Gunn; G.O. Hughes; F.H. Tyree; and W. Wilson. 1965. "Mortality of gas workers with special reference to cancers of the lung and bladder, chronic bronchitis, and pneumoconiosis." *Brit. J. Ind. Med.* 22: 1.

Donovan, D.H.; M.B. Abou-Donia; D.E. Gardner; D.L. Coffin; C. Roe; R. Ehrlich; and D.B. Menzel. 1976. "Effects of long-term low-level exposures of nitrogen dioxide on enzymatic indicators of damage." *The Pharmacol.* 18: 244 (abstract).

Douglas, J.W.B., and R.E. Waller. 1966. "Air pollution and respiratory infection in children." *Brit. J. Prev. and Soc. Med.* 20: 1.

Downing, P.B., and W.D. Watson, Jr. 1974. "The economics of enforcing air pollution controls." *J. Envir. Econ. and Manag.* 1: 219-36.

Drazen, J.M.; S.H. Loring; and R.H. Ingram, Jr. 1976. "Localization of airway constriction using gases of varying density and viscosity." *J. Appl. Physiol.* 41: 396-99.

Dungsworth, D.L. 1976. "Short-term effects of ozone on lungs of rats, mice, and monkeys." *Environ. Health Persp.* 16: 179 (abstract).

Dungsworth, D.L.; W.L. Castleman; C.K. Chow; P.W. Mellick; M.G. Mustafa; B. Tarkington; and W.S. Tyler. 1975. "Effect of ambient levels of ozone on monkeys." *Proc. Fed. Amer. Soc. Exp. Biol.* 34: 1670-74.

Durham, W.E. 1974. "Air pollution and student health." *Arch. Envir. Health* 28: 241-54.

Dzubay, T.G. 1979. "Chemical element balance method applied to dichotomous sampler data." Symposium on Aerosols: Anthropogenic and Natural Sources of Transport. *Annals of N.Y. Acad. of Sci.*, January, pp. 9-12.

Eastcott, D.F. 1956. "The epidemiology of lung cancer in New Zealand." *Lancet* 1: 37.

Eatough, D.J., and A.V. Colucci. 1975. "Determination and Possible Public Health Impact of Transition Metal Sulfate Aerosol Species." Report prepared for EPRI, December.

Ehrlich, R.; J.G. Findlay; J.D. Fenters; and D.E. Gardner. 1977. "Health effects of short-term inhalation of nitrogen dioxide and ozone mixtures." *Environ. Res.* 14: 223-31.

Eisenbud, M. 1952. "Retention, distribution and elimination of inhaled particulates." *Arch. Ind. Hyg. and Occ. Med.* 6: 214.

Eisenbud, M., and H.G. Petrow. 1964. "Radioactivity in Atmospheric effluents of power plants that used fossil fuel." *Science* 144: 288.

Emerson, P.A. 1973. "Air pollution, atmospheric conditions and chronic airways obstruction." *J. Occ. Med.* 15: 635.

EPA. 1970. "Air Quality Control Criteria for Sulphur Oxides." Environmental Protection Agency, Report AP-50. U.S. Government Printing Office.

_____. 1971. "Air quality criteria for nitrogen oxides." Washington, D.C.: AP-84 U.S. EPA.

_____. 1973a. "National Air Quality levels and trends in total suspended particulates and sulfur dioxide determined by data in the national air surveillance network." Office of Air Water Programs. Unpublished.

_____. 1973b. "Use of Supplementary Control Systems and Implementation of Secondary Standards, EPA proposed rule." *Federal Register*, September, 25697.

_____. 1976. "Environmental Protection Agency Monitoring and Air Quality Trends Report." EPA-450/1-76-001.

_____. 1977a. "National air quality and emission trend report, 1977." Office of Air and Water Management, EPA-450/1-77-002.

_____. 1977b. "Quality Assurance Handbook for Air Quality Pollution Measurement Systems, Vol. II. Ambient Air Specific Measurements." EPA-600/4-77-027a.

_____. 1978a. "Health Effects of Short-Term Exposures to Nitrogen Dioxide." Office of Research and Development.

_____. 1978b. "Air Quality Criteria for Ozone and other photochemical oxidants." EPA-600/8-78-004.

_____. 1978c. "National Air Quality, Monitoring and Emission Trends Report." EPA-450/2-78-052-December.

_____. 1978d. "Health Assessment Document for Polycyclic Organic Matter." Environmental Criteria and Assessment Office, External Review Draft No. 1.

_____. 1979. "Suspended particulate matter: A report to Congress." Research Triangle Park, North Carolina: U.S. EPA, Office of Environmental Criteria and Assessment.

EPRI (Electric Power Research Institute). 1976. "Sulfur oxides: Current status of knowledge." Report prepared by Greenfield, Attaway, and Tyler.

_____. 1978. "Cleaning the air." *EPRI Journal* 3 (October).

Eriksson, E. 1963. "The yearly circulation of sulfur in nature." *J. Geophys. Res.* 68: 4001-4008.

Envir. Sci. and Tech. 8: 1060-61 (editorial). 1974. "How does France handle her cleanup?"

Evelyn, J. 1661. *Fumiogorum, or the inconvenience of the Aer and Smoake of London Dissipated. Together with some remedies humbly proposed.* 2nd printing, 1772, quoted in Pitts and Metcalf, 1969.

Fairbairn, A.S., and D.D. Reid. 1958. "Air pollution and other local factors in respiratory disease." *Brit. J. Prev. and Soc. Med.* 12: 94.

Fairchild, G.A.; J. Roan; and J. McCarroll. 1972. "Atmospheric pollutants and the pathogenesis of viral respiratory function. Sulfur dioxide and influenza infection in mice." *Arch. Envir. Health* 25: 174–82.

Fancher. 1977. Private communication, Commonwealth Edison Corporation of Chicago.

FCLE. 1972a. Federal Common Law and the Environment, Texas v. Pankey. 2ELR 10174.

_____. 1972b. Federal Common Law and the Environment, Illinois v. Milwaukee. 2ELR 10169.

Federal Register. 1971. "National Ambient Air Quality Standards." 36: 8165.

_____. 1973. "National Ambient Air Quality Standards." 38: 25678.

_____. 1976. "Stack Height Increase Guidelines." 41: 7451.

_____. 1977. "Saccharin and its salts." 42: 200001.

Fennelly, P.F. 1975. "Primary and secondary particulates as pollutants: A literature review." *J. Air Poll. Contr. Assoc.* 25: 697–704.

Ferin, J., and L.J. Leach. 1973. "The effect of SO_2 on lung clearance of TiO_2 particles in rats." *Amer. Ind. Hyg. Assoc. J.* 34: 260–63.

Ferris, B.G., Jr. 1971. "Report on SO_2 to the World Health Organization." New York: United Nations.

_____. 1978. "Health effects of exposure to low levels of regulated air pollutants." *J. Air Poll. Cont. Assoc.* 28: 482–97.

Ferris, B.G., Jr., and D.O. Anderson. 1962. "The prevalence of chronic respiratory disease in a New Hampshire town." *Amer. Rev. Resp. Disease* 86: 165.

_____. 1964. "Epidemiological studies related to air pollution: A comparison of Berlin, New Hampshire and Chilliwack, British Columbia." *Proc. Roy. Soc. Med.* 57: 979.

Ferris, B.G., Jr., and F. Speizer. 1979. "Epidemiologic studies on health effects of SO_2 in sulfur oxides." National Academy of Sciences, Committee on Sulfur Oxides, Assembly of Life Sciences, National Research Council.

Ferris, B.G., Jr.; I.T.T. Higgins; M.W. Higgins; and J.M. Peters. 1973. "Chronic nonspecific respiratory disease in Berlin, New Hampshire, 1961–1967. A follow-up study." *Amer. Rev. Resp. Dis.* 107: 110–22.

Ferris, B.G., Jr.; H. Chen; S. Puleo; and R.L.H. Murphy. 1976. "Chronic nonspecific respiratory disease in Berlin, New Hampshire, 1967–1973. A further follow-up study." *Amer. Rev. Resp. Dis.* 113: 475–85.

Finch, S.J., and S.C. Morris. 1977. "Consistency of reported health effects of air pollution." Upton, New York: Brookhaven National Laboratory Report BNL 21808–R.

Finklea, J.F., et al. 1974. "Health consequences of sulfur oxides: A report from CHESS, 1970–71." Research Triangle Park, North Carolina: U.S. Environmental Protection Agency, EPA-650/1-74-004.

Finklea, J.F.; G.G. Akland; W.C. Nelson; R.I. Larsen; D.B. Turner; and W.E. Wilson. 1975. "Health Effects of Increasing Sulfur Oxides Emissions." Research Triangle Park, North Carolina: National Environmental Research Center, Office of Research and Development.

Finney, B.C.; R.E. Blanco; R.C. Dahlman; G.S. Hill; F.G. Kitts; R.E. Moore; and J.P. Witherspoon. 1977. "Correlation of radioactive waste treatment costs and the environmental waste treatment costs and the environmental impact of waste effluents in the nuclear fuel cycle-reprocessing light-water reactor fuel." Oak Ridge, Tennessee: Oak Ridge National Laboratory, ORNL/NUREG/TM-6.

Firket, J. 1931. "The cause of the symptoms found in the Meuse Valley during the fog of December, 1930." *Bull. Roy. Acad. Med. Belgium* 11: 683-741.

_____. 1936. "Fog along the Meuse Valley." *Trans. Faraday Soc.* 32: 1107-97.

Fletcher, C.M.; R. Peto; C. Tinker; and F.E. Speizer. 1976. *The Natural History of Chronic Bronchitis and Emphysema.* Oxford: Oxford University Press.

Foerster, T. von. 1978. "On the Use of Wood as an Energy Resource in the State of Maine." Report of the Environmental Policy Center, Harvard University.

Folinsbee, L.J.; F. Silverman; and R.J. Shephard. 1975. "Exercise responses following ozone exposure." *J. Appl. Physiol.* 38: 996-1001.

Frank, N.R.; M.O. Amdur; and J.L. Whittenberger. 1964. "A comparison of the acute effects of SO_2 administered alone or in combination with NaCl particles on the respiratory mechanics of healthy adults." *Int. J. Air Wat. Poll.* 8: 125.

Frank, N.R.; M.O. Amdur; J. Worcester; and J.L. Whittenberger. 1962. "Effects of acute controlled exposure to SO_2 on respiratory mechanics in healthy male adults." *J. Appl. Physiol.* 17: 252-58.

Frank, N.R.; R.E. Yoder; J.D. Brain; and E. Yokoyama. 1969. "SO_2 (35S labeled) absorption by the nose and mouth under conditions of varying concentration and flow." *Arch. Envir. Health* 18: 315-22.

Frederick, E.R. 1979. "Fabric filtration for fly-ash control." *J. Air Poll. Cont. Assoc.* 29: 81-85.

Freeman, A.M. III, and R.H. Haveman. 1972. "Residual charges for pollution control: A policy evaluation." *Science* 177: 322.

Friberg, L. 1950. "Health hazards in the manufacture of alkaline accumulators with special reference to chronic cadmium poisoning." *Acta. Med. Scand.* 138 (Suppl): 240.

Friberg, L., and R. Cederlöf. 1978. "Late effects of air pollution with special reference to lung cancer." *Envir. Health Persp.* 22: 45-66.

Friedrich, J.L., and R.H. Pai. 1977. "A primer of design considerations for western coal." *Heat Engineering* 48 (July-September).

Friend, J.P. 1973. "The global sulfur cycle." In *Chemistry of the Lower Atmosphere*, S.I. Rasool, ed. New York: Plenum Press.

Fry, F.A., and A. Black. 1973. "Regional deposition and clearance of particles in the human nose." *J. Aersol Sci.* 4: 113.

Gardner, D.E.; F.J. Miller; E.J. Blommer; and D.L. Coffin. 1977. "Relationship between nitrogen dioxide concentrations, time and level of effect using an animal infectivity model." Proc. Int. Conf. Photochem. Oxidant Poll. and Cont. Vol. I. EPA-600/3-77-001a: 513-525.

Gardner, D.E.; E.A. Pfitzer; R.T. Christian; and D.L. Coffin. 1971. "Loss of protective factor for alveolar macrophages when exposed to ozone." *Arch. Intern. Med.* 127: 1078-84.

Gaut, N.E. 1975. Personal communication, Environmental Research and Technology, Inc., Concord, Massachusetts.

Georgia. 1906. Georgia v. Tennessee Copper Co., 206 U.S. 238.

Giacomelli-Maltoni, G.; C. Melandri; V. Pordi; and G. Tarroni. 1972. "Deposition efficiency of monodisperse particle in human respiratory tract." *Amer. Ind. Hyg. Assoc. J.* 33: 603-10.

Giddens, W.E., Jr., and G.A. Fairchild. 1972. "Effects of sulfur dioxide on the nasal mucosa of mice." *Arch. Envir. Health* 25: 166.

Gifford, F.A. 1961. "Use of routine meteorological observations for estimating atmospheric dispersion." *Nuclear Safety* 2: 47-51.

Glasson, W.A., and C.S. Tuesday. 1970. "Hydrocarbon reactivities in atmospheric photooxidation of nitric-oxide." *Envir. Sci. and Tech.* 4: 916-24.

Goldsmith, B.J., and J.R. Mahoney. 1978. "Implications of the 1977 Clean Air Act Amendments for stationary sources." *Envir. Sci. and Tech.* 12: 144-49.

Goldsmith, J.R., and L.T. Friberg. 1977. "Effects on human health." In *Air Pollution*, A. Stern, ed. New York: Academic Press.

Goldsmith, J.R., and J. Nadel. 1969. "Experimental exposure of human subjects to ozone." *J. Air Poll. Cont. Assoc.* 19: 329-30.

Goldstein, E.; M.C. Eagle; and P.D. Hoeprich. 1973. "Effects of nitrogen dioxide on pulmonary bacterial defence mechanisms." *Arch. Envir. Health* 26: 202-204.

Goldstein, I.F., and G. Block. 1974. "Asthma and air pollution in two inner city areas in New York City." *J. Air Poll. Cont. Assoc.* 24: 665-70.

Goldstein, I.F., and L. Landowitz. 1977. "A critique of the relationship of air pollution to mortality." *J. Occ. Med.* 19: 375.

Golledge, A.H., and A.J. Wicken. 1964. "Local variations on the incidence of lung cancer and bronchitis mortality." *Med. Off.* 112: 273-77.

Gore, A.T., and C.W. Shaddick. 1958. "Atmospheric pollution and mortality in the county of London." *Brit. J. Soc. Med.* 12: 104-13.

Granat, L.; H. Rodhe; and R.O. Hallberg. 1976. "The global sulfur cycle." In "Nitrogen, Phosphorous and Sulphur—Global Cycles," SCOPE Report 7, B.H. Svensson and R. Soderlund, eds. *Ecol. Bull.* 22: 89-134.

Greenburg, L., and F. Field. 1962. "Area meterology—A component of air pollution studies." *Arch. Envir. Health* 4: 477-86.

Greenburg, L.; M. B. Jacobs; D. M. Drolette; F. Field; and M. M. Braverman. 1962. "Report on an air pollution incident in New York City, November, 1953." *Pub. Health Rep.* 77: 7–16.

Greenburg, L.; C. L. Erhardt; F. Field; and J. I. Reed. 1965. "Air pollution incidents and morbidity studies." *Arch. Envir. Health* 10: 351.

Gregor, J. J. 1976. "Mortality and air quality, the 1968–72 Allegheny County experience." Report No. 30. University Park, Pa.: Center for the Study of Environmental Policy, Pennsylvania State University.

Gregory, J. 1970. "The influence of climate and atmospheric pollution on exacerbations of chronic bronchitis." *Atmos. Envir.* 4: 453.

Hackney, J. D.; W. S. Linn; and K. A. Bell. 1978. "Experimental studies of human health effects of sulfur oxides." *Bull. N. Y. Acad. Med.* 54: 1177–85.

Hackney, J. D.; W. S. Linn; R. D. Buckley; E. E. Pedersen; S. K. Karuza; D. C. Law; and O. A. Fischer. 1975a. "Experimental studies on human health effects of air pollutants, I. Design considerations." *Arch. Envir. Health* 30: 373–78.

Hackney, J. D.; W. S. Linn; J. G. Mohler; E. E. Pedersen; P. Breisacher; and A. Russo. 1975b. "Experimental studies on human health effects of air pollutants II. Four-hour exposure to ozone alone and in combination with other pollutant gases." *Arch. Envir. Health* 30: 379–84.

Hackney, J. D.; W. S. Linn; D. C. Law; S. K. Karuza; H. Greenberg; R. D. Buckley; and E. E. Pedersen. 1975c. "Experimental studies on human health effects of air pollutants. III. Two-hour exposure to ozone alone and in combination with other pollutant gases." *Arch. Envir. Health* 30: 385–90.

Hackney, J. D.; W. S. Linn; J. G. Mohler; and C. R. Collier. 1977a. "Adaptation to short-term respiratory effects of ozone in men exposed repeatedly." *J. Appl. Physiol.: Resp. Envir. Exer. Phys.* 43: 82–85.

Hackney, J. D.; W. S. Linn; S. K. Karuza; R. D. Buckley; D. C. Law; D. V. Bates; M. Hazucha; L. D. Pengelly; and F. S. Iverman. 1977b. "Effects of ozone exposure in Canadians and Southern Californians." *Arch. Envir. Health* 32: 110–16.

Haenszel, W.; D. B. Loveland; and M. G. Sirken. 1962. "Lung cancer mortality as related to residence and smoking histories: 1. white males." *J. Nat. Can. Inst.* 28: 947–1001.

Haenszel, W.; B. C. Marcus; and E. G. Zimmerer. 1956. "Cancer morbidity in rural and urban Iowa." Public Health Mono. No. 37, Pub. Health Serv. Pub. No. 462. U. S. Government Printing Office.

Hagstrom, R. M.; H. A. Sprague; and E. Landau. 1967. "The Nashville air pollution study Pt. VII: Mortality from cancer in relation to air pollution." *Arch. Envir. Health* 15: 237.

Hakkarinen, C. 1975. "Fate of atmospheric sulfur and nitrogen from fossil fuels." *Abs. Pap. Amer. Chem. Soc.* 169: 30.

Hamilton, L. D. 1979. Private communication.

Hammer, D.I.; V. Hasselblad; and B. Portnoy. 1974. "Los Angeles student nurse study. Daily symptoms reporting and photochemical oxidants." *Arch. Envir. Health* 28: 255–60.

Hammer, D.I.; J.J. Miller; A.G. Stead; and C.G. Hayes. 1976. "Air pollution and childhood lower respiratory disease. I. Exposure to sulfur oxides and particulate matter in New York, 1972." In *Clinical Implications of Air Pollution Research*, A.J. Finkel and W.C. Duel, eds., Ch. 24. Acton, Massachusetts: Publishing Sciences Group.

Hammer, D.I.; B. Portnoy; F.W. Massey; W.S. Wayne; T. Oelsner; and P.F. Wehrle. 1965. "Los Angeles air pollution and respiratory symptoms—relationship during a selected 28 day period." *Arch. Envir. Health* 10: 475–80.

Hammond, E.C. 1966. "Smoking in relation to death rates of one million men and women." In *Epidemiologic Approaches to the Study of Cancer and Other Chronic Diseases, National Cancer Institute Monograph*, W. Haenszel, ed., pp. 127–204. Bethesda, Maryland: U.S. Public Health Service.

_____. 1972. "Smoking habits and air pollution in relation to lung cancer." In *Environmental Factors in Respiration*, K. Lee, ed., ch. 12. New York: Academic Press.

Hammond, E.C., and D.H. Horn. 1958. "Smoking and death rates: Report on forty-four month follow-up of 187,783 men." *JAMA* 166: 1294–1308.

Hammond, E.C.; J.J. Selikoff; P.L. Lawther; and H. Seidman. 1976. "Inhalation of benzopyrene and cancer in man." *Ann. N.Y. Acad. Sci.* 271: 116–24.

Harting, F.H., and W. Hesse. 1879. "Der Lungen Krebs, die Bergkraukheit in den Schneeberger." *Gruben. Viertjhrs. Ger. Med. Off. Sanit.* 30: 296; 31: 102; 31: 313.

Hatch, T.F. 1961. "Distribution and deposition of inhaled particles in the respiratory tract." *Bact. Rev.* 25: 237.

Hattori, S. 1973. "Alterations of broncho-alveolar system by polluted air: Experimental consideration." *Clinician* 219: 408 (in Japanese).

Hattori, S., and K. Takemura. 1974. "Ultrastructural changes in the bronchiolar alveolar system caused by air pollution and smoking." *J. Clin. Electron Microsc. Soc. Japan* 6: 350 (in Japanese).

Hattori, S.; R. Tateishi; T. Horai; and T. Nakajima. 1972. "Morphological changes in the bronchial alveolar system of mice following continuous exposure to NO_2 and CO." *J. Jap Soc. Chst. Dis.* 10: 16–22 (in Japanese).

Haury, G.; S. Jordan; and C. Hofmann. 1978. "Experimental investigation of the aerosol-catalyzed oxidation of SO_2 under atmospheric conditions." In *Sulfur in the Atmosphere*, R.B. Husar, J.P. Lodge, Jr., and D.J. Moore, eds. Oxford: Pergamon Press.

Hausknecht, R. 1959. "Air Pollution: Effects Reported by California Residents." Sacramento, California: California State Department of Public Health.

_____. 1962a. "Experiences of a respiratory disease panel selected from a representative sample of the adult population." *Amer. Rev. Resp. Dis.* 86: 858–66.

_____. 1962b. "Population surveys in Air Pollution Research." Berkeley, California: Bureau of Chronic Diseases, California State Department of Public Health.

Hawthorne, Sir W. 1978. "Energy and environment, conflict or compromise." Trueman Wood Lecture. *J. Roy. Soc. Arts*, July.

Hazucha, M., and D.V. Bates. 1975. "Combined effect of ozone and sulfur dioxide on human pulmonary function." *Nature* 257: 50-51.

Hazucha, M.; C. Parent; and D.V. Bates. 1977. "Development of ozone tolerance in man." Proc. Intern. Conf. Photochem. Oxid. Poll. and its control. Research Triangle Park, North Carolina: EPA-600/3-77-001a: 527-541.

Heimann, H. 1970. "Episodic air pollution in metropolitan Boston: A trial epidemiological study." *Arch. Envir. Health* 20: 230.

Herington, C.F. 1920. *Powdered Coal as a Fuel*. Princeton, New Jersey: Van Nostrand-Reinhold.

Heuss, J.M., and W.A. Glasson. 1968. "Hydrocarbon reactivity and eye irritation." *Envir. Sci. and Tech.* 2: 1109-16.

Heuss, J.M.; G.J. Nebel; and J.M. Colucci. 1971. "National air quality standards for automotive pollutants—A critical study." *JAPCA* 21: 535-44.

HEW. 1969. "Air Quality Criteria for Particulate Matter." Washington, D.C.: National Air Pollution Control Association AP-49.

_____. 1970a. "Air Quality Criteria for Sulfur Oxides." Washington, D.C.: National Air Pollution Control Association AP-50.

_____. 1970b. "Air Quality Criteria for Photochemical Oxidants." Washington, D.C.: National Air Pollution Control Association AP-63.

_____. 1971. "Air Quality Criteria for Nitrogen Oxides." Washington, D.C.: National Air Pollution Control Association AP-84.

Heyder, J., and C.N. Davies. 1971. "The breathing of half micron aerosols, III. Dispersion of particles in the respiratory tract." *J. Aerosol Sci.* 2: 437.

Hidy, G.M., and C.S. Burton. 1974. "Atmospheric Aerosol Formation by Chemical Reactions." Unpublished.

Hidy, G.M.; E.Y. Tong; and P.K. Mueller. 1977. "Design of the Sulphate Regional Experiment (SURE)." Available from G. Hidy, ERT Westlake, California, or from EPRI, Menlo Park, California.

Higgins, I.T.T. 1966. "Air pollution and chronic respiratory disease." *ASHRAE Journal* 8: 37.

_____. 1974. "Trends in respiratory cancer mortality in the U.S., England, and Wales." *Arch. Envir. Health* 28: 121-29.

_____. 1976. "Epidemiological evidence on the carcinogenic risk of air pollution." *Environmental Pollution and Carcinogenic Risk* (Insem Symp. Ser.) 52: 41-52.

Higginson, J. 1960. "Population studies on cancer action." *Acta. Un. Inst. Cancer* 16: 1667.

_____. 1969. *Present Trends in Cancer Epidemiology*, Canadian Cancer Conf. 83. Oxford: Pergamon Press.

_____. 1975. "Cancer Etiology and Prevention." In *Persons at High Risk of Cancer: An Approach to Cancer Etiology and Control*, J. Fraumeni, ed. New York: Academic Press.

Higginson, J., and O.M. Jensen. 1977. "Epidemiological review of lung cancer in man." In *Air Pollution and Cancer in Man*. Lyon, France: International Agency for the Research of Cancer Pub. No. 16.

Higginson, J., and C.S. Muir. 1973. "Epidemiology of Cancer." In *Cancer Medicine*, J.F. Holland and E. Frei III, eds., pp. 241–306. Philadelphia: Lea and Febriger.

Hill, A.B. 1965. "The environment and diseases: Association and causation." *Proc. Roy. Soc. Med., Sec. Occup. Med.* 58: 272.

Hirsch, J.A.; E.W. Sweson; and A. Wannen. 1975. "Tracheal mucous transport in beagles after long-term exposure to 1 rpm sulfur dioxide." *Arch. Envir. Health* 30: 249–53.

Hitchcock, D.R. 1976. "Atmospheric sulfates from biological sources." *J. Air Poll. Cont. Assoc.* 22: 210–15.

Hitosugi, M. 1968. "Epidemiological study of lung cancer with special reference to the effect on air pollution and smoking habit." *Bull. Inst. Pub. Health* (Japan) 17: 237.

Hoffman, E.F.; and A.G. Gilliam. 1954. "Lung cancer mortality." *Pub. Health Rep.* 69: 1033–42.

Holland, G.J.; D. Benson; A. Bush; G.Q. Rich; and R.P. Holland. 1968. "Air pollution stimulation and human performance." *Amer. J. Pub. Health* 58: 1684.

Holland, W.W.; and D.D. Reid. 1965. "The urban factor in chronic bronchitis." *Lancet* 1: 445.

Holland, W.W.; and R.W. Stone. 1965. "Respiratory disorders in U.S. East Coast telephone men." *Amer. J. Epid.* 82: 92.

Holland, W.W.; C.C. Spicer; and J.M.G. Wilson. 1961. "Influence of the weather on respiratory and heart disease." *Lancet* 2: 338.

Holland, W.W.; D.D. Reid; R. Seltser; and R.W. Stone. 1965. "Respiratory disease in England and the United States." *Arch. Envir. Health* 10: 338.

Holland, W.W.; H.S. Kasap; J.R.T. Colley; and W. Cormack. 1969. "Respiratory symptoms and ventilatory function: A family study." *Brit. J. Prev. and Soc. Med.* 23: 77.

Hoskins, W.L. 1971. "Let the pricing system provide the incentive." *Indust. Water Engineering* 8 (April): 8–10.

Hounan, R.F.; A. Black; and M. Walsh. 1969. "Deposition of aerosol particles in the nasopharyngeal region of the human respiratory tract." *Nature* 221: 1254.

Hrubec, Z.; R. Cederlöf; L. Friberg; R. Horton; and G. Ozolins. 1973. "Respiratory symptoms in twins." *Arch. Envir. Health* 27: 189–95.

Huber, T.E.; S.W. Joseph; E. Knoblock; P.L. Redfearn; and J.A. Karakawa. 1954. "New Environmental Respiratory Disease (Yokohama Asthma)." *Arch. Ind. Hyg. Occup. Med.* 10: 399.

Hull, A.P. 1971. "Radiation in perspective—some comparisons of environmental risks from nuclear fuel and fossil fueled power plants." *Nuc. Safety News* 12, no. 3: 185.

———. 1974. "Environmental Radiation Dose Criteria and Assessment—Pathway Modeling and Surveillance." *Nuc. Safety News* 21: 491–95.

Hurst, D.J. 1970. "Effects of ozone acid hydrolases of the pulmonary alveolar macrophage." *J. Reticuloendothel. Soc.* 8: 288–300.

Hurwitz, H. 1979. "Radon in homes—A challenge to regulatory constancy." Schnectady, New York: G.E. Co., Inc.

Husar, R.B.; J.P. Lodge; and D. Moore, eds. 1978. *Sulfur in the Atmosphere.* Oxford: Pergamon Press.

Hasar, R.B.; D.E. Patterson; J.D. Husar; N.V. Gillani; and W.E. Wilson, Jr. 1978. "Sulfur budget of a power plant plume." *Atmos. Envir.* 12: 549–68.

IARC. 1977. "Air Pollution and Cancer in Man." Lyon, France: International Agency for Research on Cancer, Publication No. 16.

ICAS. 1974. "Report of the ad hoc panel on the present interglacial." Washington, D.C.: National Science Foundation, Federal Council for Science and Technology Interdepartmental Committee for Atmospheric Science, 18b–FY75.

ICRP (International Commission on Radiological Protection). 1977. Reports 22 and 26. London and New York: Pergamon Press.

Ishikawa, S.; D.H. Bowden; V. Fisher; and J.P. Wyatt. 1969. "The 'Emphysema Profile' in two midwestern cities in North America." *Arch. Envir. Health* 18: 660.

Islam, M.S.; E. Vastag; and W.T. Ulmer. 1972. "Sulphur-dioxide induced bronchial hyperactivity against acetulcholine." *Int. Arch. Arbeitsmed* 29: 221–32.

IWCI (International Workshop on Climate Issues). 1978. "International Perspectives on the Study of Climate and Society." A Report of the International Workshop on Climate Issues. Washington, D.C.: National Academy of Sciences.

Jackson, A.C.; P.J. Gulesian, Jr.; and J. Mead. 1975. "Glottal aperture during panting with voluntary limitation of tidal volume." *J. Appl. Physio.* 39: 834–36.

Jackson, A.C.; J.P. Butler; E.J. Millet; F.G. Hoppins, Jr.; and S.V. Dawson. 1977. "Airway geometry by analysis of acoustic pulse response measurements." *J. Appl. Physio.* 43: 523–36.

JASON. 1979. "The long-term impact of atmospheric carbon dioxide on climate." Arlington, Virginia: JASON Technical Report for the U.S. DOE JRS-78-07 by SRI International.

Jaworowski, Z.; J. Bilkiewicz; E. Dobosz; D. Grybowska; L. Kownacka; and Z. Wronski. 1974. "Radiation hazards to the population resulting from conventional and nuclear powered electric production." In "Environmental Surveillance Around Nuclear Installations," SM-180/20, Vol. I, pp. 403-12. Vienna: International Atomic Energy Agency.

Joensuu, O. 1971. "Fossil fuels as a source of mercury pollution." *Science* 172: 1027-28.

Johnson, N.R.; J.G. Mohler; and B.W. Armstrong. 1971. "Specific conductance in human subjects acutely exposed to 0.6 to 0.8 ppm ozone." In *Proc. Second Int. Clean Air Conf.* H.M. Englund and W.T. Berry, eds., pp. 195-99. New York: Academic Press.

Johnson, W.B.; D.E. Wolf; and R.L. Mancuso. 1978. "Long-term regional patterns and transfrontier exchanges of airborne sulfur pollution in Europe." *Atmos. Envir.* 12: 511-27.

Junge, C.E. 1963. *Air Chemistry and Radioactivity.* New York: Academic Press.

Kagawa, J., and T. Toyama. 1975. "Photochemical air pollution: Its effects on respiratory function of elementary school children." *Arch. Envir. Health* 30: 117-22.

Kagawa, J.; T. Toyama; and M. Nakaza. 1976. "Pulmonary function tests in children exposed to air pollution." In *Clinical Implications of Air Pollution Research,* A.J. Finkel and W.C. Duel, eds., pp. 305-20. Acton, Massachusetts: Publishing Sciences Group.

Kawai, M.; H. Amamoto; and K. Harada. 1967. "Epidemiological study of occupational lung cancer." *Arch. Envir. Health* 14: 859.

Keeling, C.D. 1973. "The carbon dioxide cycle: Reservoir models to depict the exchange of atmospheric carbon dioxide with the oceans and land plants." In *Chemistry of the Lower Atmosphere,* S.I. Rasool, ed. New York: Plenum Press.

Keeling, C.D.; R.B. Bacastow; A.E. Bainbridge; A. Ekdahl, Jr.; R. Guenther; and S. Waterman. 1976a. "Atmospheric carbon dioxide variations at Mauna Lua Observatory, Hawaii." *Tellus* 28: 538-51.

Keeling, C.D.; J.A. Adams, Jr.; A. Ekdahl, Jr.; and R. Guenther. 1976b. "Atmospheric carbon dioxide variations at the South Pole." *Tellus* 28: 552-64.

Kellogg, W.W.; R.D. Cadle; E.R. Allen; A.Z. Lazarus; and E.A. Martell. 1972. "The sulfur cycle: Man's contributions are compared to natural sources of sulfur in the atmosphere and oceans." *Science* 175: 587-96.

Kenline, P.A. 1966. "October 1963 New Orleans asthma study." *Arch. Envir. Health* 12: 295.

Kennecott Copper. 1972. Kennecott Copper v. EPA, 462 F 2d 846, 3 ERC 1682 2 ELR 20166. U.S. Court of Appeals D.C. Circuit 231.

Kleinfeld, M.; C. Giel; and I.R. Tabershaw. 1957. "Health hazards associated with inert-gas-shielded metal arc welding." *Arch. Ind. Health* 15: 27-31.

Kletz, T.A. 1977. "The risk equations: what risks should we run." *New Scientist* (May): 320.

Kneese, A.V., and B.T. Bower. 1968. *Managing Water Quality*. Baltimore: Johns Hopkins University Press.

Knelson, J.H., and R.E. Lee. 1977. "Oxides of nitrogen in the atmosphere. Origin, fate and public health implication." *Ambio* 6: 126-30.

Knelson, J.H.; M.L. Peterson; G.M. Goldstein; D.E. Gardner; and C.G. Hayes. 1974. "Health Effects of Oxidant Exposures: A Research Progress Report. Report on UC-ARB Conference—Technical Bases for control strategies of photochemical oxidant: current status and priorities in research." Riverside: University of California Statewide Air Pollution Research Center.

Korfmacher, W.A.; D.F.S. Natusch; and E. Wehry. 1977. Unpublished results cited in Natusch (1978).

Koshal, R.K., and M. Koshal. 1973. "Environments and urban mortality, an econometric approach." *Envir. Poll.* 4: 247-59.

Kramer, B.M. 1976. "Economics, technology and the Clean Air Amendments of 1970, the first six years." *Ecology Law Quarterly* 6: 161-230.

Kreyberg, L. 1956. "Occurrence and aetiology of lung cancer in Norway and in the light of pathological anatomy." *Brit. J. Prev. and Soc. Med.* 10: 145.

Lambelet, J.C. 1972. "Recent controversies over environmental policies." Private report, Economics Department, University of Pennsylvania.

Lambert, P.M., and D.D. Reid. 1970. "Smoking, air pollution and bronchitis in Britain." *Lancet* 1: 853.

Lammers, B.; R.F. Schilling; and J. Walford. 1964. "A study of byssinosis, chronic respiratory symptoms, and ventilatory capacity in England and Dutch cotton workers, with special reference to atmospheric pollution." *Brit. J. Ind. Med.* 21: 124.

Larsen, R.I. 1969. "A new mathematical model of air pollution concentration averaging time and frequency." *J. APCA* 19: 24-30.

_____. 1971. *A Mathematical Model for Relating Air Quality Measurements to Air Quality Standards*. Research Triangle Park, North Carolina: U.S. EPA AP-89.

Laskin, S.; M. Kuschner; and R.T. Drew. 1970. "Studies in pulmonary carcinogenesis." In *Inhalation Carcinogenesis*, M.G. Hanna, Jr., P. Nettesheim, and J.R. Gilbert, eds. Oak Ridge, Tennessee: U.S. AEC Symposium Series 18.

Last, J.A., and C.E. Cross. 1978. "A new model for health effects of air pollutants. Evidence for synergistic effects of mixtures of ozone and sulfuric acid aerosols on rat lungs." *J. Lab. Clin. Med.* 91: 328-39.

Last, J.A.; M. Jennings; and C.E. Cross. 1977. "Mucous glyoprotein secretion by tracheal explants from rats exposed to ozone." *Fed. Proc.* 36: 413.

Lave, L.B., and L.C. Freeburg. 1973. "Health effects of electricity generation from coal, oil and nuclear fuel." *Nuc. Safety* 14: 409-28.

Lave, L.B., and E. Seskin. 1970. "Air Pollution and Human Health." *Science* 69: 723–33.

———. 1971. "Health and Air Pollution." *Swed. J. Econ.* 73: 76–95.

———. 1972. "Air pollution, climate and home heating: their effects on U.S. mortality rates." *Amer. J. Pub. Health* 62: 909.

———. 1973. "An analysis for the association between U.S. mortality and air pollution." *J. Amer. Stat. Assoc.* 68: 284–90.

———. 1977. *Air Pollution and Human Health.* Baltimore: Johns Hopkins University Press.

Lavery, T.F. 1978. "Regional Transport and Photochemical model of atmospheric sulfates." Downsview, Ontario: NATP/CCMS Ninth International Meeting on Air Pollution Modelling and Its Application.

Lavery, T.F.; G.M. Hidy; R. Baskett; and J. Thrasher. 1978. "Occurrence of long range transport of sulphur-oxides in Northeastern United States." *B. Am Meteor.* 59: 1239.

Lawther, P.J. 1958. "Climate, air pollution and chronic bronchitis." *Proc. Roy. Soc. Med.* 51: 262.

———. 1963. "Compliance with the Clean Air Act: medical aspects." *J. Instit. Fuel* 36: 341–44.

Lawther, P.J., and E. Waller. 1978. "Trends in urban air pollution in the United Kingdom in relation to lung cancer mortality." *J. Envir. Health Persp.* (Feb): 71–73.

Lawther, P.J.; B.T. Commins; and R.G. Waller. 1965. "A study of the concentration of aromatic hydrocarbons in gas works retort houses." *Brit. J. Ind. Med.* 22: 13.

Lawther, P.J.; R.E. Waller; and M. Henderson. 1970. "Air pollution and exacerbations of bronchitis." *Thorax* 25: 525–39.

Lawther, P.J.; A.J. MacFarland; R.W. Waller; and A.G.F. Brooks. 1975. "Pulmonary function and sulfur dioxide: Some preliminary findings." *Envir. Res.* 10: 355–69.

Lebowitz, M.D., and G.A. Fairchild. 1973. "The effects of sulfur dioxide and A_2 influenza virus on pneumonia and weight reduction in mice: An analysis of stimulus–response relationships." *Chem. Biol. Interact* 7: 317–26.

Lebowitz, M.D.; P. Bendheim; G. Cristea; D. Markowitz; J. Misiaszele; M. Staniec; and D. Van Wyck. 1974. "The effect of air pollution and weather on lung function in exercising children and adolescents." *Amer. Rev. Resp. Dis.* 109: 262–63.

Lee, B.K., and J.F. Fraumeni. 1969. "Arsenic and respiratory cancer in man: An occupational study." *J. Nat. Can. Inst.* 42: 10.

Lee, R.E., Jr., and S. Goranson. 1976. "National Air Surveillance Cascade Impactor Network. III. Variations in size of airborne particulate matter over three-year period." *Envir. Sci. and Tech.* 10: 1022.

Lee, R.E., Jr.; and D.J. VonLehnden. 1973. "Trace metal pollution in the environment." *J. Air Poll. Cont. Assoc.* 23: 853.

Lee, R.E., Jr.; R.K. Patterson; and J. Wagman. 1968. "Particle-size distribution of metal components in urban air." *Envir. Sci. and Tech.* 2: 288-90.

Lee, R.E., Jr.; S.S. Goranson; R.E. Enrione; and G.B. Morgan. 1972. "National Air Surveillance Cascade Impactor Network. II. Size distribution measurements of trace metal components." *Envir. Sci. and Tech.* 6: 1025-30.

Van der Lende, R.; J.P.M. DeKroon; G.J. Tammerling; B.F. Visser; K. DeVries; J. Wever-Hess; and N.G.M. Orie. 1972. "Prevalence of chronic nonspecific lung disease in a nonpolluted and an air polluted area of the Netherlands." In *Ecology of Chronic Nonspecific Respiratory Diseases.* Warsaw: Panstwowy Zaklad Wydawnictw Lekavskich.

Van der Lende, R.; C. Huygen; E. Jansen-Koster; S.K. Nijpstra; R. Peset; B.F. Visser; E.H.E. Wolfs; and N.G.M. Orie. 1974. "A temporary decrease in the ventilatory function of an urban population during an acute increase in air pollution." *Bull. Physio. Path. Resp.* 11: 31-43.

Leonard, A.G.; D. Crowley; and J. Belton. 1950. "Atmospheric pollution in Dublin during the years 1944-1950." *Roy. Dub. Soc. Sci. Proc.* 25: 166-67.

Levin, M.L.; W. Haenszel; B.E. Carrol; D.R. Gerhardt; V.H. Handy; and S.C. Ingraham. 1960. "Cancer incidence in urban and rural areas of New York State." *J. Nat. Can. Inst.* 24: 1243-57.

Lewis, R.; M.M. Gilkeson; and R.O. McCaldin. 1962. "Air pollution and New Orleans asthma." *Pub. Health Rep.* 77: 947.

Lewis, T.R.; W.J. Moorman; W.F. Lundmann; and K.I. Campbell. 1973. "Toxicity of long-term exposure to oxides of sulfur." *Arch. Envir. Health* 26: 16-21.

Lieben, J., and H. Pistawka. 1967. "Mesothelioma and asbestos exposure." *Arch. Envir. Health* 14: 559-631.

Likens, G.E. 1976. "Acid Precipitation." *Chem. Eng. News* 54: 29-44.

Likens, G.E., and F.H. Bormann. 1974. "Acid rain: A serious regional environmental problem." *Science* 184: 1171-79.

Lindberg, W. 1968. "Der Alminnilige luft forurensing i norse" ("General air pollution in Norway"). Oslo: Utgit av Royksderadit (Smoke Damage Council).

Linn, W.S.; J.D. Hackney; E.E. Pedersen; P. Breisacker; J.V. Patterson; C.H. Mulry; and J.F. Coyle. 1976. "Respiratory function and symptoms in urban office workers in relation to oxidant air pollution exposure." *Amer. Rev. Resp. Dis.* 114: 477-83.

Linzon, S.N. 1971. "Economic effects of sulfur dioxide on forest growth." *J. Air Poll. Cont. Assoc.* 21: 81-86.

Lipfert, F.W. 1977. "The association of air pollution with human mortality: multiple regression results for 136 U.S. cities, 1969." Presented at the 70th annual conference of the Air Pollution Control Association, Toronto, Canada, June 22-24.

_____ . 1978. "The association of human mortality with air pollution." Cincinnati, Ohio: Union Graduate School Thesis.

Lippmann, M. 1970a. "'Respirable' dust sampling." *Amer. Ind. Hyg. Assoc. J.* 31: 138-59.

_____ . 1970b. "Deposition and clearance of inhaled particles in the human nose." *Ann. Otol.* 70: 519.

Lippmann, M., and R.E. Albert. 1969. "The effect of particle size on the regional deposition of inhaled aerosols in the human respiratory tract." *Am. Ind. Hyg. Assoc. J.* 30: 257.

Lippmann, M.; R.E. Albert; and H.T. Peterson, Jr. 1971. "The regional deposition of aerosols in man." In *Inhaled Particles, III,* W.H. Walton, ed., I: 105. Surrey, England: Unwin Brothers.

Little, J.B.; E.P. Radford, Jr.; and R.B. Holtzman. 1967. "Polonium 210 in bronchial epithelium of cigarette smokers." *Science* 155: 606.

Lloyd, J.W. 1971. "Long term mortality of steel workers v. respiratory cancer in coal oven workers." *J. Occup. Med.* 13: 53.

Loudon, R.G., and J.F. Kilpatrick. 1969. "Air pollution, weather, and cough." *Arch. Envir. Health* 18: 641.

Louisville Courier-Journal. 1931. Meuse Valley, Belgium air pollution episode. Reprinted in *Literary Digest,* January 10, 1931.

Lowrance, W.W. 1976. *Of Acceptable Risk — Science and Determination of Safety.* Los Altos, California: Kaufman.

Ludwig, J.H.; G.B. Morgan; and T.B. McMullen. 1971. "Trends in Urban Air Quality." In *Man's Impact on Climate,* W.H. Matthew, W.W. Kellogg, and G.O. Robinson, eds., pp. 321-38. Cambridge, Massachusetts: MIT Press.

Lundgren, D.A., and H.J. Paulus. 1975. "The mass distribution of large atmospheric particles." *J. Air Poll. Control Assoc.* 25: 1227.

Lunn, J.E.; J. Knowelden; and A.J. Handyside. 1967. "Patterns of respiratory illness in Sheffield infant school children." *Brit. J. Prev. and Soc. Med.* 21: 7.

Lunn, J.E.; J. Knowelden; and J.W. Roe. 1970. "Patterns of respiratory illness in Sheffield junior school children. A follow-up study." *Brit. J. Prev. and Soc. Med.* 24: 223.

Lynn, D.A.; G.L. Deane; R.G. Galkiewicz; R.M. Bradway; and F.A. Record. 1976. "National assessment of the urban particulate problem, Vol. 1. Summary of National." EPA-950/3-76-024, July 1976.

Lyons, W.A., and R.B. Husar. 1976. "Visible images detect a synoptic-scale air pollution episode." *Mon. Weather Rev.* 104: 1623-26.

MacFarland, H.N.; E.C. Ulrich; A. Martin; A. Krumm; W.M. Busey; and Y. Alarie. 1971. "Chronic exposure of cynomolgus monkeys to fly ash." In *Inhaled Particles III,* I: 313-27. Surrey, England: Unwin Brothers.

Mage, D.T., and W.R. Otto. 1978. "Refinements on the lognormal probability model for analysis of aerometric data." *J. Air Poll. Cont. Assoc.* 28: 796-97.

Makino, K., and I. Mizoguchi. 1975. "Symptoms caused by photochemical smog." *Jap. J. of Pub. Health* 22: 421–30.

Maltoni, C. 1977. "Recent findings on the carcinogenicity of chlorinated olefins." *Envir. Health Persp.* 21: 1.

Manabe, S., and R.T. Wetherald. 1975. "The effects of doubling the CO_2 concentration in the climate of a general circulation model." *J. Atom Sci.* 32: 3–15.

Mancuso, T.F. 1970. "Relation of duration of employment and prior respiratory illness to respiratory cancer among beryllium workers." *Envir. Res.* 3: 251–75.

Mancuso, T.F., and E.J. Coulter. 1958. "Cancer mortality among native white, foreign-born white, and nonwhite male residents of Ohio: Cancer of the lung, larynx, bladder, and central nervous system." *J. Nat. Can. Inst.* 20: 79–105.

Mandi, A.; E. Galambos; G. Galgocy; M. Szabo; and K. Koller. 1974. "Relationship between lung function values and air pollution data in Budapest school children." *Pneumologie* 150: 217.

Manzhenko, E.G. 1966. "The effect of atmospheric pollution on the health of children." *Hygiene and Sanit.* 31, no. 10: 126.

Marland, G., and R.M. Rotty. 1977. "The Question Mark over Coal." Third IIASA Conference on Energy Resources, Moscow.

Martens, A., and W. Jacobi. 1973. "Die in-vivo Bestimmung der Aerosolteilchendeposition in Atemtrakt bei Mund-bzw. Naseatmung." In *Aerosole in Physik, Medizin und Technik*, V. Bohlay, ed., pp. 117–21. Bad Soden, West Germany: Gesselschaft fur Aerosolforschung.

Martin, A.E., and W.H. Bradley. 1960. "Mortality, fog and atmospheric pollution: an investigation during winter of 1958–59." *Monthly Bull. Min. Health* (London) 19: 56–72.

Martin, J.E.; E.D. Harward; and D.T. Oakley. 1969. "Comparison of radiation from fossil fuel and nuclear power plants." U.S. Public Health Service, Department of H.E.W. Reprinted in *Environmental Effects of Producing Electric Power*. Committee Report, Hearings of Joint Commitee on Atomic Energy. 91st Congress. Government Printing Office. 773–809.

MASN (Massachusetts Air Surveillance Network). 1970. Unpublished data on sulfate concentrations. Data collected by the Massachusetts Department of Environmental Quality Engineering within the Executive Secretary's Office of Environmental Affairs, Boston, Massachusetts.

McBride, J.P.; R.E. Moore; J.P. Witherspoon; and R.E. Blanco. 1978. "Radiological impact of airborne effluents of coal and nuclear plants." *Science* 202: 1045–58.

McCarroll, J. 1979. "Health effects associated with increased use of coal." Paper presented at the 72nd Annual meeting of the Air Pollution Control Association, Cincinnati, Ohio.

McCarroll, J., and W. Bradley. 1966. "Excess mortality as an indicator of health effects of air pollution." *Amer. J. Pub. Health* 56: 1933–42.

McDonald, A.D.; A. Harper; O.A. El Attar; and J.C. McDonald. 1970. "Epidemiology of primary malignant mesothelial tumors in Canada." *Cancer* 26: 914–19.

McEwen, J.; A. Finlayson; A. Muir; and A.A.M. Gibson. 1970. "Mesothelioma in Scotland." *Brit. Med. J.* 4: 575–78.

McJilton, C.; R. Frank; and R. Charlson. 1973. "Role of relative humidity in the synergistic effect of a sulfur dioxide-aerosol mixture in the lungs." *Science* 182: 503.

McMillian, R.S.; D.H. Wiseman; B. Hanes; and P.F. Wehrle. 1969. "Effects of oxidant air pollution on peak expiratory flow rates in Los Angeles school children." *Arch. Envir. Health* 18: 941.

Mellick, P.W.; D.L. Dungsworth; L.W. Schwartz; and W.S. Tyler. 1977. "Short-term morphologic effects of high ambient levels of ozone on lungs of rhesus monkeys." *Lab. Invest.* 36: 82–90.

Melville, G.N. 1970. "Changes in specific airway conductance in healthy volunteers following nasal and oral inhalation of SO_2." *West Ind. Med. J.* 19: 231–35.

Mendelsohn, R., and G. Orcutt. 1978. "An empirical analysis of air pollution dose response curves." New Haven, Connecticut: Institute for Social and Policy Studies, Working Paper 800.

Menzel, D.B. 1976. "The role of free radicals in the toxicity of the air pollutants (nitrogen oxides and ozone)." In *Free Radicals in Biology*, W. Pryor, ed., pp. 181–201. New York: Academic Press.

Menzel, D.B.; M.D. Abou-Donia; C.R. Roe; R. Ehrlich; D.E. Gardner; and D.L. Coffin. 1977. "Biochemical indices of nitrogen dioxide intoxication of guinea pigs following low-levels long-term exposure." *Proc. Intern. Conf. on Photochem. Oxidant Poll. and Its Control*, B. Dimitriades, ed., vol. II: 577–587. Research Triangle Park, North Carolina: U.S. EPA–600/3–77–001b.

Meselson, M., and K. Russell. 1977. "Comparison of carcinogenic and mutagenic potency." Proc. of the Cold Springs Harbor Conf. on Origins of Human Cancer, New York.

Meyer, G.W. 1976. "Environmental respiratory disease (Tokyo–Yokohama asthma): the case for allergy." In *Clinical Implications of Air Pollution Research*, A.J. Finkel and W.C. Duel, eds., pp. 177–82. Acton, Massachusetts: Publishing Sciences Group.

Meyers, R.E.; R.T. Cederwall; L.I. Kleinman; S.E. Schwartz; and M. McCoy. 1978. "Constraints on coal utilization with respect to air pollution production and transport over long distances: summary." Brookhaven, New York: Brookhaven National Laboratory. Draft report.

Michel, F.B.; J.P. Marty; L. Quet; and P. Cour. 1977. "Penetration of inhaled pollen into the respiratory tract." *Amer. Rev. Resp. Dis.* 115: 609–16.

Michie. 1974. Michie v. Great Lakes Steel Div., 495 F. 2d 213 (6th Cir.).

Miller, F.J. 1977. "A mathematical model for transport and removal of ozone in mammalian lungs." Thesis, North Carolina State University (Raleigh), Department of Statistics.

Miller, F.J.; J.W. Illing; and D.E. Gardner. 1978. "Effect of urban ozone levels on laboratory induced respiratory infections." *Toxicol. Lett.* 2: 163–69.

Miller, F.J.; D.E. Gardner; J.A. Graham; R.E. Lee; W.E. Wilson; and J.D. Bachmann. 1979. "Size considerations for establishing a standard for inhalable particles." *J. Air Poll. Cont. Assoc.* 29: 610–15.

Mills, C.A. 1957. "Respiratory and cardiac deaths in Los Angeles smog." *Amer. J. Med. Sci.* 233: 379–86.

Mills, M.T., and A.A. Hirata. 1978. "A multi-scale transport and dispersion model for local and regional scale sulfur dioxide/sulfur concentrations." Downsview, Ontario: NATO/CCMS Ninth International Meeting on Air Pollution Modeling and Its Application.

Ministry of Public Health. 1954. "Mortality and morbidity during the London fog of September, 1952." London: Reports on Public Health and Related Subjects No. 95.

Mishan, E.J. 1976. *Cost-Benefit Analysis.* New York: Praeger Publishers. Especially chapters on "Loss of life and limb" and "Measuring pollution damage."

MIT. 1970. *Man's Impact on the Global Environment. Assessment and Recommendations for Action.* Report of the Study of Critical Environmental Problems (SCEP). Cambridge, Massachusetts: MIT Press.

Mitchell, J.M., Jr. 1961. "Recent secular changes of global temperature." *Ann. N.Y. Acad. of Sci.* 95: 235.

Moeller, D.W., and D.W. Underhill. 1976. "Study of the effects of building materials on population dose equivalents." Boston, Massachusetts: Harvard School of Public Health Report for EPA.

Mohler, J.G.; J.P. Butler; J.D. Hackney; C.R. Collier; D.C. Law; and W.S. Linn. 1978. "Redistribution of specific ventilation in humans exposed to ozone: a new nitrogen model." Downey, California: Report from Rancho Los Amigos Hospital.

Montgomery, T.L.; J.M. Keavitt; T.L. Crawford; and F.E. Gartrell. 1973. "Controlling ambient SO_2." *Journal of Metals* 25: 35–41.

Morgan, M.G.; S.C. Morris; A.K. Meir; and D.L. Shenk. 1978a. "A probabilistic methodology for estimating air pollution effects from coal-fired power plants." *Energy Systems and Policy* 2: 287–310.

_____. 1978b. "Sulfur control in coal-fired power plants: a probabilistic approach to policy analysis." *JAPCA* 28: 993–97.

Morgan, M.S.; J.Q. Koenig; D.S. Covert; and R. Frank. 1977. "Acute effects of inhaled SO_2 combined with hydroscopic aerosol in healthy man." *Amer. Rev. Resp. Dis.* 115: (4, pt. 2 abstract): 231.

Morrow, P.E. 1972. "Theoretical and experimental models for dust deposition and retention in man." Rochester, N.Y.: University of Rochester UR-3490-169.

_____. 1975. "An evaluation of recent nox toxicity data and an attempt to derive an ambient air standard for nox by established toxicological procedures." *Envir. Res.* 10: 92-112.

Mostardi, R.A., and D. Leonard. 1974. "Air pollution and cardiopulmonary function." *Arch. Envir. Health* 29: 325-28.

Mostardi, R.A., and R. Martell. 1975. "The effects of air pollution on pulmonary functions in adolescents." *Ohio J. Sci.* 75: 65-69.

Motley, H.L., and H.W. Phelps. 1964. "Pulmonary function impairment produced by atmospheric pollution." *Dis. Chest* 45: 154-62.

Motley, H.L.; R.H. Smart; and C.I. Leftwich. 1959. "Effect of polluted Los Angeles air (SMOG) on lung volume measurements." *JAMA* 171: 1469-77.

Mueller, P.K., and M. Hitchcock. 1969. "Air quality criteria toxicological appraisal for oxidants, nitrogen oxides and hydrocarbons." *JAPCA* 19: 670-76.

Mustafa, M.G., and S.D. Lee. 1976. "Pulmonary biochemical alterations resulting from ozone exposure." *Ann. Occ. Hyg.* 19: 17-26.

Mustafa, M.G.; S.M. Macres; B.K. Tarkington; C.K. Chow; and M.Z. Hussain. 1975. "Lung superoxide dismutase (SOD): stimulation by low level ozone exposure." *Clin. Res.* 23 (Abst): 138.

Nadel, J.A. 1977. "Autonomic control of airway smooth muscle and airway secretions." *Am. Rev. Resp. Dis.* 115: 117-26.

Nadel, J.A.; H. Salem; B. Tamplin; and Y. Tokiwa. 1965. "Mechanism of bronchoconstriction during inhalation of sulfur dioxide." *J. Appl. Physio.* 20: 164.

Nakajima, T.; S. Kusomoto; C. Chen; and K. Okamoto. 1969. "Effects of prolonged continuous exposure of nitrogen dioxide in the quantity of reduced glutathione in lungs of mice and their histopathological changes." In *Appendix: Effects of Nitrite and Nitrate on the Glutathione Reductase.* Osaka, Japan: Osaka Pretectual Pub. Health Inst. Reports, Labor and Sanitation Ser. No. 7: 35-41 (in Japanese).

Nakamura, K. 1964. "Response of pulmonary airway resistance by interaction of aerosols and gases in different physical and chemical nature." *Jap. J. Hyg.* 19: 322-33 (in Japanese). Translation by EPA. Research Triangle Park, North Carolina: APTIC No. 11425.

NAPCA. 1968. "Report for consultation on the metropolitan Boston intrastate air quality control region." Washington, D.C.: Consumer Protection and Environmental Health Service, National Air Pollution Control Administration.

NAS. 1972. *Particulate Polycyclic Organic Matter.* Washington, D.C.: National Academy of Science Committee on Biological Effects of Atmospheric Pollutants.

_____. 1973a. *Man, Materials and Environment.* Washington, D.C.: National Academy of Sciences, National Academy of Engineering: A Report for the National Commission on Materials Policy.

_____. 1973b. "Proceedings of the Health Effects of Air Pollutants." NAS–NRC October 3–5. Prepared for U.S. Senate Committee on Public Works.

_____. 1974a. "Air Quality and Automobile Emission Control." Washington, D.C.: National Academy of Sciences, National Academy of Engineering, Summary Report, No. 93–124, vol. I.

_____. 1974b. "How Safe is Safe? – The Design of Policy on Drugs and Food Additives." Washington, D.C.: National Academy of Sciences.

_____. 1975. "Air Quality and Stationary Source Emission Control." Washington, D.C.: National Academy of Sciences, National Academy of Engineering Commission on Natural Resources Report prepared for the U.S. Senate Committee on Public Works, Ser. No. 94–4.

_____. 1977a. *Airborne Particles.* Washington, D.C.: National Academy of Sciences, National Research Council, Subcommittee on Airborne Particles, Committee on Medical and Biological Effects of Environmental Pollutants.

_____. 1977b. *Nitrogen Oxides.* Washington, D.C.: National Academy of Science Committee on Medical and Biological Effects of Environmental Pollutants.

_____. 1977c. *Ozone and Other Photochemical Oxidants.* Washington, D.C.: National Academy of Science Committee on Medical and Biological Effects on Environmental Pollutants.

_____. 1978. *Sulfur Oxides.* National Academy of Science Committee on Sulfur Oxides Board of Toxicological and Environmental Health Hazards Draft Report.

_____. 1979a. *Airborne Particles.* National Academy of Sciences Committee on Airborne Particles, Committee on Medical and Biological Effects of Environmental Pollutants. Baltimore: University Park Press.

_____. 1979b. *Sulfur Oxides.* Washington, D.C.: National Academy of Science, National Research Council Committee on Sulfur Oxides, Board on Toxicological and Environmental Health Hazards, Assembly of Life Sciences.

Natusch, D.F.S. 1978. "Potentially carcinogenic species emitted to the atmosphere by fossil fuel-power plants." *Envir. Health Persp.* 22: 79–90.

Natusch, D.F.S.; J.R. Wallace; and C.A. Evans. 1974. "Toxic trace elements: preferential concentrations in respirable particles." *Science* 183: 79–90.

NEEPC (New England Energy Policy Council). 1975. Recommendation to the New England States. Boston, Massachusetts: Center for Energy Policy.

New Jersey. 1931. New Jersey v. New York City. Quoted in 2 ELR 10174.

New York. 1921. New York v. New Jersey and Passiac Valley Sewage Commissioners, 256 U.S. 296, October term, 1920.

Newhouse, M.L., and H. Thompson. 1965. "Mesothelioma of pleora and peritoneum following exposure to asbestos in the London area." *Brit. J. Ind. Med.* 22: 261-69.

vonNieding, G. 1978. "Effects of NO_2 on chronic bronchitis." *Conference on Pollutants and High Risk Groups*, E.J. Calabrese, ed. Washington, D.C.: EPA.

vonNieding, G., and H. Krekeler. 1971. "Pharmakologische Beeinflussung der akuten NO_2 —wirkung auf die lungenfunktion von Gesunden und Kraken mit einer chronischen Bronchitis." *Intern. Arch. Arbeitsmed.* 29: 55-63.

vonNieding, G.; H. Krekeler; R. Fuchs; M. Wagner; and K. Koppenhagen. 1973. "Studies of the effects of NO_2 on lung function: influence on diffusion, perfusion and ventilation in the lungs." *Int. Arch. Arbeitsmed.* 31: 61.

vonNieding, G.; M. Wagner; H. Löllgen; and H. Krekeler. 1977. "Zur Akuten Wirkung von ozon auf die lungenfunktion des menschen." Presented at the VDI Commission Colloquium on Ozone and Related Substances in Photochemical Smog. Dusseldorf, West Germany, VDI-Verlag GmbH. Report 270, pp. 123-29.

Nishiwaki, Y.; Y. Tsunetoshi; T. Shimizu; M. Ueda; N. Nakayama; H. Takahashi; A. Ichinosawa; S. Kajihara; A. Ohshino; M. Ogino; K. Sakaki. 1971. "Atmospheric contamination of industrial areas including fossil-fuel power stations, and a method of evaluating possible effects on inhabitants." Vienna: IAEA-SM-146/16.

Nochumson, D. 1978. "Models for the Long Distance Transport of Atmospheric Sulfur Oxides." Cambridge, Massachusetts: Harvard University Thesis.

NRC. 1975. "Nuclear Regulatory Commission decision in the "As Low as Practicable" hearing, RM-50-1.

NRDC. 1974. National Resources Defense Council v. Environmental Protection Agency, No. 27-2402 (5th Circ. February 8), 4 ELR 20204.

OECD (Organization for Economic Cooperation and Development). 1974. Seminar Problems in Transfrontier Pollution. Paris, January.

_____. 1977. "The Organization for Economic Cooperation and Development Programme on long range transport of air pollution: Measurements and findings." Paris.

Oeschger, H.; U. Siegenthaler; U. Schotterer; and A. Gugelmann. 1975. "A box diffusion model to study the carbon dioxide exchange in nature." *Tellus* 27: 168-92.

Orehek, J., and P. Gayard. 1976. "Les testes de provocation bronchique nonspecifiques dans l'asthme." *Bull. Europ. Physiopath. Resp.* 12: 565-98.

Orehek, J.; J.P. Massari; P. Gayard; C. Grimaud; and J. Charpin. 1976. "Effect of short-term, low-level nitrogen dioxide exposure on bronchial sensitivity of asthmatic patients." *J. Clin. Invest.* 57: 301-307.

Oshima, H.; M. Imai; and T. Kawagishi. 1972. "A study on air pollution effects on the respiratory symptoms in the city of Yokkaichi." *Mie Igaku* 16: 25.

Oshima, Y.; T. Ishizaki; T. Miyamoto; T. Shimizu; T. Shida; and J. Kabe. 1964. "Air pollution and respiratory diseases in the Tokyo-Yokohama area." *Amer. Rev. Resp. Dis.* 90: 572.

OTA. 1979. *The Direct Use of Coal: Prospects and Problems of Production and Combustion, and Appendices.* Office of Technology Assessment, U.S. Government Printing Office Doc. 052-003-00664-2.

Pace, T.G., and E.L. Mayer. 1979. "Preliminary characterization of inhalable particulates in urban areas." 72nd Annual Conf. of the Air Poll. Control Assoc., Cincinnati, Ohio, June 24-29.

Palmes, E.D., and M. Lippmann. 1977. "Influence of respiratory air space dimensions on aerosol deposition." In *Inhaled Particles,* W.H. Walton, ed., I: 127-36. London: Pergamon Press.

Pasquill, F. 1961. "The estimation of the dispersion of windborn materials." *Meteorology Mag.* 90: 1963.

_____. 1962. *Atmospheric Diffusion.* Princeton, New Jersey: Van Nostrand-Reinhold.

Pattle, R.E., and H. Collumbine. 1956. "Toxicity of some atmospheric pollutants." *Brit. Med. J.* 2: 913-16.

Peacock, P.R., and J.B. Spence. 1967. "Incidence of lung tumors in LX mice exposed to (1) free radicals, (2) SO_2." *Br. J. Cancer* 21: 606-18.

Pearlman, M.E.; J.F. Finklea; J.P. Creason; C.M. Shy; M.M. Young; and R.J.M. Horton. 1971. "Nitrogen dioxide and lower respiratory illness." *Pediatrics* 47: 391.

Pedersen, E.; K. Magnus; T. Mork; A. Hougen; and F. Bjelke. 1969. "Lung cancer in Finland and Norway—An epidemiological study." *Acta. Pathol. Biol. Scan.* Suppl: 199.

Pemberton, J. 1961. "Air pollution as a possible cause of bronchitis and lung cancer." *J. Hyg. Epid. Micro. and Immun.* 5: 189.

Perera, F.P., and A.K. Ahmed. 1978. "Respirable Particles: Impact of Airborne Fine Particulates on Health and Environment." New York, New York: Natural Resources Defense Council Report.

Permutt, S. 1977. Private communication.

Petering, D.H., and N.T. Shih. 1975. "Biochemistry of bisulfite-sulfur dioxide." *Envir. Res.* 9: 55-65.

Petr, B., and P. Schmidt. 1967. "Der Einfluss der durch Schwefeldioxid und nitrose gase verunreinigten atmosphäre auf den gesandeitszustand der kinder." *Z. Ges. Hyg.* 13: 34-38.

Petrilli, F.L.; G. Agnese; and S. Kanitz. 1966. "Epidemiologic studies of air pollution effects in Genoa, Italy." *Arch. Envir. Health* 12: 733.

Phelps, H.W. 1961. "Pulmonary function studies used to evaluate air pollution asthma disability." *Military Med.* 126: 282.

_____. 1965. "Follow-up studies in Tokyo-Yokohama respiratory disease." *Arch. Envir. Health* 10: 143.

Phelps, H.W., and S. Koike. 1962. "Tokyo–Yokohama asthma." *Amer. Rev. Resp. Dis.* 86: 55.

Phelps, H.W.; G.W. Sobel; and N.E. Fisher. 1961. "Air pollution asthma among military personnel in Japan." *JAMA* 175: 990.

Pike, M.C.; R.J. Gordon; B.E. Henderson; H.R. Merck; and J. SooHoo. 1975. "Air Pollution." In *Persons at High Risk of Cancer, An Approach to Cancer Etiology and Control,* J. Fraumeni, ed. New York: Academic Press.

Pitts, J.N., and R.L. Metcalf, eds. 1969. *Advances in Environmental Science,* vol. 1, p. 8. New York: Wiley–Intersciences.

Plantagenet, E. 1307. "Letter to the Sheriff of Surrey." Quoted and translated by Sir W. Hawthorne, in "Energy and environment, conflict or compromise," Trueman Wood Lecture, *J. Roy. Soc. Arts,* July 1978.

Pochin, E.E. 1975. "The acceptability of risk." *Brit. Med. Bull.* 31: 184–90.

Prindle, R.A.; G.W. Wright; R.O. McCaldin; S.C. March; T.C. Lloyd; and W.E. Bye. 1963. "Comparison of pulmonary functions and other parameters in two communities with widely different air pollution levels." *J. Pub. Health* 53: 200.

Pszenny, A.P. 1978. "Chemical characteristic of rainfall in a Boston suburb." Chestnut Hill, Massachusetts: Boston College M.S. thesis.

Rall, D.P. 1974. "Review of the health effects of sulfur oxides." *Envir. Health Persp.* 8: 97–121.

———. 1978. "Health and environmental effects of increased use of coal utilization." *Federal Register* 43, no. 10: 2229.

Rawls, J. 1957. "Justice as fairness." *J. Phil.* 54: 653.

Redmond, C.K.; R.B. Strobino; and R.H. Cypress. 1976. "Cancer experience among coke by-product workers." *Ann. N.Y. Acad. Sci.* 271: 102–15.

Reid, D.D. 1956. "Symposium: chronic bronchitis." *Proc. Roy. Soc. Med.* 49: 767.

———. 1958. "Environmental factors in respiratory disease." *Lancet* 1: 1289–94.

Reid, D.D., and A.S. Fairbairn. 1958. "The natural history of chronic bronchitis." *Lancet* 1: 1147.

Reid, D.D.; D.O. Anderson; B.G. Ferris, Jr.; and C.M. Fletcher. 1964. "An Anglo–American comparison of the prevalence of bronchitis." *Brit. Med. J.* 2: 1487.

Remmers, J.E., and O.J. Balchum. 1965. "Effects of Los Angeles urban air pollution upon respiratory function of emphysematous patients. The effect of the microenvironment on patients with chronic respiratory disease." Presented at the annual conference of the Air Pollution Control Association, Pittsburgh, Pennsylvania.

Renzetti, N.A., and V. Gobran. 1957. "Studies of Eye Irritation Due to Los Angeles Smog." San Marino, California: Air Pollution Foundation Report.

Revelle, R., and H.E. Seuss. 1957. "Carbon dioxide exchange between the atmosphere and the ocean, and the question of an increase in atmospheric carbon dioxide during the past decades." *Tellus* 9: 18.

Richardson, N.A., and W.C. Middleton. 1958. "Evaluation of filter removing irritants from polluted air." *Heating, Piping and Air Conditioning* 30: 147.

Ritter, L.E., and P.A. Podolick. 1973. "Determining an equitable surcharge for industrial wastes." *Public Works,* July.

Roberts, E.F. 1975. *Land Use Planning, Cases and Materials.* New York: Matthew Behder.

Roberts, M. 1978. "The political economy of implementation of the Clean Air Act and stationary sources." In *Approaches to Controlling Air Pollution,* A.F. Friedlander, ed. Cambridge, Massachusetts: MIT Press.

Robinson, E., and R.C. Robbins. 1968. "Sources, abundance and fate of gaseous atmospheric pollutants." Menlo Park, California: SRI International PR-6755.

_____. 1970. "Gaseous sulfur pollutants from urban and natural sources." *JAPCA* 20: 233-35.

Robinson. 1875. Robinson, William B. and others v. Baugh, John B. 31 Michigan 290.

Rodhe, H. 1972. "A study of the sulfur budget for the atmosphere over northern Europe." *Tellus* 24: 128-38.

_____. 1978. "Budgets and turnover times of atmospheric sulfur compounds." *Atmos. Envir.* 12: 671-80.

Rogers, T., and C.C. Gamertsfelder. 1971. "Proceedings for Environmental Effects of Nuclear Power Stations." Vienna: International Atomic Energy Agency.

Rokaw, S.N., and F. Massey. 1962. "Air pollution and chronic respiratory disease." *Amer. Rev. Resp. Dis.* 86: 703.

Rokaw, S.N.; H.E. Swann, Jr.; P.L. Keenan; and J.R. Phillips. 1968. "Human exposure to single pollutants: NO_2 in a controlled environmental facility." Presented at the 9th Air Pollution Medical Research Conference, Denver, Colorado.

Ross, F.F.; and T.T. Frankenberg. 1971. "What sulphur dioxide problem?" *Combustion,* August, p. 6.

Roth, H.D.; J.R. Viren; and A.V. Colucci. 1977. "An evaluation of CHESS New York asthma data, 1970-71, Vol. I, Findings and Supporting Tables." Menlo Park, California: EPRI Report EA-50.

Rotty, R.M. 1973. "Global production of CO_2 from fossil fuels and possible changes in the world's climate." *Mech. Eng.* 96: 67.

_____. 1976. "Global carbon dioxide production from fossil fuels and cement, A.D. 1950-A.D. 2000." Office of Naval Research Conference on the Fate of Fossil Fuel Carbonates, Honolulu, Hawaii, January 19-23.

_____ . 1977. "Uncertainties associated with future atmospheric CO_2 levels." Oak Ridge, Tennessee: Institute for Energy Analysis, ORAU/IEA (O)-16.

Rudolf, G., and J. Heyder. 1974. "Deposition of aerosol particles in the human nose." In *Aerosole in Naturwissenschaft, Medizin und Technik,* V. Bohlau, ed. Bad Soden, West Germany, Gesellschaft fur Aerosolforschung.

Sackner, M.A., and M. Reinhardt. 1977. "Effect of microaerosols of sulfate particulate matter on tracheal velocity in conscious sheep." *Amer. Rev. Resp. Dis.* 115 (4, pt. 2): A241.

Sackner, M.A.; R.D. Dougherty; and G.A. Chapman. 1976. "Effect of inorganic nitrate and sulfate salts on cardio-pulmonary function." *Amer. Rev. Resp. Dis.* 113: A89.

Sackner, M.A.; M. Reinhardt; and D. Ford. 1977. "Effect of sulfuric acid mist on pulmonary function in animals and man." *Amer. Rev. Resp. Dis.* (Abstract) 115 (4, pt. 2): 240.

Sackner, M.A.; D. Ford; R. Fernandez; E.D. Michaelson; R.M. Schreck; and A. Warner. 1977. "Effect of sulfate aerosols in cardio-pulmonary function of normal humans." *Amer. Rev. Resp. Dis.* (Abstract) 115: 240.

Samuelsen, G.S.; R.E. Rasmussen; B.K. Nair; and T.T. Crocker. 1978. "Novel culture and exposure system for measurement of effects of airborn pollutants on mammalian cells." *Envir. Sci. and Tech.* 12: 426–30.

Sato, S.; M. Kawakami; S. Maeda; and T. Takishima. 1976. "Scanning electron microscopy of the lung of vitamin E deficient rats exposed to low concentrations of ozone." *Am. Rev. Resp. Dis.* 113: 809–21.

Sawicki, F. 1972. "Air pollution and prevalence of chronic nonspecific respiratory diseases." In *Ecology of Chronic Nonspecific Respiratory Diseases.* Panstwowy Zaklad Wydawnictw Lekavskich, Warsaw, Poland.

_____ . 1977. "Chemical composition and potential genotoxic aspects of polluted atmosphere." In *Air Pollution and Cancer in Man,* U. Mohr, D. Schmahl, and L. Tomatis, eds. Lyon, France: IARC Scientific Publications No. 160.

Sawicki, F., and P.S. Lawrence. 1977. "Chronic non-specific respiratory disease in the city of Cracow." Report of a 5-year follow-up study among adult inhabitants. Warsaw, Poland: National Institute of Hygiene.

Sawicki, R.; W.C. Elbert; T.R. Hauser; R.T. Fox; and T.W. Stanley. 1960. "Benzo (α) pyrene content of the air of American communities." *Amer. Ind. Hyg. Assoc. J.* 21: 443.

Schimmel, H., and L. Greenberg. 1973. "A study of the relation of air pollution to mortality, New York City 1963–1964." *J. Air Poll. Cont. Assoc.* 22: 606.

Schimmel, H., and L. Jordan. 1977. "The relation of air pollution to mortality: New York City, 1963–72. II. Refinements in methodology and data analysis." Scientific Papers absts of the International Epidemiological Association (London). International Scientific Meeting, San Juan, Puerto Rico.

Schimmel, H., and T.J. Murawski. 1975. "SO$_2$ —Harmful pollutant or air quality indicator?" *JAPCA* 25: 739-40.

——. 1976. "The relation of air pollution to mortality." *J. Occ. Med.* 18: 316.

Schlesinger, R.B.; M. Lippmann; and R.E. Albert. 1978. "Effects of short-term exposure to sulfuric acid and ammonium sulfate aerosols upon bronchial airway function in donkeys." *Amer. Ind. Hyg. Assoc. J.* 39: 275-86.

Schneider, S.H. 1975. "On carbon dioxide–climate confusion." *J. of Atmos. Sci.* 32: 2060.

Schneiderman, M.A., and D.L. Levin. 1972. "Trends in lung cancer. Mortality, incidence, diagnosis, treatment, smoking, and urbanization." *Cancer* 30: 1320-25.

Schoettlin, C.E., and E. Landau. 1961. "Air pollution and asthmatic attacks in the Los Angeles area." *Pub. Health Rep.* 76: 545-48.

——. 1962. "The effect of air pollution on elderly males." *Amer. Rev. Resp. Dis.* 86: 878.

Schrenk, H.M.; H. Heinmann; G.D. Clayton; W.M. Gafafer; and K. Wexler. 1949. "Air pollution in Donora, Pennsylvania. Epidemiology of the unusual fog episode of October, 1948." Washington, D.C.: HEW: Pub. Health Serv. Bull. No. 306.

Schumacher, E. 1973. *Small is Beautiful.* New York: Harper and Row.

Schwartz, L.W. 1976. "Comparison of the effects of ozone and oxygen on lungs of rats." *Envir. Health Persp.* 16: A179.

Schwartz, L.W.; D.L. Dungsworth; M.G. Mustafa; B.K. Tarkington; and W.S. Tyler. 1976. "Pulmonary responses in rats to ambient levels of ozone. Effect of seven day intermittent or continuous exposure." *Lab. Invest.* 34: 565-78.

Schwing, R., and R. McDonald. 1976. "Measures of association of some air pollutants, natural ionizing radiation, and cigarette smoking with mortality rates." *The Science of the Total Environment* 5: 139-69.

Scott, J.A. 1963. "The London Fog of December, 1962." *Med. Off.* (London) 109: 250-52.

Selikoff, I.J., and E.C. Hammond. 1964. "Asbestos Exposure and Neoplasia." *JAMA* 188: 22-26.

Selikoff, I.J.; J. Churg; and E.C. Hammond. 1965. "Relation between exposure to asbestos and neoplasia." *New Eng. J. Med.* 272: 560-65.

Selikoff, I.J.; E.C. Hammond; and J. Churg. 1968. "Asbestos exposure, smoking, and neoplasia." *JAMA* 204: 106-12.

Selikoff, I.J.; E.C. Hammond; and H. Seidman. 1973. In *Biological Effects of Asbestos*, P. Bogovski, J.C. Gilson, V. Timbrell, and J.C. Wagner, eds., pp. 209-16. Lyon, France: IARC Science Publication No. 8.

SGR. 1979. "Smoking and health: A report of the Surgeon General's Office." Washington, D.C.: DHEW Publication 79-50066.

Shabad, L.M. 1977. "The carcinogenicity of automobile exhausts, from data obtained in the U.S.S.R." In *Air Pollution and Cancer in Man.* Lyon, France: IARC Scientific Publications No. 160.

Shalamberidze, O.P. 1967. "Reflex effects of mixtures of sulphur and nitrogen dioxide." *Hyg. Sanit.* 32, no. 7: 10.

Shapiro, R.; V. DiFate; and M. Welcher. 1974. "Deanimation of cytosine derivatives by bisulfite. Mechanism of the reaction." *J. Amer. Chem. Soc.* 96: 906–12.

Sherwin, R.P., and D.A. Carlson. 1973. "Protein content of lung lavage fluid of guinea pigs exposed to 0.4 ppm nitrogen dioxide." *Arch. Envir. Health* 27: 90–93.

Sherwin, R.P.; J.B. Margolick; and E.A. Aguilar. 1974. "Acid phophotase intensity equilibrium fractions of the lungs of guinea pigs exposed to 0.4 ppm nitrogen dioxide." *Fed. Proc.* 33: A633.

Shrad, L.M. 1977. "The carcinogenicity of automobile exhausts from data obtained in the U.S.S.R. In *Air Pollution and Cancer in Man.* Lyon, France: IARC Scientific Publication No. 160.

Shy, C.M. 1973. "Human health consequences of nitrogen dioxide: a review." *Proc. Conf. on Health Effects of Air Pollutants* 15: 363–405.

_____. 1976. "Lung cancer and the urban environment." In *Clinical Implications of Air Pollution Research*, A.J. Finkel and W.C. Duel, eds. Acton, Massachusetts: Publishing Sciences Group.

Shy, C.M.; J.P. Creason; M.E. Pearlman; K.E.McClain; F.B. Benson; and M.M. Young. 1970a. "The Chattanooga school children study: Effects of community exposure to nitrogen dioxide. II. Incidence of acute respiratory illness." *JAPCA* 20: 582.

_____. 1970b. "The Chattanooga school children study: Effects of community exposure to nitrogen dioxide. I. Methods, Description of Pollutant Exposure and Results of Ventilatory Testing." *JAPCA* 20: 539–82.

Shy, C.M.; J.R. Goldsmith; J.D. Hackney; M.D. Lebowitz; and D.B. Menzel. 1978. *The Health Effects of Air Pollution.* New York: American Medical Association American Thoracic Society Medical Section.

Shy, C.M.; V. Hasselblad; R.M. Burton; C.J. Nelson; and A. Cohen. 1973a. "Air pollution effects on ventilatory function of U.S. school children, results of studies in Cincinnati, Chattanooga and New York." *Arch. Envir. Health* 27: 124.

Shy, C.M.; L. Niemeyer; L. Truppi; and T. English. 1973b. "Re-evaluation of the Chattanooga school-children studies and the health criteria for NO_2 exposures, in-house technical report." Research Triangle Park, North Carolina: EPA NERC.

Sierra Club. 1972. Sierra Club v. Ruckelhause (EPA), Affirmed by D.C. Court of Appeals 2 ELR 20656. Affirmed by U.S. Supreme Court Fri. v. Sierra Club, 1973 412 US 541.

Silverman, F. 1978. "Asthma and respiratory irritants (ozone). In *Conference on Pollutants and High Risk Groups*, E.J. Calabrese, ed. Washington, D.C.: EPA.

Silverman, F.; L.J. Follinsbee; J. Barnard; and R.J. Shepard. 1976. "Pulmonary function changes in ozone—interaction of concentration and ventilation." *J. Appl. Phys. Resp.* 41: 859.

Silverman, L.P. 1973. "The determinants of daily emergency admissions to hospitals." Washington, D.C.: Public Research Institute of the Center for Naval Analyses.

Sim, V.M., and R.E. Pattle. 1957. "Effect of possible smog irritants on human subjects." *JAMA* 165: 1908-13.

Sinnet, P.F., and H.M. Whyte. 1973. "Epidemiological studies in a highland population of New Guinea: Environment, culture and health status." *Human Ecology* 1: 245-77.

Skalpe, I.O. 1964. "Long-term effects of sulphur dioxide exposure in pulp mills." *Brit. J. Ind. Med.* 21: 69.

Slade, D.H., ed. 1968. *Meterology and Atomic Energy*. Washington, D.C.: U.S. Atomic Energy Commission report number TID: 24190.

Smith, T.J., and H.J. Paulus. 1971. "An epidemiological study of atmospheric pollution and bronchial asthma attacks." Paper presented at the 64th annual meeting of the Air Pollution Control Association, Atlantic City.

Smith, V.K. 1976. *The Economic Consequences of Air Pollution*. Cambridge, Massachusetts: Ballinger Publishing Company.

Snell, R.E., and P.C. Lucksinger. 1969. "Effects of sulfur dioxide on expiratory flow rates and total respiratory resistance in normal human subjects." *Arch. Envir. Health* 18: 693-98.

Solow, R.M. 1971. "The economist's approach to pollution and its control." *Science* 173: 498.

South Terminal Corporation. 1974. South Terminal Corporation v. EPA, 504 F. 2d 646, 675.

Speizer, F.E., and B.G. Ferris, Jr. 1963. "The prevalence of chronic non-specific respiratory disease in road tunnel employees." *Amer. Rev. Resp. Dis.* 88: 204.

_____ . 1973a. "Exposure to automobile exhaust. I. Prevalence of respiratory symptoms and disease." *Arch. Envir. Health* 26: 313.

_____ . 1973b. "Exposure to automobile exhaust. II. Pulmonary function measurement." *Arch. Envir. Health* 26: 319.

Speizer, F.E., and N.R. Frank. 1966. "The uptake and release of SO_2 by the human nose." *Arch. Envir. Health* 12: 725-28.

Speizer, F.E.; B.G. Ferris, Jr.; Y.M. Bishop; and J.D. Spengler. 1979. "Health effects of indoor nitrogen dioxide exposures." Paper presented at the Symposium on Health Effects of Nitrogen Oxides, AECS/CJS, Chemical Congress 1979, American Chemical Society Japan, Honolulu, Hawaii.

Spengler, J.D. 1979. "Comments on 'Atmospheric dispersion modeling: a critical review' by D.B. Turner." *JAPCA* 29: 9.

Spengler, J.D.; D.W. Dockery; and W.A. Turner. 1979. "Indoor-outdoor air pollution relationships: results of indoor-outdoor monitoring program of Harvard Six City Study." 72nd Annual Meeting, Air Pollution Control Association, Cincinnati, Ohio, June 24-29.

Spicer, W.S., Jr., and D.H. Kerr. 1966. "Variations in respiratory function." *Arch. Envir. Health* 12: 217.

_____. 1970. "Effects of environment on respiratory function, Part III." *Arch. Envir. Health* 21: 635.

Spicer, W.S., Jr.; W.A. Reinke; and D.H. Kerr. 1966. "Effects of environment upon respiratory function. II. Daily studies in patients with chronic obstructive lung disease." *Arch. Envir. Health* 13: 753.

Spodnick, M.J., Jr.; G.D. Cushman; D.H. Kerr; R.W. Blide; and W.S. Spicer, Jr. 1966. "Effects of environment on respiratory function: weekly studies on young male adults." *Arch. Envir. Health* 13: 243.

Spotnitz, M. 1965. "The significance of Yokohama asthma." *Amer. Rev. Resp. Dis.* 92: 371-75.

Stahley, S. 1976. Personal communication.

Starr, C. 1969. "Social Benefit v. Technological Risk." *Science* 165: 1231.

Statistical Abstract of the United States. 1978. Washington, D.C.: U.S. Government Printing Office.

Stebbings, J.H.; D.G. Fogelman; K.E. McLain; and M.C. Townsend. 1976. "Effect of the Pittsburgh air pollution episode upon pulmonary function in school children." *Amer. J. Epidem.* 104: 344.

Stensland, G.J., and R.G. dePena. 1973. "The influence of meteorological parameters on the concentration of contaminants in rainfall due to the washout of hydroscopic particles." Pennsylvania State University Report.

Stephens, R.J.; M.F. Sloan; M.J. Evans; and G. Freeman. 1974. "Alveolar Type 1 cell response to exposure to 0.5 ppm ozone for short periods." *Exp. Molec. Pathol.* 20: 11-23.

Sterling, T.D.; J.J. Phair; and S.V. Pollack. 1967. "Urban hospital morbidity and air pollution: a second report." *Arch. Envir. Health* 15: 362-74.

Sterling, T.D.; J.J. Phair; S.V. Pollack; D.A. Schumsky; and I. DeGroot. 1966. "Urban morbidity and air pollution: A first report." *Arch. Envir. Health* 13: 158-70.

Stern, A.C., ed. 1977. *Air Pollution.* New York: Academic Press.

Stocks, P. 1958. "Report on cancer in North Wales and Liverpool region in British Empire campaign." Supplement to 1957 annual report.

_____. 1966. "Recent epidemiological studies of lung cancer mortality, cigarette smoking, and air pollution, with discussion of a new hypothesis of causation." *Brit. J. Cancer* 20: 595.

Stocks, P., and M.J. Campbell. 1959. "Lung cancer death rates among non-smokers and pipe and cigarette smokers: evaluation in relation to air pollution by benzo (α) pyrene and other substances." *Brit. Med. J.* 2: 923.

Stokinger, H.E. 1972. "Concepts of thresholds in standard setting: analysis of concept and its application to industrial air limits." *Arch. Envir. Health* 25: 153.

Stuart, B.O. 1976. "Deposition and clearance of inhaled particles." *Envir. Health Persp.* 16: 41.

Stuiver, M. 1978. "Atmospheric carbon dioxide and carbon reservoir changes." *Science* 199: 253.

Stumphius, J., and P.B. Meyer. 1968. "Asbestos bodies and mesothelioma." *Ann. Occup. Hyg.* 11: 283–93.

Suess, H.E. 1955. "Radio carbon concentration in modern wood." *Science* 122: 416.

Sultz, H.A.; J.G. Feldman; E.R. Schlesinger; and W.E. Mosher. 1970. "An effect of continued exposure to air pollution on the childhood allergic disease." *Amer. J. Pub. Health* 60: 891.

Sutton, O.G. 1932. "A theory of eddy diffusion in the atmosphere." *Proc. Roy. Soc. A.* 135: 143.

Swanson, V.E. et al. 1976. U.S. Geological Survey Open Field Report 76–468 (draft).

Takahashi, H. 1970. *J. Jap. Soc. Air Poll.* 5: 131 (Abstract in Japanese).

Terrill, J.G., Jr.; E.D. Harward; and I.P. Leggett, Jr. 1967. "Environmental aspects of nuclear and conventional power plants." *Ind. Med. Surg.* 36: 412.

TGLD (Task Group on Lung Dynamics), Committee II–ICRP. 1966. "Deposition and retention models for internal dosimetry of the human respiratory tract." *Health Physics* 12: 173–207.

Thibodeau, L.; R. Reed; and Y. Bishop. 1979. A review of 'Air Pollution and Human Health' by Lester B. Lave and Eugene P. Seskin, Department of Biostatistics, Harvard School of Public Health. Submitted for publication.

Thomas, H.V.; P.K. Mueller; and R. Wright. 1967. "Response of rat lung mast cells to nitrogen dioxide inhalation." *JAPCA* 17: 33–35.

Toyama, T. 1964. "Air pollution and health effects in Japan." *Envir. Health* 8: 153.

Toyama, T., and K. Nakamura. 1964. "Synergistic response of hydrogen perocide aerosols and sulfur dioxide to pulmonary airway resistance." *Ind. Health* 2: 34–45.

Travis, C.C. et al. 1979. "A radiological assessment of radon–222 released from uranium mills and other natural and technologically enhanced sources." Oak Ridge, Tennessee: Oak Ridge Nat. Lab. NUREG/CR–0573 ORNL/NUREG–55.

Turner, D.B. 1969. *Workbook of Atmospheric Dispersion Estimates.* Washington, D.C.: U.S. DHEW.

_____. 1979. "Atmospheric dispersion modeling: a critical review." *JAPCA* 29: 502.

Turner, J. 1972. "Handbook of Atmospheric Dispersion Estimates." Washington, D.C.: U.S. Environmental Protection Agency AP-26.

TVA (Tennessee Valley Authority). 1974. "Summary of Tennessee Valley Authority atmospheric dispersion modeling." Conference on the TVA Experience at the Institute of Applied Systems Analysis, Schloss Laxenberg, Austria.

U.S. House. 1976. "Community Health and Environmental Surveillance System (CHESS): An Investigative Report." U.S. House of Representatives, November.

U.S. Senate. 1970. "Report accompanying 1970 Clean Air Amendments, 91st Congress." Washington, D.C.: U.S. Government Printing Office Report No. 91-1196.

Union Electric. 1976. Union Electric Co. v. EPA, 427 U.S. 246.

Ury, H. 1968. "Photochemical air pollution and automobile accidents in Los Angeles." *Arch. Envir. Health* 17: 334–42.

Ury, H., and A.C. Hexter. 1969. "Relating photochemical pollution to human physiological reactions under controlled conditions. Statistical procedures." *Arch. Envir. Health* 18: 473–80.

Ury, H.; N.M. Perkins; and J.R. Goldsmith. 1972. "Motor vehicle accidents and vehicular pollution in Los Angeles." *Arch. Envir. Health* 25: 314–22.

Versailles Borough. 1935. Versailles Borough v. McKeesport Coal Coke Co., 83 *Pittsburgh Legal Journal* 379–84.

Vigdortschik, N.A.; E.C. Andreeva; I.L. Mattusswitch; M.M. Nikulina; L.M. Fromina; and V.A. Stritter. 1937. "The symptomatology of chronic poisoning with oxides of nitrogen." *J. Ind. Hyt. Toxicol.* 19: 461–73.

Viren, J.J. 1978. "Cross-sectional estimates of mortality due to fossil fuel pollutants: A case for spurious association." Washington, D.C.: Report for the Division of Policy Analysis, U.S. Department of Energy.

Voisin, C.; C. Aerts; E. Jakubezak; J.L. Houdret; and A.B. Tonnel. 1977. "Effects of nitrogen dioxide on alveolar macrophages surviving in the gas phase." *Bull. Europ. Physiopath. Resp.* 13: 137.

Wagner, J.C.; C.A. Slegges; and P. Marchand. 1960. "Diffuse pleural mesothelioma and asbestos exposure in the northwest cape province." *Brit. J. Ind. Med.* 17: 260–71.

Waller, R.E. 1971. "Air pollution and community health." *Journal of the Royal College of Physicians* (London) 5: 362.

Waller, R.E., and P.J. Lawther. 1955. "Some observations on London fog." *Brit. Med. J.* 2: 1356.

_____. 1957. "Further observations on London fog." *Brit. Med. J.* 4: 1473.

Watanabe, H. 1965. "Air pollution and its health effects in Osaka, Japan." Paper presented at the 58th Annual Conference of the Air Pollution Control Association, Toronto, Canada.

Watanabe, S.; F. Frank; and E. Yokoyama. 1973. "Acute effects of ozone in lungs of cats." *Amer. J. Rev. Resp. Dis.* 108: 1141-51.

Wayne, W.S.; P.F. Wehrle; and R.E. Carroll. 1967. "Oxidant air pollution and athletic performance." *JAMA* 199: 901-904.

Weir, F.W., and P.A. Bromberg. 1973. "Effects of sulfur dioxide on human subjects exhibiting peripheral airway impairment." Amer. Petroleum Inst. Proj. CAWC S-15.

_____. 1975. "Effects of sulfur dioxide on healthy and peripheral airway impaired subjects." *Recent Advancements in the Assessment of the Health Effects of Environmental Pollutants* 4: 1989-2004.

Weiss, E. 1974. "Sulphur pollution and its mitigation." Appendix to Wilson and Chen (1974).

Welch, R.E. 1974. Private communication. American Smelting and Refining Company, Tacoma, Washington.

West, J.B. 1977. *Pulmonary Pathophysiology.* Baltimore: William and Wilkens.

West, P.W. 1968. "Chemical analysis of inorganic particulate pollutants." In *Air Pollution,* A.C. Stern, ed., vol. II. New York: Academic Press.

Whitby, K.T., and B. Cantrell. 1975. "Atmospheric aerosols—characteristics and measurements." International Conference on Environmental Sensing and Assessment, IEEE, Las Vegas, Nevada.

Wicken, A.J. 1966. "Environmental and personal factors in lung cancer and bronchitis mortality in Northern Ireland 1960-1962." London: Tobacco Research Council.

Wilson, D.G., and P.W. Chen. 1975. "An Analysis of Emission Charges as a Method of Reducing Sulfur Pollution." BNL Report No. 21654.

Wilson, R. 1975. "Examples in risk benefit analysis." *Chem. Tech.* 5: 604-607.

_____. 1978. "Risks caused by low-levels of pollution." *Yale J. Biol. and Med.* 51: 37-51.

Wilson, R., and W.J. Jones. 1974. *Energy, Ecology and the Environment.* New York: Academic Press.

Winkelstein, W., Jr. 1970. "Utility or futility or ordinary mortality statistics in the study of air pollution effects." Proc. of the 6th Berkeley Symposium on Mathematical Statistics and Probability, Berkeley, California.

Winkelstein, W., Jr., and S. Kantor. 1969. "Respiratory symptoms and air pollution in an urban population of northeastern United States." *Arch. Envir. Health* 18: 760.

Winkelstein, W., Jr.; W.S. Kantor; E.W. Davis; C.S. Maneri; and W.W. Mosher. 1967. "The relationship of air pollution and economic status of total mortality and selected respiratory mortality in men. I. Suspended particulates." *Arch. Envir. Health* 14: 162-71.

_____. 1968. "The relationship of air pollution and economic status to total mortality and selected respiratory system mortality in men. II. Oxides of sulfur." *Arch. Envir. Health* 16: 401-405.

Wolf, P.C. 1974. "Systems for continuous stack-gas monitoring. Part I." *Mechanical Engineering* 96 (4).

Wong, C.S. 1978. "Atmospheric input of carbon dioxide from burning wood." *Science* 200: 19.

Woodwell, G.M.; R.H. Whittaker; W.A. Reiners; G.E. Likens; C.C. Delwiche; and D.B. Botkin. 1978. "The biota and the world carbon budget." *Science* 199: 141.

Woodwell, M., and E.V. Pecan, eds. 1973. "Carbon in the Biosphere. Proceedings of the 24th Brookhaven Symposium in Biology." Upton, New York: USAEC CONF-720510.

World Health Organization. 1960. World Health Organization Technical Report Ser. 192.

_____. 1962. "Epidemiology of Air Pollution." Geneva: WHO Public Health Paper No. 15.

Wynder, E.L., and D.A. Hoffman. 1965. "Some laboratory and epidemiological aspects of air pollution carcinogenesis." *JAPCA* 15: 155-59.

Wyzga, R.E. 1978. "The effect of air pollution on mortality—A consideration of distributed lag models." *J. Amer. Stat. Assoc.* 73: 463-72.

Yeager, K. 1975. "Stacks vs. Scrubbers." Research Progress Report FF-3. Menlo Park, California: EPRI 7.

Yoshida, K.; H. Oshima; and M. Imai. 1964. "Air pollution in Yokkaichi area with special regards to the problem of 'Yokkaichi asthma.'" *Ind. Health* 2: 87-94.

_____. 1966. "Air pollution and asthma in Yokkaichi." *Arch. Envir. Health* 13: 763.

Yoshida, K.; Y. Takatsuka; M. Kitabatake; H. Oshima; and M. Imai. 1969. "Air pollution and its health effects in Yokkaichi area: Review of Yokkaichi asthma." *Mie Med. J.* 18: 195-209.

Yoshii, M.; J. Nonoyama; H. Oshima; H. Yamagiwa; and S. Takeda. 1969. "Chronic pharyngitis in air-polluted districts of Yokkaichi in Japan." *Mie Med. J.* 19: 17-27.

Young, W.A.; D.B. Shaw; and D.V. Bates. 1963. "Pulmonary function in welders exposed to ozone." *Arch. Envir. Health* 7: 337-40.

Zabezinsky, M.A. 1968. "City air pollution with carcinogenic substances released by motor car engines." *Vop. Onkol.* 14: 87.

Zarkower, A. 1972. "Alterations in antibody response induced by chronic inhalation of SO and carbon." *Arch. Envir. Health* 25: 45-50.

Zeidberg, L.D.; R.A. Prindle; and E. Landau. 1961. "The Nashville air pollution study Part I. Sulfur dioxide and bronchial asthma." *Amer. Rev. Resp. Dis.* 84: 489-503.

_____. 1964. "The Nashville air pollution study. III. Morbidity in relation to air pollution." *Amer. J. Pub. Health* 54: 85-97.

Zeidberg, L.D.; R.J.M. Horton; E. Landau; R.M. Hagstrom; and H.A. Sprague. 1967a. "The Nashville air pollution study. V. Mortality from diseases of the respiratory system in relation to air pollution." *Arch. Envir. Health* 15: 214–24.

_____. 1967b. "The Nashville air pollution study. VI. Cardiovascular disease mortality in relation to air pollution." *Arch. Envir. Health* 15: 225–36.

ABBREVIATIONS

AQCR	air quality control region
BACT	best available control technology
B α P	benzo (α) pyrene
$CaCO_3$	calcium carbonate
CAMP	Continuous Air Monitoring Project
CaO	calcium oxide
$(CH_3)_2 S$	dimethyl sulfide
CO	carbon monoxide
CO_2	carbon dioxide
COH	coefficient of haze
EPA	U.S. Environmental Protection Agency
ESP	electrostatic precipitator
FGD	flue gas desulfurization
$H_2 S$	hydrogen sulfide
$H_2 SO_4$	sulfuric acid
ICS	intermittent control system
MWe	megawatts electric

NAAQS	national ambient air quality standards
NADB	National Aerometric Data Bank
NAMS	National Air Monitoring Station
NASN	National Air Surveillance Network (EPA)
$Na_2 S_2 O_5$	sodium metabisulfite
NaOH	sodium hydroxide
$Na_2 CO_3$	sodium carbonate
$NH_4 HSO_4$	ammonium bisulfate
$NH_4 OH$	ammonium hydroxide
$(NH_4)_2 SO_4$	ammonium sulfate
NO_2	nitrogen dioxide
NO_x	nitrogen oxides
O_3	ozone
OH	hydroxyl radical
PAH	polycyclic aromatic hydrocarbons
PAQS	primary air quality standard
POM	particulate organic matter
ppm	parts per million
PSD	prevention of significant deterioration
RACT	reasonably available control technology
rem	roentgen-equivalent-man
SLAMS	State and Local Air Monitoring Station
SMS	Synchronous Meteorological Satellite
SMSA	standard metropolitan statistical area
SMR	standard mortality ratio
SO_2	sulfur dioxide
SO_x	sulfur oxides
Sv	severt (unit of International Commission of Radiological Protection)
TiO_2	titanium oxide
TSP	total suspended particulates
$Zn(NH_4)_2 (SO_4)_2$	zinc ammonium sulfate

GLOSSARY

acute/chronic: acute—diseases or responses with short and generally severe course (often due to high pollutant concentrations); chronic—diseases linger a long time (often due to low, continuous pollutant concentrations)

aerosol: a suspension of fine liquid or solid particles in a gas

albedo: the portion of the sun's light incident upon the earth which is reflected from it

ambient level: the level (of pollutant) in the general environment

anthropogenic: of human origin

asphyxiation: the action of suffocation by depriving tissue of oxygen

bronchitis: inflammation of the bronchial tubes

bronchoconstriction: reduced caliber of the bronchial tubes by contraction of the bronchial walls

busbar costs: cost of electricity at the generating station

carcinogenic: cancer causing

chronic: see acute

combustion: the process of burning

edema: a swelling due to effusion of watery fluid into intercellular spaces

377

epidemiology: the study of disease cause and control by observation of human populations

false negative results: results which show no effect when one is there

fly-ash: small solid ash particles from the noncombustible portion of fuel that are small enough to escape with the exhaust gases

harvesting: the sudden death or disease of people who have been rendered sensitive by previous exposures or existing disease

haze: obscuration of the earth's atmosphere, often over large areas

homeostasis: a tendency toward stability of the body

hyperplasia: abnormal multiplication of the normal cells in tissue

hypertrophy: excessive enlargement of a part or an organ

in vivo: within living organisms

in vitro: outside of living organisms (in glass)

isopleth: lines on a graph connecting points of constant value; e.g., isopleths of visibility are lines of equal visibility

lung: alveoli—small sacs through which gas exchange takes place
bronchi—branching air passages in the lungs
bronchioli—final branching of the bronchi
cilia—small hairlike processes that move mucus in the
bronchi
epithelium—the lining cells of the lung
macrophage—a wandering cell that injests (phagocytizes)
foreign particles

mass-median diameter: the diameter that divides the mass distribution of an aerosol in half

measurements: ng-μg-mg-μm-ppm, etc. relationships—
1 kilometer (km) = 1000 meters (m) = 10^5 cm = 10^6 mm = 10^9 μm
1 metric tonne = 1 thousand (10^3) kilograms (kg) = 1 million
(10^6) grams (g) = 10^9 milligrams (mg) = 10^{12} micrograms
(μg) = 10^{15} nanogram (ng)
1 part per million (ppm) of a pollutant usually indicates 1 gram
weight of the pollution to every million grams weight of air
(or water)
1 μg/m^3 (1 microgram in every cubic meter)

meteorology: a study of the weather and processes of the atmosphere

model: Gaussian distribution model (normal statistical)—(after Gauss a famous 19th century scientist) is expressed by the formula:

$$\frac{1}{\sigma_x \sqrt{2\pi}} \; exp \; \frac{(x - \overline{x})^2}{2\sigma_x^2}$$

$\overline{x}$ is the mean, σ_x is the standard deviation. It is sometimes called the normal distribution. A Gaussian air dispersion model is one in which the air is assumed to spread according to such a distribution and described by the two parameters $\overline{x}$ and σ_x.

proportional: two quantities are proportional to each other when if one doubles in size, the other does also. In disease, the death rate is proportional to pollutant concentration if it doubles when the concentration doubles.

morbidity: sickness

mortality: death

necrosis: death of tissue

particles: airborne particles can be natural, caused by stirring of soil dusts, or anthropogenic. They vary in size from coarse (diameter $>$ 3 μm) and fine ($<$ 3 μm). Sometimes inhalable or respirable is used to describe those particles ($<$ 2 μm) which can be inhaled through the nose and enter the lungs.

plume: the cloud of steam or smoke that comes from a chimney stack and blows downwind

point source: a source of pollutant which is in one place, such as a power plant

pollutant: any material entering the environment that has undesired effects

radionuclides: radioactive elements. These may be subdivided into natural radionuclides such as radium or uranium which are normally present in the earth, or artificial radionuclides which are not normally present (or normally present in very small amounts) and are produced by nuclear fission.

sampler: dichotomous—a particle capture device that separates particles into two size categories: ($< 2\ \mu$m, $> 2\ \mu$m and $< 15\ \mu$m) high-volume—a particle capture device that samples for all suspended aerosols or TSP

sink: a place where pollutants can "disappear," for example the sea

smelter: a works where ore is melted to extract the metal

smoke: the visible aerosol that results from incomplete combustion

source: a place where pollutants are emitted—for example a chimney stack

synergism: where the combined action of two or more pollutants is greater than the sum of individual actions

threshold: a pollutant concentration below which no deleterious effect occurs; this concept is increasingly doubted

trace: a very small amount of a material

tonne: a metric ton--1000 kilograms (see measurements)

topography: the detailed delineation of the geographic features of a locality

toxicology: the study of toxic materials

INDEX

ABOUT THE AUTHORS

Richard Wilson is a professor of physics at Harvard University. In the last ten years he has been involved with environmental concerns. He helped to start the Energy and Environmental Policy Center in 1973 and served as its director from 1975 through 1978. He has written many articles on environmental risks.

John D. Spengler is an Associate Professor of Environmental Health at the Harvard School of Public Health. For several years he has directed air monitoring and exposure assessment for a large epidemiologic study in six U.S. cities. As part of this work, he has been measuring the indoor/outdoor concentrations as well as the personal exposures of gases and particles. This work has lead to the physical and chemical characterization of respirable particles. This research has meshed in recent years with the public policy issues surrounding air quality standards, coal utilization, and indoor air pollution. Dr. Spengler chaired a National Academy of Sciences study of indoor air pollution.

Steven D. Colome is an Assistant Professor in the Social Ecology Program at the University of California, Irvine. He holds degrees from Stanford and Harvard Universities. His research interests include the physical characterization and health effects of atmospheric particles. He has consulted for state and federal agencies on the health and environmental effects of air pollutants and is interested in methods for conveying scientific uncertainty in the development of public policy.

David Gordon Wilson was involved in the turbomachinery and power-generation industry before joining the MIT faculty where he is a professor of mechanical engineering. In the last ten years he has been particularly concerned with the technology of and the externalities from power generation and the treatment of wastes. He has served on several legislative commissions and has drafted legislation concerning, particularly, energy conservation and materials policy.